Snoring and Obstructive Sleep Apnea

Hsueh-Yu Li
Editor

Snoring and Obstructive Sleep Apnea

Evolution and Innovation

Springer

Editor
Hsueh-Yu Li
Department of Otolaryngology
Chang Gung Memorial Hospital, Chang Gung University
Taoyuan, Taiwan

ISBN 978-981-95-2647-5 ISBN 978-981-95-2648-2 (eBook)
https://doi.org/10.1007/978-981-95-2648-2

This Springer imprint is published by the registered company Springer Nature Singapore Pte Ltd.
The registered company address is: 152 Beach Road, #21-01/04 Gateway East, Singapore 189721, Singapore

Foreword

Obstructive sleep apnea syndrome (OSAS) is a clinical entity known and described in the ancient times. But "The symptoms and characteristics of sleep apnoea syndrome—excessive daytime sleepiness, loud snoring, restless and non-restorative sleep—are so impressive that it is difficult to understand why its true recognition was delayed until the 1970s" (Peretz Lavie 2008). The key of modern sleep surgery was the demonstration ulmost in the same years by Dr. Khulo in Freiburg and Dr. Lugaresi, my Mentor, in Bologna that repetitive mechanical upper airways obstructions are responsible for OSAS. And even more that was feasible to address obstruction by a surgical by-pass of the collapsed sites (tracheostomy). It was the seed for the development of more targeted and better accepted surgical procedures. UPPP was introduced in the USA as an alternative to permanent tracheostomy by Fujita in 1981 (K. Yaremchuk, 2017). It was the dawn of the modern sleep surgery. In this new, stimulating but challenging field ENT Surgeons play a key role for many reasons. The most common referral symptom (snoring) bring the patient more usually to the otolaryngologysts. The complex anatomy of upper airways is mastered by ENT specialists better than by any other medical or surgical specialists. Awake and Sedation Endoscopy are the mainstay of the best treatment, and again are a routine experience in most of the ENT Clinics. From the scientific point of view just today PubMed reports more than 8000 papers using as key word "obstructive sleep apnea syndrome surgery" and more than 1300 items using "snoring surgery." It is a boundless ocean of new data in which navigation is very difficult. But the book you have just started to read now may be a very useful compass for you for many reasons. First of all for the experience, competence, and international reputation of the Editorial Board. All of them are well known in the exclusive worldwide Community of experienced Sleep Surgeons. For that reason, all of them with their relative chapters will serve at the best in the challenging task of orienting you of a growing number of news in diagnosis and therapy of sleep disordered breathing. The general organization of the monograph first of all offers you a really comprehensive overview of the most updated tools for the diagnosis, ranging from the most conventional sleep studies to the snoring sound analysis and the promising application of ultrasounds. A special place is reserved to DISE. The core of the book is centered around the incredible variety of surgical solutions, all discussed in detail. The classic nose, palate, hypopharyngeal, and multilevel as well skeletal surgery are followed by the most interesting new entries like robotic surgery and hypoglossal nerve stimulation. Conservative treatments are covered as well (CPAP, drugs, positional, myofunctional, weight loss, oral appliances). Last but not least, according to the most modern trend of sleep surgery, a special discussion is reserved to the crucial perspective of gender and age-related sleep medicine and of the most recently described clinical entity (COMISA). Take your time and enjoy the lecture of this really useful monography. It is produced by the generous effort of a group of great professionals to whom I extend my thanks for their commitment and the compliments for the remarkable result.

Faenza, Italy
2025

Claudio Vicini

Preface

The field of sleep medicine has undergone a remarkable transformation over the past few decades. What was once viewed primarily as a benign nuisance—snoring—is now recognized as a potentially serious health condition with systemic consequences, particularly when associated with obstructive sleep apnea (OSA). The textbook *Snoring and Obstructive Sleep Apnea: The Evolution and Innovation* was conceived to encapsulate this transformation, documenting the rapid advances in understanding, diagnosing, and treating sleep-disordered breathing.

This volume highlights not only the scientific evolution of sleep medicine but also the innovations that are reshaping clinical care. From the paradigm shift in diagnostics—moving from traditional laboratory polysomnography to wearable home sleep apnea testing, interventional drug-induced sleep endoscopy, and ultrasound-based obstruction localization—to the reinvention of sleep surgery, this book captures a dynamic and multidisciplinary field in motion.

Surgical management of OSA is no longer a one-size-fits-all intervention. The emergence of novel techniques—such as palatal hybrid surgery, transoral robotic surgery, and hypoglossal nerve stimulation—has enabled precision treatment tailored to specific patterns of anatomical collapse. Evidence-based innovations, including echo-conducted tongue assessment and loco-sonic snoring source identification, have redefined the very concepts that guide clinical decisions. The focus has shifted from merely enlarging the airway to restoring function, improving sleep architecture, and ultimately enhancing quality of life.

Beyond airway management, sleep surgery now intersects with broader domains of health. It plays an increasingly recognized role in addressing comorbidities such as cardiovascular disease, neurocognitive dysfunction, and metabolic syndromes. By integrating surgical care with holistic and preventive medicine, we stand at the frontier of a new era—where treating sleep disorders contributes to long-term systemic wellness.

The chapters in this textbook are written by leading experts who have pioneered many of these advances. Their contributions reflect a commitment to both academic rigor and clinical relevance, offering practical insights for specialists, general practitioners, and researchers alike.

It is with great enthusiasm that I present this work to the global sleep medicine community. May it serve as a catalyst for continued innovation, deeper understanding, and improved care for patients affected by snoring and obstructive sleep apnea.

Taoyuan, Taiwan

Hsueh-Yu Li

Contents

Pathophysiology of Obstructive Sleep Apnea (OSA): From Animal to Human

1

Fang-Chia Chang and Nai-Chieh Liu

Contents

F.-C. Chang (✉) · N.-C. Liu
School of Veterinary Medicine, National Taiwan University, Taipei, Taiwan
e-mail: fchang@ntu.edu.tw; naichiehliu@ntu.edu.tw

H.-Y. Li (ed.), *Snoring and Obstructive Sleep Apnea*,
https://doi.org/10.1007/978-981-95-2648-2_1

Abstract

Obstructive Sleep Apnea (OSA) is a prevalent sleep disorder characterized by repeated upper airway obstructions during sleep, leading to intermittent hypoxia and sleep fragmentation. This chapter examines the pathophysiology of OSA, focusing on the interplay of anatomical vulnerabilities, neuromuscular deficits, and systemic responses. Animal models, including rodents, rabbits, dogs, and non-human primates, are instrumental in advancing our understanding of OSA mechanisms and therapeutic strategies. Rodent models replicate key features of OSA, such as oxidative stress and cardiovascular dysfunction, while rabbit models offer insights into upper airway dynamics. Larger mammals like dogs and primates provide translational relevance due to their anatomical and physiological similarities to humans. Naturally occurring OSA in brachycephalic dogs mirrors human disease, offering unique opportunities for studying disease progression and testing interventions. These findings elucidate the importance of animal models in bridging basic research with clinical applications to improve OSA management.

Keywords

Obstructive Sleep Apnea (OSA) · Pathophysiology · Animal models · Chronic intermittent hypoxia · Brachycephalic obstructive airway syndrome (BOAS) · Translational research · Neuromuscular control · Cardiovascular complications

1.1 Introduction

Obstructive sleep apnea (OSA) is a prevalent sleep disorder marked by repeated upper airway obstruction during sleep, causing intermittent hypoxia (IH) and sleep fragmentation [1]. This leads to serious health issues like cardiovascular diseases, metabolic disorders, and increased accident risk from daytime sleepiness [2, 3]. Affecting 9–38% of adults globally [4]. OSA burdens healthcare systems and is linked to hypertension, coronary artery disease, stroke, and type 2 diabetes [2, 5]. Its pathophysiology involves anatomical vulnerabilities, neuromuscular deficits, and systemic responses to hypoxemia. Animal models and human studies elucidate these mechanisms, guiding therapeutic strategies [2, 6–8]. This chapter aims to comprehensively explore OSA's underlying mechanisms, systemic consequences, and evolving therapeutic approaches informed by this expanding body of evidence.

1.2 Animal Models for Human OSA Study

Animal models are vital for studying OSA under controlled conditions. Rodents, rabbits, dogs, and non-human primates are commonly used.

1.2.1 Rodent Models

Rodents, especially mice and rats, are cost-effective and genetically manipulable. Chronic intermittent hypoxia (CIH) mimics OSA's hypoxic episodes, inducing oxidative stress, inflammation, and sympathetic overactivation [9–11]. CIH increases reactive oxygen species (ROS), causing endothelial dysfunction and vascular remodeling [12, 13] and impairs hypoglossal nerve function, raising airway collapsibility [6, 14]. These mirror human OSA deficits [8, 14].

1.2.2 Rabbit Models

Rabbits' upper airway anatomy resembles humans', making them useful for studying OSA dynamics. Paralyzing the genioglossus with botulinum toxin increases apnea–hypopnea index (AHI), reflecting human OSA mechanisms [7, 15].

1.2.3 Larger Mammals: Dogs and Non-Human Primates

Dogs and primates model complex OSA responses. Primates under CIH show glucose metabolism and insulin resistance changes akin to human OSA [16]. English bulldogs, a natural OSA model, exhibit sleep patterns and oxygen desaturation similar to humans [17, 18]. Serotonin antagonists (e.g., ondansetron, trazodone) and modafinil show therapeutic potential [19–21]. However, bulldogs' daytime obstruction from brachycephalic obstructive airway syndrome (BOAS) and the different sleep positions between dogs and humans limit their human OSA mimicry [22, 23].

1.3 Naturally Occurring Obstructive Sleep Apnea in Dogs

1.3.1 Relevance of Canine Models

Dogs serve as unique animal models for studying OSA due to their natural predisposition to OSA-like conditions arising from anatomical factors. Brachycephalic breeds, such as bulldogs and pugs, are prone to structural abnormalities like compressed nasal passages and elongated soft palates, which contribute to upper airway obstruction during sleep. Research has shown that these breeds experience significantly higher rates of sleep-related breathing disorder (SRBD), including obstructive events and snoring, compared to non-brachycephalic breeds, highlighting their translational relevance in understanding OSA [24–26].

The study of naturally occurring OSA in dogs allows for the investigation of disease mechanisms without artificial induction, closely mimicking the spontaneous development and progression of OSA seen in humans.

1.3.2 Canine Brachycephalic Obstructive Airway Syndrome Versus Human OSA

Dogs are obligate nasal breathers during sleep, and severe nasal or nasopharyngeal obstruction may cause them to wake and breathe through their mouths [27]. Chronic cases lead to adaptations like maintaining an elevated chin or sleeping with mouths open, resulting in sleep-related breathing disorder (SRBD), which can escalate to choking fits and severe sleep apnea [17, 27–30]. Breeds most affected include French bulldogs, English bulldogs, pugs, Chihuahuas, Pomeranians, and Cavalier King Charles Spaniels, all classified as brachycephalic and often suffering from BOAS.

BOAS is a common condition in these breeds, resulting from selective breeding for traits that create anatomical imbalances. This leads to upper airway obstruction due to oversized soft tissues relative to the bony structure. Lesions associated with BOAS include an elongated and thickened soft palate, macroglossia, hypoplastic larynx and trachea, nasal septal deviation, laryngomalacia, hypertrophied and aberrant nasal turbinates, and stenotic nares and nasal vestibules (Fig. 1.1) [31–34]. Sleep exacerbates obstruction as muscle relaxation occurs, and OSA—a subtype of SRBD—has been observed in dogs with BOAS. Approximately 56% of these dogs exhibit signs of SRBD, with 48% showing detectable sleep apnea [27].

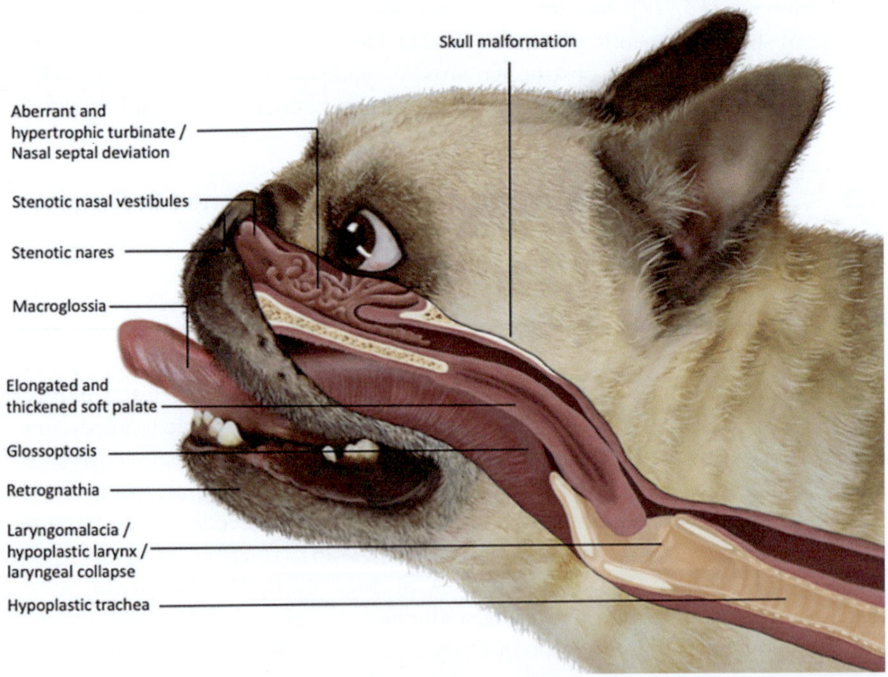

Fig. 1.1 Anatomical lesions of brachycephalic obstructive airway syndrome

In addition, obesity and comorbid cardiac diseases are prevalent among affected dogs, leading to intrathoracic pressure changes and potential heart failure. Research on metabolic and hematological changes in brachycephalic dogs suggests parallels with human OSA, including hypomagnesemia [35], hypercoagulability [36], hypertension, lower arterial oxygen partial pressure (PaO2), and elevated carbon dioxide partial pressure (PaCO2) [37]. Although chronic airway obstruction is more pronounced in these dogs, studying the long-term effects of BOAS and OSA could provide valuable insights into human OSA.

1.3.3 Diagnostic Tests for Canine OSA

The gold standard for diagnosing OSA in humans is in-hospital polysomnography (PSG). Although similar techniques have been developed for canine research [17, 19, 22, 38], PSG is not widely used in clinical settings for dogs due to compliance challenges and instrumentation limitations. Standard human PSG setups involve multiple leads and a nasal flow detector, but analogous equipment for dogs often results in only partial recordings [17].

Recently, a modified canine PSG system has been validated for use with client-owned dogs, offering non-invasive surface electrodes for electroencephalogram (EEG), electromyogram (EMG), and electrooculogram (EOG) monitoring. However, this system does not record nasal airflow, limiting its utility for comprehensive OSA studies [39, 40].

Whole-body barometric plethysmography (WBBP) has emerged as a valuable non-invasive technique for assessing respiratory function in small animals, including during sleep. A case series demonstrated WBBP's ability to detect sleep apnea in Cavalier King Charles Spaniels, with observed respiratory patterns resembling those seen in humans (Fig. 1.2) [30].

A recent study from Finland adapted a portable neckband device to measure obstructive respiratory event index (OREI) and snore percentage in dogs. However, it lacks critical features such as EEG and respiratory flow measurements, raising concerns about its accuracy compared to PSG [41]. This highlights the urgent need for reliable diagnostic tests specifically for canine OSA.

Imaging techniques are used to evaluate upper airway anatomy rather than diagnose OSA. While cephalometric analysis and ultrasonography are effective in humans, they are challenging in dogs. Instead, awake video fluoroscopy and computed tomography (CT) under anesthesia are employed to identify obstructed sites dynamically and statically. Liu et al. noted that extubated CT scanning mimicking natural sleep positions is vital for detecting lesions causing obstruction during sleep, particularly in dogs with positional pharyngeal collapse [42].

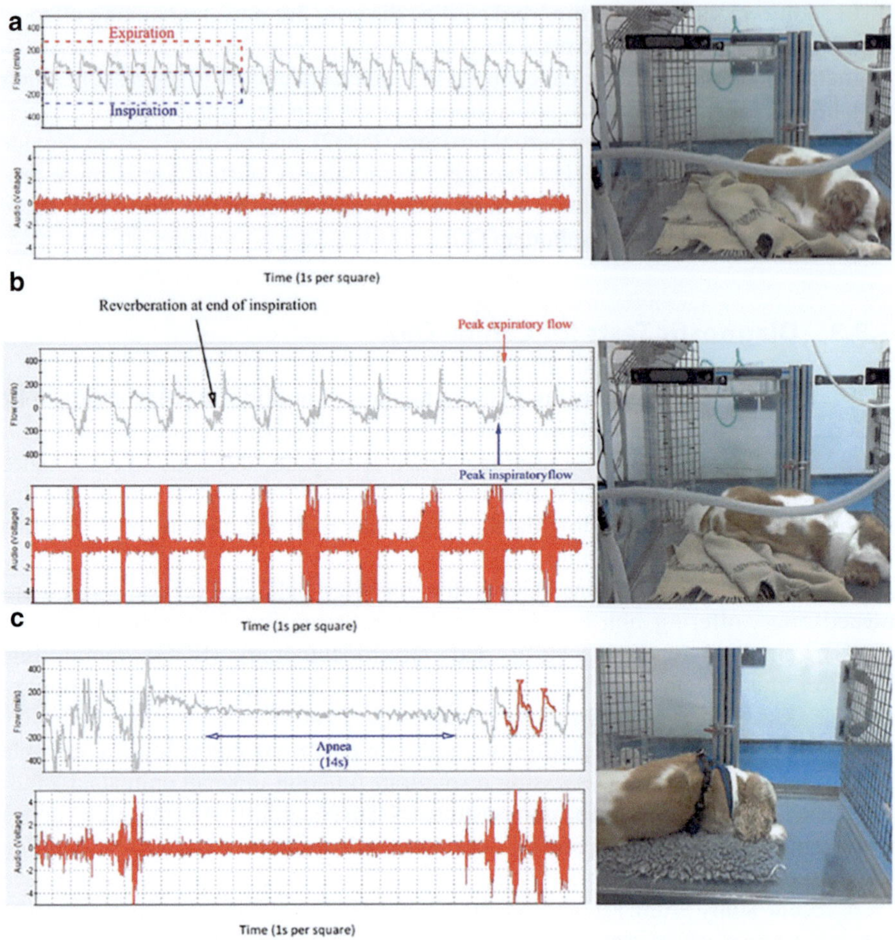

Fig. 1.2 Examples of using whole-body barometric plethysmography (WBBP) detecting obstructive sleep apnea (OSA) in dogs. (**a**) Awake respiratory trace without respiratory noise; (**b**) obstructive respiratory trace during sleep with high peak expiratory flows and inspiratory noise (snoring); (**c**) a 14-second apnea period during sleep. Time is marked as one second per square on the graphs. (Figure adapted from Hinchliffe et al. [30]. Result section, Fig. 1.1)

1.3.4 Pathophysiological Features

In brachycephalic dogs with naturally occurring OSA, researchers have observed many hallmarks of the disorder:

Intermittent Hypoxia Similar to humans with OSA, affected dogs experience repeated episodes of oxygen desaturation during sleep. Studies using brachycephalic breeds have confirmed increased rates of obstructive respiratory events and hypoxemia during sleep, as seen in human OSA patients [43, 44].

Cardiovascular Changes Studies have documented elevated blood pressure levels and signs of left ventricular hypertrophy in brachycephalic dogs with severe airway obstruction. Additionally, BOAS has been associated with a hypercoagulable state, similar to cardiovascular complications observed in human OSA [35, 44].

Neuromuscular Deficits Reduced activity of pharyngeal dilator muscles has been noted during sleep phases prone to airway collapse. Morphological abnormalities in the upper airway muscles of brachycephalic breeds, such as English Bulldogs, further impair their activity and contribute to airway obstruction during sleep [42, 43].

These findings highlight the translational potential of canine models for studying both the natural history and therapeutic interventions for OSA.

1.3.5 Implications for Human Studies

The parallels between naturally occurring canine OSA and human disease provide valuable insights into shared pathophysiological mechanisms. For example:

Anatomical Vulnerabilities Brachycephalic dogs exhibit anatomical vulnerabilities similar to craniofacial abnormalities in humans, such as retrognathia and maxillary deficiency. Structural features like shortened muzzles and altered skull shapes lead to significant airway obstruction during respiration, akin to conditions in humans with craniofacial anomalies [45–47]. These anatomical changes contribute to health issues, including obstructive sleep apnea, by creating a mismatch between airway size and surrounding soft tissues. In these dogs, the shortening of the facial skeleton without a proportional reduction in soft tissues results in airway obstruction and conditions like BOAS. This mismatch increases airway resistance and negative pressure, exacerbating respiratory dysfunction [46, 48].

Therapeutic Testing Canine models are increasingly recognized for their utility in testing novel therapies for OSA. These models facilitate the evaluation of interventions such as hypoglossal nerve stimulation and pharmacological agents targeting neuromuscular control, leveraging anatomical and physiological similarities between dogs and humans to assess treatment efficacy. Research has shown that canine models can effectively mimic human OSA conditions, enabling the exploration of pulmonary remodeling and other complications associated with the disorder [49]. Additionally, studies confirm that these models reproduce key characteristics of human OSA, including apneas and hypersomnolence, validating their use in therapeutic assessments [41].

Canine models have also been instrumental in evaluating surgical interventions and pharmacological treatments for OSA. They have been used to assess the effectiveness of continuous positive airway pressure (CPAP) treatments and other therapeutic approaches, contributing to a better understanding of how these interventions can improve patient's outcomes [50, 51]. The ability to conduct such research in canine subjects underscores the importance of these models in advancing therapeutic options for OSA.

Longitudinal Studies Longitudinal studies are valuable for understanding OSA as they allow researchers to observe disease progression over time in real-world conditions. Unlike induced animal models, naturally occurring canine OSA provides a more authentic representation of how the condition develops, enabling comprehensive evaluations of environmental and genetic factors influencing OSA. Research highlights the significance of these studies in revealing the natural history of OSA and its long-term health implications, emphasizing the need for continuous monitoring [52, 53]. Additionally, longitudinal investigations using canine models can yield critical data on the effectiveness of therapeutic interventions and their impact on disease trajectory, further supporting their role in translational research [54].

1.4 Pathophysiological Consequences of OSA

1.4.1 Intermittent Hypoxia and Oxidative Stress

Intermittent hypoxia (IH) is a significant pathophysiological consequence of OSA, resulting from repeated airway obstruction during sleep. IH induces oxidative stress through the overproduction of ROS, which damage cellular components and impair normal physiological functions. Research links IH to systemic inflammation and metabolic dysfunction; for example, IH leads to a pro-inflammatory phenotype in adipose tissue, correlating with insulin resistance and metabolic disorders associated with OSA [55]. Additionally, IH activates pathways that increase ROS production, contributing to cellular injury in various tissues [56]. This oxidative stress is crucial in developing cardiovascular and metabolic complications in OSA patients, as it can lead to endothelial dysfunction, a factor in cardiovascular diseases like hypertension and ischemic heart disease [57, 58].

In rodent models exposed to IH, elevated ROS levels have been associated with endothelial dysfunction and vascular remodeling. Studies show that IH exposure significantly increases ROS production, impairing endothelial function and vascular health, evidenced by reduced vasodilatory responses [59, 60]. In humans, biomarkers of oxidative stress such as 8-isoprostane and malondialdehyde are consistently elevated in OSA patients [61], indicating that oxidative damage contributes to vascular dysfunction and broader systemic issues, including metabolic and neurocognitive impairments. Antioxidant therapies like N-acetylcysteine (NAC) have shown promise in reducing oxidative stress markers and improving endothelial function in early studies [62].

1.4.2 Systemic Inflammation

IH triggers systemic inflammation, which plays a key role in mediating comorbidities associated with OSA. Inflammatory cytokines such as tumor necrosis factor-alpha (TNF-α), interleukin-6 (IL-6), and interleukin-1 beta (IL-1β) are upregulated in both animal models and human patients with OSA. This inflammation is primarily due to chronic IH, which induces oxidative stress and activates inflammatory pathways, contributing to cardiovascular and metabolic complications [61, 63].

These pro-inflammatory mediators worsen endothelial dysfunction, insulin resistance, and adipose tissue inflammation, creating a feedback loop that accelerates disease progression. Animal studies have shown that inhibiting these cytokines can alleviate OSA-related metabolic and cardiovascular issues; for instance, blocking IL-6 signaling in rodents reduces vascular inflammation and improves insulin sensitivity [64]. Translational research supports these findings, revealing elevated serum levels of TNF-α and IL-6 correlate with OSA severity and treatment response [65].

1.4.3 Cardiovascular Dysfunction

OSA significantly impacts the cardiovascular system through mechanisms such as sympathetic overactivation, oxidative stress, and systemic inflammation. During apneic episodes, sympathetic nerve activation increases vasoconstriction and blood pressure, contributing to cardiovascular strain [66]. The cyclical hypoxia and reoxygenation in OSA patients generate oxidative stress, damaging vascular endothelium and promoting inflammation [67, 68]. This interplay exacerbates existing cardiovascular conditions and elevates the risk of developing new diseases related to OSA.

Repeated hypoxemia activates the sympathetic nervous system, leading to sustained increases in blood pressure and heart rate variability. In animal models, CIH induces hypertension, ventricular hypertrophy, and arrhythmias, mirroring clinical observations in humans [69]. Studies show that OSA is an independent risk factor for cardiovascular diseases, including hypertension, atrial fibrillation, and heart failure [70, 71].

CPAP therapy effectively reduces blood pressure and improves cardiac function, demonstrating the reversibility of these changes with treatment [72]. However, long-term outcomes often depend on patient adherence to therapy, which remains a significant challenge [73].

1.4.4 Neurocognitive Impairments

Neurocognitive deficits are common in OSA, resulting from sleep fragmentation, intermittent hypoxia, and reduced cerebral perfusion. Patients often report memory impairments, decreased attention span, and executive function difficulties [74]. Animal studies indicate that CIH leads to hippocampal neuronal apoptosis and impaired synaptic plasticity, providing a mechanistic basis for cognitive deficits in humans [75, 76].

Neuroimaging studies reveal significant structural and functional changes in the hippocampus and prefrontal cortex, areas critical for memory and cognition [77, 78]. These alterations emphasize the need for early intervention to prevent irreversible neurocognitive decline. Changes in the hippocampus and prefrontal cortex can lead to cognitive impairments if not addressed promptly [79]. Techniques like fMRI have been essential in mapping these changes, highlighting how early brain modifications can impact cognitive development over time [77, 78]. The evidence underscores the importance of timely therapeutic strategies to mitigate potential long-term cognitive deficits.

1.5 Therapeutic Approaches

1.5.1 CPAP

CPAP remains the gold standard for treating OSA by maintaining airway patency throughout the respiratory cycle. Numerous studies demonstrate CPAP's efficacy in reducing the apnea–hypopnea index, improving sleep quality, and mitigating cardiovascular and metabolic consequences. Randomized controlled trials indicate that CPAP reduces daytime sleepiness and lowers blood pressure in hypertensive patients [80–82].

Despite these benefits, CPAP adherence is often suboptimal, with many patients discontinuing therapy due to discomfort and inconvenience. Efforts to improve adherence include developing more comfortable masks, pressure titration algorithms, behavioral interventions, and novel CPAP devices with advanced technologies such as auto-titration and heated humidification [83, 84].

1.5.2 Pharmacological Therapies

The search for pharmacological alternatives to CPAP has led to exploring several therapeutic targets. These include antioxidant therapies like NAC and ascorbic acid, which aim to counteract oxidative stress and enhance vascular function. Additionally, anti-inflammatory agents such as TNF-α inhibitors and IL-6 antagonists are being investigated for their potential to reduce systemic inflammation and improve metabolic outcomes. Emerging therapies targeting central respiratory regulation include orexin receptor antagonists and serotonin reuptake inhibitors, which modulate respiratory drive and neuromuscular control. Although these pharmacological approaches are still in early development, they represent promising options for patient's intolerant to CPAP or other device-based therapies [85, 86].

1.5.3 Surgical Device-Based Interventions

Surgical interventions such as uvulopalatopharyngoplasty (UPPP) and maxillomandibular advancement (MMA) are important options for patients with anatomical contributors to OSA [87, 88]. MMA has demonstrated high success rates, with studies showing an average airway volume increase of approximately 43.75% and a significant reduction in the AHI [89, 90]. This procedure is also associated with improved oximetric indicators and overall quality of life for patients with OSA [91].

Hypoglossal nerve stimulation has emerged as a minimally invasive alternative, electrically stimulating the hypoglossal nerve to activate the genioglossus muscle and prevent airway collapse during sleep. Clinical trials indicate meaningful reductions in AHI and improvements in sleep quality, with a favorable safety profile, making it an appealing option for those that are intolerant to CPAP therapy [92, 93].

1.5.4 Lifestyle Modifications

Lifestyle changes, including weight loss, positional therapy, and alcohol avoidance, are crucial in managing OSA. Weight loss has been shown to reduce the AHI and improve symptoms in obese patients, with dietary-induced weight loss recommended by the American Academy of Sleep Medicine as a key treatment option [94, 95]. Positional therapy, which involves avoiding supine sleeping positions, can significantly reduce airway obstruction in patients with positional OSA [96, 97]. While these interventions are beneficial, they often require sustained behavioral changes and multidisciplinary support for long-term success.

1.5.5 Medical and Surgical Interventions for Canine OSA

Multilevel upper airway corrective surgery is recommended for treating BOAS in dogs with OSA. Surgical interventions typically include procedures such as palatoplasty, partial tonsillectomy, laryngeal ventriculectomy, and turbinate tissue reduction [29, 98]. Pohl et al. reported a significant reduction in sleep apnea incidence from 55% to 3% following such surgeries in pugs and French bulldogs [29].

A case series on Cavalier King Charles Spaniels undergoing upper airway corrective surgery for elongated soft palates and hypertrophied turbinates showed successful resolution of OSA in all cases [30]. Additionally, two single-case reports highlighted the use of Ondansetron in a Chihuahua and a pug with persistent OSA symptoms after surgery; both dogs responded well to the medication, achieving near-complete symptom resolution [99, 100].

Experimental studies have shown that combining trazodone and L-tryptophan can significantly reduce sleep-disordered breathing in English bulldogs [20]. Comorbidities such as hypothyroidism, cardiovascular diseases, obesity, rhinitis, and gastrointestinal disorders are frequently observed in dogs with OSA, necessitating concurrent management of these conditions. Further research into clinical presentations and breed-specific treatment protocols is essential for advancing precision treatment approaches for OSA in dogs.

1.6 Conclusion

OSA is a complex disorder characterized by intricate pathophysiological mechanisms and diverse systemic consequences. Research advancements, particularly from animal models and human studies, have clarified the interplay of anatomical, neuromuscular, inflammatory, and metabolic factors involved in OSA. These insights have facilitated the development of diagnostic tools and therapeutic interventions. While CPAP remains the cornerstone treatment, emerging pharmacological therapies, surgical techniques, and innovative devices hold promise for more personalized management approaches.

Despite these advancements, significant challenges remain. These include improving treatment adherence, identifying biomarkers for disease phenotyping, and addressing the global burden of undiagnosed cases. Future research should aim to integrate basic science with clinical insights to effectively tackle this pervasive disorder and improve outcomes for millions affected.

The shared characteristics of OSA in both humans and dogs underscore the importance of precision medicine, particularly in focusing on treatable traits within this heterogeneous disease [101]. However, the genetic complexity of brachycephalic dogs complicates their role as direct experimental models for human OSA. Instead, adopting a "one-medicine" approach, leveraging insights from both species, may offer greater benefits for advancing understanding and treatment options for OSA in both humans and dogs.

References

1. Franklin KA, Lindberg E. Obstructive sleep apnea is a common disorder in the population—a review on the epidemiology of sleep apnea. J Thorac Dis. 2015;7(8):1311–22.
2. Ryan S. Mechanisms of cardiovascular disease in obstructive sleep apnoea. J Thorac Dis. 2018;10(Suppl 34):S4201–11.
3. Dempsey JA, Veasey SC, Morgan BJ, O'Donnell CP. Pathophysiology of sleep apnea. Physiol Rev. 2010;90(1):47–112.
4. Peppard PE, Young T, Barnet JH, Palta M, Hagen EW, Hla KM. Increased prevalence of sleep-disordered breathing in adults. Am J Epidemiol. 2013;177(9):1006–14.
5. Parish JM, Somers VK. Obstructive sleep apnea and cardiovascular disease. Mayo Clin Proc. 2004;79(8):1036–46.
6. Kim LJ, Freire C, Fleury Curado T, Jun JC, Polotsky VY. The role of animal models in developing pharmacotherapy for obstructive sleep apnea. J Clin Med. 2019;8(12):2049.
7. Lee MC, Lee CH, Hong SL, Kim SW, Lee WH, Lim JY, Joe S, Yoon IY, Kim JW. Establishment of a rabbit model of obstructive sleep apnea by paralyzing the genioglossus. JAMA Otolaryngol Head Neck Surg. 2013;139(8):834–40.
8. Pham LV, Schwartz AR. The pathogenesis of obstructive sleep apnea. J Thorac Dis. 2015;7(8):1358–72.

9. Garvey JF, Taylor CT, McNicholas WT. Cardiovascular disease in obstructive sleep apnoea syndrome: the role of intermittent hypoxia and inflammation. Eur Respir J. 2009;33(5):1195–205.
10. Prabhakar NR, Peng YJ, Nanduri J. Hypoxia-inducible factors and obstructive sleep apnea. J Clin Invest. 2020;130(10):5042–51.
11. Turnbull CD. Intermittent hypoxia, cardiovascular disease and obstructive sleep apnoea. J Thorac Dis. 2018;10(Suppl 1):S33–9.
12. Arnaud C, Billoir E, de Melo Junior AF, Pereira SA, O'Halloran KD, Monteiro EC. Chronic intermittent hypoxia-induced cardiovascular and renal dysfunction: from adaptation to maladaptation. J Physiol. 2023;601(24):5553–77.
13. Janaszak-Jasiecka A, Siekierzycka A, Płoska A, Dobrucki IT, Kalinowski L. Endothelial dysfunction driven by hypoxia—the influence of oxygen deficiency on NO bioavailability. Biomolecules. 2021;11(7):982.
14. Lv R, Liu X, Zhang Y, Dong N, Wang X, He Y, Yue H, Yin Q. Pathophysiological mechanisms and therapeutic approaches in obstructive sleep apnea syndrome. Signal Transduct Target Ther. 2023;8(1):218.
15. Yu MS, Jung NR, Choi KH, Choi K, Lee BJ, Chung YS. An animal model of obstructive sleep apnea in rabbit. Laryngoscope. 2014;124(3):789–96.
16. Martins FO, Sacramento JF, Olea E, Melo BF, Prieto-Lloret J, Obeso A, Rocher A, Matafome P, Monteiro EC, Conde SV. Chronic intermittent hypoxia induces early-stage metabolic dysfunction independently of adipose tissue deregulation. Antioxidants. 2021;10(8):1233.
17. Hendricks JC, Kline LR, Kovalski RJ, O'Brien JA, Morrison AR, Pack AI. The English bulldog: a natural model of sleep-disordered breathing. J Appl Physiol. 1987;63:1344–50.
18. Brooks D, Horner RL, Kozar LF, Render-Teixeira CL, Phillipson EA. Obstructive sleep apnea as a cause of systemic hypertension. Evidence from a canine model. J Clin Invest. 1997;99(1):106–9.
19. Veasey SC, Chachkes J, Fenik P, Hendricks JC. The effects of ondansetron on sleep-disordered breathing in the English bulldog. Sleep. 2001;24(2):155–60.
20. Veasey SC, Fenik P, Panckeri K, Pack AI, Hendricks JC. The effects of trazodone with L-tryptophan on sleep-disordered breathing in the English bulldog. Am J Respir Crit Care Med. 1999;160(5 Pt 1):1659–67.
21. Panckeri KA, Schotland HM, Pack AI, Hendricks JC. Modafinil decreases hypersomnolence in the English bulldog, a natural animal model of sleep-disordered breathing. Sleep. 1996;19(8):626–31.
22. Hendricks JC, Kovalski RJ, Kline LR. Phasic respiratory muscle patterns and sleep-disordered breathing during rapid eye movement sleep in the English bulldog. Am Rev Respir Dis. 1991;144(5):1112–20.
23. Hendricks JC, Petrof BJ, Panckeri K, Pack AI. Upper airway dilating muscle hyperactivity during non-rapid eye movement sleep in English bulldogs. Am Rev Respir Dis. 1993;148(1):185–94.
24. Horner RL, Brooks D, Kozar LF, Leung E, Hamrahi H, Render-Teixeira CL, Makino H, Kimoff RJ, Phillipson EA. Sleep architecture in a canine model of obstructive sleep apnea. Sleep. 1998;21(8):847–58.
25. Mondino A, Delucchi L, Moeser A, Cerdá-González S, Vanini G. Sleep disorders in dogs: a pathophysiological and clinical review. Top Companion Anim Med. 2021;43:100516.
26. O'Donnell CP, King ED, Schwartz AR, Robotham JL, Smith PL. Relationship between blood pressure and airway obstruction during sleep in the dog. J Appl Physiol. 1994;77(4):1819–28.
27. Roedler FS, Pohl S, Oechtering GU. How does severe brachycephaly affect dog's lives? Results of a structured preoperative owner questionnaire. Vet J. 2013;198(3):606–10.
28. Hendricks JC. Brachycephalic airway syndrome. Vet Clin North Am Small Anim Pract. 1992;22(5):1145–53.
29. Pohl S, Roedler FS, Oechtering GU. How does multilevel upper airway surgery influence the lives of dogs with severe brachycephaly? Results of a structured pre- and postoperative owner questionnaire. Vet J. 2016;210:39–45.

30. Hinchliffe TA, Liu N-C, Ladlow J. Sleep-disordered breathing in the cavalier King Charles spaniel: a case series. Vet Surg. 2019;48(4):497–504.
31. Aron DN, Crowe DT. Upper airway obstruction. General principles and selected conditions in the dog and cat. Vet Clin North Am Small Anim Pract. 1985;15(5):891–917.
32. Torrez CV, Hunt GB. Results of surgical correction of abnormalities associated with brachycephalic airway obstruction syndrome in dogs in Australia. J Small Anim Pract. 2006;47:150–4.
33. Riecks TW, Birchard SJ, Stephens JA. Surgical correction of brachycephalic syndrome in dogs: 62 cases (1991–2004). J Am Vet Med Assoc. 2007;230:1324–8.
34. Dupre G, Heidenreich D. Brachycephalic syndrome. Vet Clin North Am Small Anim Pract. 2016;46(4):691–707.
35. Mellema MS, Hoareau GL. Hypomagnesemia in brachycephalic dogs. J Vet Intern Med. 2014;28(5):1418–23.
36. Hoareau G, Mellema M. Pro-coagulant thromboelastographic features in the bulldog. J Small Anim Pract. 2015;56(2):103–7.
37. Hoareau GL, Jourdan G, Mellema M, Verwaerde P. Evaluation of arterial blood gases and arterial blood pressures in brachycephalic dogs. J Vet Intern Med. 2012;26(4):897–904.
38. Kis A, Szakadát S, Kovács E, Gácsi M, Simor P, Gombos F, Topál J, Miklósi A, Bódizs R. Development of a non-invasive polysomnography technique for dogs (Canis familiaris). Physiol Behav. 2014;130:149–56.
39. Reicher V, Kis A, Simor P, Bódizs R, Gombos F, Gácsi M. Repeated afternoon sleep recordings indicate first-night-effect-like adaptation process in family dogs. J Sleep Res. 2020;29(6):e12998.
40. Mondino A, Ludwig C, Menchaca C, Russell K, Simon KE, Griffith E, Kis A, Lascelles BDX, Gruen ME, Olby NJ. Development and validation of a sleep questionnaire, SNoRE 3.0, to evaluate sleep in companion dogs. Sci Rep. 2023;13(1):13340.
41. Niinikoski I, Himanen SL, Tenhunen M, Lilja-Maula L, Rajamäki MM. Description of a novel method for detection of sleep-disordered breathing in brachycephalic dogs. J Vet Intern Med. 2023;37(4):1475–81.
42. Liu N-C, Riggs J, Ladlow J, Owen L, editors. The effect of head position during computed tomography on nasopharyngeal patency in brachycephalic dogs with and without obstructive sleep disorders. European College of Veterinary Surgeons Annual Scientific Meeting. Porto: European College of Veterinary Surgeons; 2022.
43. Barker DA, Tovey E, Jeffery A, Blackwell E, Tivers MS. Owner reported breathing scores, accelerometry and sleep disturbances in brachycephalic and control dogs: a pilot study. Vet Rec. 2021;189(4):e135.
44. Crane C, Rozanski EA, Abelson AL, deLaforcade A. Severe brachycephalic obstructive airway syndrome is associated with hypercoagulability in dogs. J Vet Diagn Invest. 2017;29(4):570–3.
45. di Virgilio F, Guelette V, Prata D, Boulouha L, Santos M, Belluzzi E, Caraty J, Bongartz A, Deneuche A. Comparative histological analysis of the alae nasi in French bulldogs with brachycephalic obstructive airway syndrome and unaffected non-brachycephalic dogs. Vet Rec. 2023;193(4):e3058.
46. Packer RM, Hendricks A, Tivers MS, Burn CC. Impact of facial conformation on canine health: brachycephalic obstructive airway syndrome. PLoS One. 2015;10(10):e0137496.
47. Packer RM, Tivers MS. Strategies for the management and prevention of conformation-related respiratory disorders in brachycephalic dogs. Vet Med (Auckl). 2015;6:219–32.
48. Meola SD. Brachycephalic airway syndrome. Top Companion Anim Med. 2013;28(3):91–6.
49. Ding X, Yu C, Liu Y, Yan S, Li W, Wang D, Sun L, Han Y, Li M, Zhang S, Yun F, Zhao H, Li Y. Chronic obstructive sleep apnea accelerates pulmonary remodeling via TGF-β/miR-185/CoLA1 signaling in a canine model. Oncotarget. 2016;7(36):57545–55.
50. Horner RL. Pathophysiology of obstructive sleep apnea. J Cardiopulm Rehabil Prev. 2008;28(5):289–98.

51. Neuzeret PC, Gormand F, Reix P, Parrot S, Sastre JP, Buda C, Guidon G, Sakai K, Lin JS. A new animal model of obstructive sleep apnea responding to continuous positive airway pressure. Sleep. 2011;34(4):541–8.
52. Bonsignore MR, Mazzuca E, Baiamonte P, Bouckaert B, Verbeke W, Pevernagie DA. REM sleep obstructive sleep apnoea. Eur Respir Rev. 2024;33(171):230166.
53. Somers VK, White DP, Amin R, Abraham WT, Costa F, Culebras A, Daniels S, Floras JS, Hunt CE, Olson LJ, Pickering TG, Russell R, Woo M, Young T. Sleep apnea and cardiovascular disease: an American Heart Association/American College of Cardiology Foundation Scientific Statement. Circulation. 2008;118(10):1080–111.
54. Chien WC, Lin CW, Liu CK, Chen SL, Chou MC, Hsu CY. The associations between polysomnographic parameters and memory impairment among patients with obstructive sleep apnea: a 10-year hospital-based longitudinal study. Biomedicine. 2023;11(2):621.
55. Murphy AM, Thomas A, Crinion SJ, Kent BD, Tambuwala MM, Fabre A, Pepin JL, Roche HM, Arnaud C, Ryan S. Intermittent hypoxia in obstructive sleep apnoea mediates insulin resistance through adipose tissue inflammation. Eur Respir J. 2017;49(4):1601731.
56. Song R, Baker TL, Watters JJ, Kumar S. Obstructive sleep apnea-associated intermittent hypoxia-induced immune responses in males, pregnancies, and offspring. Int J Mol Sci. 2024;25(3):1852.
57. Yamauchi M, Kimura H. Oxidative stress in obstructive sleep apnea: putative pathways to the cardiovascular complications. Antioxid Redox Signal. 2008;10(4):755–68.
58. Gozal D, Kheirandish-Gozal L. Cardiovascular morbidity in obstructive sleep apnea: oxidative stress, inflammation, and much more. Am J Respir Crit Care Med. 2008;177(4):369–75.
59. Chu A, Gozal D, Cortese R, Wang Y. Cardiovascular dysfunction in adult mice following postnatal intermittent hypoxia. Pediatr Res. 2015;77(3):425–33.
60. Tuleta I, França CN, Wenzel D, Fleischmann B, Nickenig G, Werner N, Skowasch D. Hypoxia-induced endothelial dysfunction in apolipoprotein E-deficient mice; effects of infliximab and L-glutathione. Atherosclerosis. 2014;236(2):400–10.
61. Lavalle S, Masiello E, Iannella G, Magliulo G, Pace A, Lechien JR, Calvo-Henriquez C, Cocuzza S, Parisi FM, Favier V, Bahgat AY, Cammaroto G, La Via L, Gagliano C, Caranti A, Vicini C, Maniaci A. Unraveling the complexities of oxidative stress and inflammation biomarkers in obstructive sleep apnea syndrome: a comprehensive review. Life (Basel). 2024;14(4):425.
62. Eisele HJ, Markart P, Schulz R. Obstructive sleep apnea, oxidative stress, and cardiovascular disease: evidence from human studies. Oxidative Med Cell Longev. 2015;2015:608438.
63. Gabryelska A, Łukasik ZM, Makowska JS, Białasiewicz P. Obstructive sleep apnea: from intermittent hypoxia to cardiovascular complications via blood platelets. Front Neurol. 2018;9:635.
64. Koritala BSC, Gaspar LS, Bhadri SS, Massie KS, Lee YY, Paulose J, Smith DF. Murine proinflammatory responses to acute and sustained intermittent hypoxia: implications for obstructive sleep apnea research. Laryngoscope. 2024;134(Suppl 4):S1–S11.
65. Kheirandish-Gozal L, Gozal D. Obstructive sleep apnea and inflammation: proof of concept based on two illustrative cytokines. Int J Mol Sci. 2019;20(3):459.
66. Shamsuzzaman AS, Gersh BJ, Somers VK. Obstructive sleep apnea: implications for cardiac and vascular disease. JAMA. 2003;290(14):1906–14.
67. Jean-Louis G, Zizi F, Clark LT, Brown CD, McFarlane SI. Obstructive sleep apnea and cardiovascular disease: role of the metabolic syndrome and its components. J Clin Sleep Med. 2008;4(3):261–72.
68. Tietjens JR, Claman D, Kezirian EJ, De Marco T, Mirzayan A, Sadroonri B, Goldberg AN, Long C, Gerstenfeld EP, Yeghiazarians Y. Obstructive sleep apnea in cardiovascular disease: a review of the literature and proposed multidisciplinary clinical management strategy. J Am Heart Assoc. 2019;8(1):e010440.
69. Takahashi K, Ueda S, Kobayashi T, Nishiyama A, Fujisawa Y, Sugaya T, Shiota S, Takahashi K, Gohda T, Horikoshi S, Suzuki Y. Chronic intermittent hypoxia-mediated renal sympathetic nerve activation in hypertension and cardiovascular disease. Sci Rep. 2018;8(1):17926.

70. Bao G, Metreveli N, Fletcher EC. Acute and chronic blood pressure response to recurrent acoustic arousal in rats. Am J Hypertens. 1999;12(5):504–10.
71. Xie A, Skatrud JB, Puleo DS, Morgan BJ. Exposure to hypoxia produces long-lasting sympathetic activation in humans. J Appl Physiol. 2001;91(4):1555–62.
72. Lui MM, Tse HF, Lam DC, Lau KK, Chan CW, Ip MS. Continuous positive airway pressure improves blood pressure and serum cardiovascular biomarkers in obstructive sleep apnoea and hypertension. Eur Respir J. 2021;58(5):2003687.
73. Franke K, Loffler KA, Nicholls SJ, Anderson CS, Cowan BR, McEvoy RD, Young AA, Verjans JW. Effects of CPAP treatment on cardiac structure and function in individuals with obstructive sleep apnea and cardiovascular disease: a prospective 6-month randomised control trial. Euro Heart J. 2020;41(Suppl. 2):ehaa946.2889.
74. Lal C, Strange C, Bachman D. Neurocognitive impairment in obstructive sleep apnea. Chest. 2012;141(6):1601–10.
75. Khuu MA, Nallamothu T, Castro-Rivera CI, Arias-Cavieres A, Szujewski CC, Garcia Iii AJ. Stage-dependent effects of intermittent hypoxia influence the outcome of hippocampal adult neurogenesis. Sci Rep. 2021;11(1):6005.
76. Yin X, Zhang X, Lv C, Li C, Yu Y, Wang X, Han F. Protocatechuic acid ameliorates neurocognitive functions impairment induced by chronic intermittent hypoxia. Sci Rep. 2015;5:14507.
77. Smith EE, Jonides J. Neuroimaging analyses of human working memory. Proc Natl Acad Sci USA. 1998;95(20):12061–8.
78. Yen C, Lin CL, Chiang MC. Exploring the frontiers of neuroimaging: a review of recent advances in understanding brain functioning and disorders. Life (Basel). 2023;13(7):1472.
79. Harnett NG, Goodman AM, Knight DC. PTSD-related neuroimaging abnormalities in brain function, structure, and biochemistry. Exp Neurol. 2020;330:113331.
80. Chen TA, Mao ST, Lin HC, Liu WT, Tam KW, Tsai CY, Kuan YC. Effects of inspiratory muscle training on blood pressure- and sleep-related outcomes in patients with obstructive sleep apnea: a meta-analysis of randomized controlled trials. Sleep Breath. 2023;27(5):1953–66.
81. Lacroix J, Tatousek J, Den Teuling N, Visser T, Wells C, Wylie P, Rosenberg R, Bogan R. Effectiveness of an intervention providing digitally generated personalized feedback and education on adherence to continuous positive airway pressure: randomized controlled trial. J Med Internet Res. 2023;25:e40193.
82. Li Z, Cai S, Wang J, Chen R. Predictors of the efficacy for daytime sleepiness in patients with obstructive sleep apnea with continual positive airway pressure therapy: a meta-analysis of randomized controlled trials. Front Neurol. 2022;13:911996.
83. Jacobsen AR, Eriksen F, Hansen RW, Erlandsen M, Thorup L, Damgård MB, Kirkegaard MG, Hansen KW. Determinants for adherence to continuous positive airway pressure therapy in obstructive sleep apnea. PLoS One. 2017;12(12):e0189614.
84. Paz Y, Mar H, Castriotta RJ. Duration of positive airway pressure adherence: how much PAP is enough? J Clin Sleep Med. 2022;18(4):969–70.
85. Aboussouan LS, Bhat A, Coy T, Kominsky A. Treatments for obstructive sleep apnea: CPAP and beyond. Cleve Clin J Med. 2023;90(12):755–65.
86. Ayas N, Pépin JL. Pharmacologic therapy for obstructive sleep apnea: are we seeing some light at the end of the tunnel? Am J Respir Crit Care Med. 2023;208(12):1263–4.
87. He M, Yin G, Zhan S, Xu J, Cao X, Li J, Ye J. Long-term efficacy of uvulopalatopharyngoplasty among adult patients with obstructive sleep apnea: a systematic review and meta-analysis. Otolaryngol Head Neck Surg. 2019;161(3):401–11.
88. Tannyhill Iii RJ. Treatment of obstructive sleep apnea using maxillomandibular advancement surgery. Compend Contin Educ Dent. 2023;44(6):326–30.
89. Furundarena-Padrones L, Cabriada-Nuño V, Brunsó-Casellas J, Garcia-Fernandez RI, Castro-Quintas S, Santos-Zorrozúa B, González-Zapico G, Calvo-Guirado JL, De-Carlos-Villafranca F. Correlation between polysomnographic parameters and volumetric changes generated by maxillomandibular advancement surgery in patients with obstructive sleep apnea: a fluid dynamics study. J Clin Sleep Med. 2024;20(3):371–9.

90. Rojo-Sanchis C, Almerich-Silla JM, Paredes-Gallardo V, Montiel-Company JM, Bellot-Arcís C. Impact of bimaxillary advancement surgery on the upper airway and on obstructive sleep apnea syndrome: a meta-analysis. Sci Rep. 2018;8(1):5756.
91. Wan CCJ, Leung YY. The role of skeletal surgical treatment in the management of obstructive sleep apnea: an overview. Front Oral Maxillofac Med. 2022;4:18.
92. Eastwood PR, Barnes M, MacKay SG, Wheatley JR, Hillman DR, Nguyên XL, Lewis R, Campbell MC, Pételle B, Walsh JH, Jones AC, Palme CE, Bizon A, Meslier N, Bertolus C, Maddison KJ, Laccourreye L, Raux G, Denoncin K, Attali V, Gagnadoux F, Launois SH. Bilateral hypoglossal nerve stimulation for treatment of adult obstructive sleep apnoea. Eur Respir J. 2020;55(1):1901320.
93. Schwartz AR, Jacobowitz O, Eisele DW, Mickelson SA, Miller MB, Oliven A, Certal V, Hopp ML, Winslow DH, Huntley TC, Nachlas NE, Pham LV, Gillespie MB, Weeks BH, Lovett EG, Shen J, Malhotra A, Maurer JT. Targeted hypoglossal nerve stimulation for patients with obstructive sleep apnea: a randomized clinical trial. JAMA Otolaryngol Head Neck Surg. 2023;149(6):512–20.
94. Bouloukaki I, Daskalaki E, Mavroudi E, Moniaki V, Schiza SE, Tsiligianni I. A dietary and lifestyle intervention improves treatment adherence and clinical outcomes in overweight and obese patients with obstructive sleep apnea: a randomized, controlled trial. Life (Basel). 2023;13(8):1755.
95. Dobrosielski DA, Papandreou C, Patil SP, Salas-Salvadó J. Diet and exercise in the management of obstructive sleep apnoea and cardiovascular disease risk. Eur Respir Rev. 2017;26(144):160110.
96. Ravesloot MJL, White D, Heinzer R, Oksenberg A, Pépin JL. Efficacy of the new generation of devices for positional therapy for patients with positional obstructive sleep apnea: a systematic review of the literature and meta-analysis. J Clin Sleep Med. 2017;13(6):813–24.
97. Wu B, Neher JO, Safranek S. Positional therapies for obstructive sleep apnea. Am Fam Physician. 2021;103(9):568–9.
98. Liu N-C, Oechtering GU, Adams VJ, Kalmar L, Sargan DR, Ladlow JF. Outcomes and prognostic factors of surgical treatments for brachycephalic obstructive airway syndrome in 3 breeds. Vet Surg. 2017;46(2):271–80.
99. Kopke MA, Wightman P, Ruaux CG. Obstructive sleep apnea in a Chihuahua successfully managed with ondansetron. Clin Case Reports. 2019;7(5):872–6.
100. Villedieu E, Robinson W, Shales C. Suspected obstructive sleep apnea successfully treated with ondansetron in a pug. J Am Anim Hosp Assoc. 2022;58(3):152–5.
101. Pevernagie DA, Gnidovec-Strazisar B, Grote L, Heinzer R, McNicholas WT, Penzel T, et al. On the rise and fall of the apnea-hypopnea index: a historical review and critical appraisal. J Sleep Res. 2020;29(4):e13066.

Home Sleep Apnea Test (HSAT) 2

Febryan Setiawan, Che-Wei Lin, Cheng-Yu Lin, and Meiho Nakayama

Contents

F. Setiawan
Department of Biomedical Engineering, College of Engineering, National Cheng Kung University, Tainan, Taiwan

C.-W. Lin
Department of Biomedical Engineering, College of Engineering, National Cheng Kung University, Tainan, Taiwan

Institute of Gerontology, College of Medicine, National Cheng Kung University, Tainan, Taiwan

Medical Device Innovation Center, National Cheng Kung University, Tainan, Taiwan

Institute of Medical Informatics, College of Electrical Engineering and Computer Science, National Cheng Kung University, Tainan, Taiwan

C.-Y. Lin (✉)
Department of Otolaryngology, National Cheng Kung University Hospital, College of Medicine, National Cheng Kung University, Tainan, Taiwan

Sleep Medicine Center, National Cheng Kung University Hospital, College of Medicine, National Cheng Kung University, Tainan, Taiwan
e-mail: yu621109@ms48.hinet.net

M. Nakayama
Meiho Sleep & Balance Clinic, Nagoya, Japan

H.-Y. Li (ed.), *Snoring and Obstructive Sleep Apnea*,
https://doi.org/10.1007/978-981-95-2648-2_2

Abstract

The use of home sleep apnea testing (HSAT) has been increasingly recognized as a practical and economical alternative to in-laboratory polysomnography (PSG) for diagnosing obstructive sleep apnea (OSA), especially in underserved settings. Challenges associated with PSG—such as high cost, limited accessibility, and patient discomfort—have driven interest in simplified, home-based approaches. Particular attention has been given to electrocardiogram (ECG)-based HSAT, which utilizes heart rate variability and advanced computational algorithms, including machine learning and deep learning, to detect apneic events, distinguish sleep stages, and identify cardiovascular comorbidities such as atrial fibrillation. Although current HSAT devices show promise, limitations in diagnostic accuracy, ease of use, and standardization remain. Recent innovations in algorithmic design, such as transformer-based and object detection models, have enhanced detection capabilities and processing efficiency. Future progress is anticipated through the integration of multimodal physiological signals, improvements in wearable technology, and validation via multicenter clinical studies, ultimately advancing HSAT's role in individualized OSA diagnosis and management.

Keywords

Home sleep apnea testing (HSAT) · Obstructive sleep apnea (OSA) · Sleep stages Atrial fibrillation · Electrocardiogram (ECG) · Artificial intelligence (AI) · Deep learning

2.1 Introduction

2.1.1 Background and Significance

Obstructive sleep apnea (OSA) affects approximately one billion people worldwide and is characterized by repeated upper airway collapse during sleep, leading to reduced or stopped airflow [1]. This condition significantly impacts public health, increasing risks of cardiovascular disease, stroke, motor vehicle accidents, and cognitive impairments, while also lowering quality of life and burdening healthcare systems [1]. OSA prevalence is higher among men, older adults, and individuals with obesity [2, 3]. Despite its global scope, including in developing nations, public awareness and diagnosis rates remain low, underscoring the need for early detection [1].

Home sleep apnea testing (HSAT) has emerged as a cost-effective, convenient alternative to in-lab polysomnography, with devices like the Nox-A1 and Watch-PAT proving effective for diagnosing OSA in high-risk populations [4, 5]. These advancements emphasize the importance of accessible diagnostic tools and highlight the need for further research into their integration into public health systems to enhance OSA management [6].

2.1.2 Limitations of Traditional Polysomnography (PSG)

While polysomnography (PSG) is the gold standard for diagnosing OSA, it has notable limitations [6], including high costs, reliance on specialized equipment and trained personnel, and reduced accessibility, particularly in rural areas. Conducting PSG in clinical settings often causes discomfort, sleep disturbances, and the "first-night effect," further complicating its use. These drawbacks, coupled with long waiting times and scalability issues, have driven the development of simpler and more affordable HSAT methods to enhance accessibility and address these limitations.

2.1.3 Current Challenges in OSA Diagnosis

The diagnosis of OSA is hindered by resource limitations [7, 8], leading to delayed identification and care disparities, especially in underserved and rural areas lacking sleep centers and specialists. Traditional PSG, while the gold standard, requires specialized facilities and trained personnel, limiting accessibility and resulting in long waiting times. To address these barriers, scalable and user-friendly diagnostic tools are needed to facilitate widespread testing and earlier detection. These innovations must balance accuracy with ease of use, making them suitable for diverse healthcare settings, including primary care, particularly in resource-constrained environments.

2.2 The Importance of HSAT

2.2.1 Overview of Sleep Study Levels

Diagnostic methods for OSA are classified into four levels based on complexity and application [9]:

1. Level 1: Comprehensive Laboratory PSG
   - Gold standard in sleep diagnostics.
   - Monitors extensive physiological signals (e.g., electroencephalogram (EEG), electrooculography (EOG), electromyography (EMG), ECG, SpO_2, respiratory, airflow, body position).
   - High accuracy but costly, time-consuming, and inconvenient.
2. Level 2: Portable Multichannel Monitoring
   - Home-based devices recording multiple parameters (e.g., SpO_2, respiratory effort, airflow, heart rate).
   - More accessible than PSG but requires professional setup and analysis.
3. Level 3: Simplified Home-Based Devices
   - Monitors fewer parameters (e.g., SpO_2, respiratory effort, airflow, heart rate, chest/abdominal movement).
   - Cost-effective and user-friendly for initial screening.
4. Level 4: Single- or Dual-Parameter Devices
   - Minimal monitoring (e.g., SpO_2, respiratory effort, heart rate, snoring sound).
   - Highly affordable, easy to use, ideal for preliminary screening.

This stratification allows tailored diagnostic approaches based on patient needs, resource availability, and healthcare settings.

2.2.2 HSAT as a Solution

HSAT provides a cost-effective and convenient alternative to traditional PSG for diagnosing OSA. Conducted at home, it eliminates the need for costly sleep labs and personnel, reducing expenses. Testing in a familiar setting enhances comfort, reduces anxiety, and improves patient compliance. HSAT also addresses accessibility challenges, such as long waiting times and limited resources, especially in underserved or remote areas. By improving affordability, comfort, and reach, HSAT plays a vital role in early OSA detection and management [10, 11].

2.2.3 Existing HSAT Devices

Current HSAT devices offer a convenient, cost-effective alternative for diagnosing OSA, monitoring events like apnea, hypopnea, oxygen saturation, and airflow. However, they often underestimate AHI, lack accuracy compared to PSG for sleep stages and arousals, and face challenges like device variability, complex setups, and inconsistent user experiences. Limited telemedicine integration and the absence of standardized clinical trial guidelines further restrict their utility. Table 2.1 highlights available HSAT devices and their key features [29, 30].

To fully realize their potential, HSAT devices require improvements in diagnostic precision, usability, telemedicine compatibility, and standardized clinical trial protocols, ensuring they can reliably and effectively support home-based OSA diagnosis.

Table 2.1 Commercially available HSAT devices

HSAT Device	Key Features
Alice PDx (Philips) [12]	A portable, clinically validated monitor for detecting sleep apnea and cardiopulmonary dysfunction, known for high diagnostic consistency, but no longer available in the U.S. market
ApneaLink Air (ResMed) [13]	Measures respiratory effort, pulse, oxygen saturation, airflow, and snoring, aiding in OSA diagnosis, though specificity could be improved
WatchPat One (Itamar Medical) [14–17]	Provides seven-channel data with comprehensive analysis and good OSA sensitivity, though specificity needs improvement
MediByte [18]	Compact and portable; measures ECG, oxygen saturation, respiratory effort, and nasal pressure with good sensitivity and specificity for AHI > 15
ApneaLink Plus [19]	Records respiratory effort, pulse, oxygen saturation, and nasal airflow with good specificity validated against polysomnography for home studies
Nox-T3 [20]	Portable OSA monitor validated for adult use in China
Rooti Rx (Rooti Labs) [21]	Compact wearable device with high-resolution ECG for diagnosing cardiac events and sleep apnea, featuring real-time event tagging for improved accuracy
BEATINFO (Singular Wings) [22]	Lightweight, biocompatible wearable monitors heart rate and physiology; waterproof, dustproof, with long battery life; integrates with BEATINFO Health apps
ApneaGraph AG200 [23]	Portable device analyzes cardiorespiratory patterns, sleep position, nasopharyngeal pressure, and temperature, detailing upper airway obstruction sites
SenseEcho [24–27]	Wearable device measuring ECG and respiratory signals; uses semi-supervised machine learning to detect signal quality and abnormalities
Nox-A1 [28]	Type 2 ambulatory device for home sleep tests; records EEG, airflow, and respiratory effort with reliable quality and high patient satisfaction, suitable for diagnosing moderate to severe OSA

2.3 ECG-Based HSAT

2.3.1 Introduction to ECG-Based HSAT

ECG-based HSAT is a promising diagnostic tool for OSA, leveraging heart rate variability (HRV) as markers of apneas, hypopneas, and physiological stress. Its non-invasive nature and compatibility with remote monitoring make it accessible for patients in mobility-limited or remote settings, providing real-time feedback [31–33].

By simultaneously evaluating cardiovascular risks, ECG-based HSAT addresses the frequent overlap between OSA and cardiovascular diseases, enabling comprehensive patient care. Research indicates that single-lead ECG signals, combined with artificial intelligence (AI) and deep learning, can achieve diagnostic accuracy comparable to traditional methods, highlighting its potential for efficient and accessible OSA diagnostics [34–36].

2.3.2 Existing ECG-Based HSAT Devices

MediByte, Rooti Rx, BEATINFO, SenseEcho, and Nox A1 are advanced HSAT devices that integrate ECG monitoring (refer to Table 2.1 for details). These devices assess both respiratory parameters, such as respiratory effort and nasal pressure, and cardiac activity, including detailed ECG signals, enabling simultaneous diagnosis of sleep apnea and cardiovascular risk evaluation. Designed for home use, they emphasize portability, user-friendliness, and AI-driven signal analysis. By combining ECG with other sensors, these devices represent a significant advancement in HSAT technology, providing comprehensive and accurate diagnostic capabilities in clinical and real-world settings.

2.3.3 Drawbacks of Existing ECG-Based HSAT Devices

ECG-based HSAT is a promising method for diagnosing OSA through HRV analysis but is less suitable for patients with severe arrhythmias, where in-lab PSG is preferred [4]. Other limitations include difficulty distinguishing OSA from central sleep apnea (CSA), signal interference, and reliance on total recording time instead of sleep time, which may underestimate AHI.

Challenges such as patient discomfort, improper electrode placement, algorithm variability, lack of standardization, high costs, and reliance on trained personnel limit the usability and accessibility of ECG-based HSAT, especially in low-resource settings. Addressing these issues by improving accuracy, comfort, standardization, and affordability is crucial for optimizing its role in OSA management [37, 38].

2.4 Research Gaps on ECG-Based HSAT

OSA is closely linked to atrial fibrillation (AF), increasing its risk and recurrence through mechanisms like intermittent hypoxia and heightened sympathetic activity [39–41]. However, current HSAT devices often fail to detect arrhythmias essential for diagnosing OSA-associated AF due to limited ECG monitoring capabilities, and their integration with CPAP data for treatment optimization remains underutilized. Future HSAT advancements should enhance ECG monitoring, automate arrhythmia detection, and support personalized management to improve OSA and AF outcomes.

ECG-based HSAT devices struggle to differentiate rapid eye movement (REM) and non-rapid eye movement (NREM) sleep, limiting detection of REM-specific OSA, which poses higher cardiovascular risks [42, 43]. Without EEG monitoring, they often underestimate OSA severity by using total recording time instead of sleep time, hindering personalized treatments like optimized CPAP settings [44]. Advancements in REM/NREM differentiation, targeted therapies, and assessing cognitive and quality-of-life impacts are needed to improve diagnostics and outcomes.

2.5 Innovation on ECG-Based HSAT

2.5.1 Algorithmic Approaches in ECG-Based HSAT

Figure 2.1 highlights AI-driven advancements in ECG-based HSAT, including machine learning, deep learning, transformers, and object detection. These approaches enable real-time detection of apnea types, REM/NREM OSA, AF-related OSA, and OSA vs. CSA, enhancing diagnostic accuracy and accessibility.

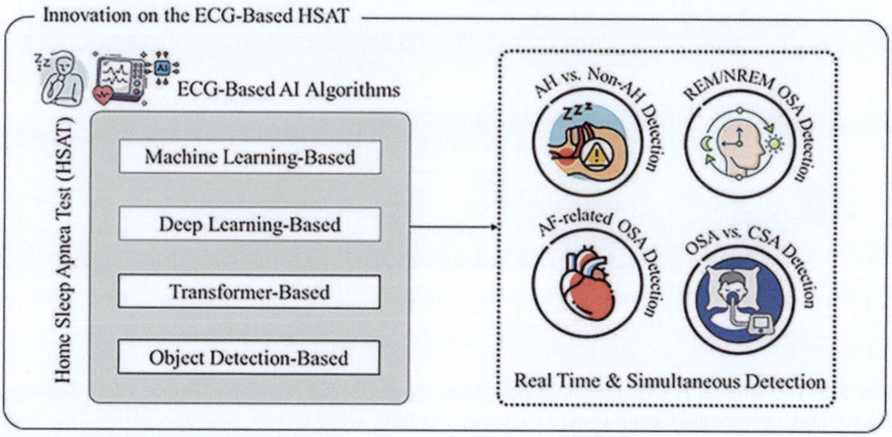

Fig. 2.1 The novel ECG-based HSAT system

2.5.2 Machine Learning-Based Methods [45]

This method enhances ECG-based OSA detection by combining high temporal resolution, advanced feature extraction, and machine learning, as shown in Fig. 2.2. Using PhysioNet Apnea-ECG database (PAED) and National Cheng Kung University Hospital Sleep Center database (NCKUHSCD) datasets, it integrates preprocessing, continuous wavelet transform (CWT), and bag-of-features (BoF), with support vector machine (SVM), *k*-nearest neighbor (KNN), and exclusive lasso (EL) classifiers achieving 91.4% accuracy and 89.8% sensitivity, outperforming previous techniques and advancing HSAT diagnostics.

2.5.3 Deep Learning-Based Methods

Recent advances in HSAT, leveraging lightweight deep learning frameworks, have improved REM/NREM sleep stage and AF-related OSA detection with high accuracy and efficiency, as shown in Fig. 2.2. Key models include

- A 1D-CNN system using single-lead ECG, EMD, and synthetic minority oversampling technique (SMOTE), achieving 91.5% accuracy, 93.4% sensitivity, and 90.5% specificity [46].

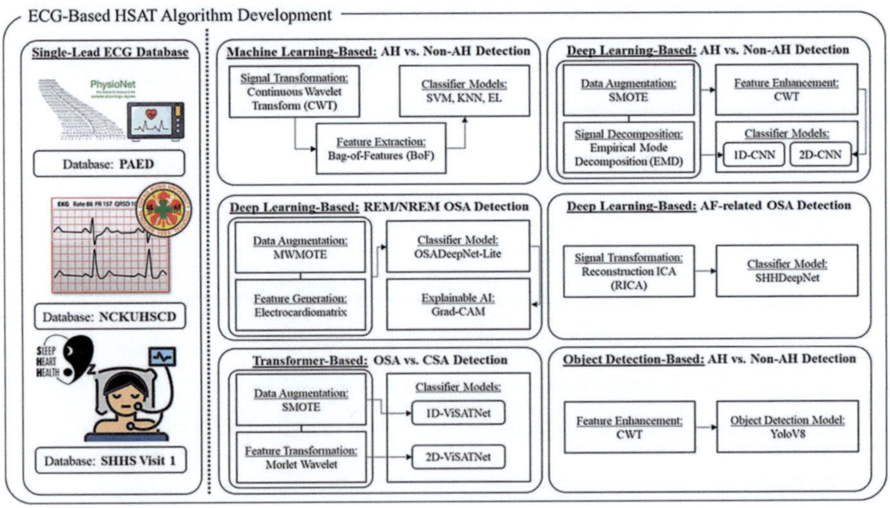

Fig. 2.2 ECG-based HSAT algorithm flowchart. (*Note:* PAED PhysioNet Apnea-ECG database, NCKUHSCD National Cheng Kung University Hospital Sleep Center database, SHHS The Sleep Heart Health Study database, SVM support vector machine, KNN *k*-nearest neighbor, SMOTE synthetic minority oversampling technique, CNN convolutional neural network, MWMOTE majority-weighted minority oversampling technique, Grad-CAM gradient-class activation mapping, ICA independent component analysis, ViSATNet vision sleep apnea transformer network)

- OSADeepNet-Lite, utilizing NCKUHSCD data with ECM visualizations and majority-weighted minority oversampling technique (MWMOTE), detecting REM/NREM-related OSA with 91.42% accuracy and processing a full night's data in under 3 minutes [47].
- SHHDeepNet, combining RICA and a DarkNet-based CNN, achieving 98.22% accuracy and an AUC of 0.9981 for AF-associated OSA detection [47].

These models offer significant potential for early cardiovascular risk assessment, PSG prioritization, and personalized OSA diagnostics.

2.5.3.1 Transformer-Based Methods [47]

Differentiating OSA from CSA is crucial for effective treatment. Figure 2.2 highlights a novel ViT-based system using single-lead ECG with CWT spectrograms to differentiate OSA from CSA. The 2D-ViSATNet model achieved 95.3% accuracy, 93.1% sensitivity, and 96.5% specificity, showcasing the potential of advanced algorithms to enhance HSAT diagnostics and personalized treatment.

2.5.3.2 Object Detection-Based Methods

The integration of YOLOv8 with CWT analysis of single-lead ECG data has been developed for event-based OSA detection, as illustrated in Fig. 2.2. Using NCKUHSCD data, it achieved F1 scores of 0.701 for hypopnea and 0.723 for apnea, with a mAP of 72.9%. Further refinements are needed to improve clinical applicability.

2.5.4 Advantages of the Algorithmic Approaches

2.5.4.1 Simultaneous Detection of REM/NREM OSA and AF-Associated OSA

Advanced ECG-based HSAT algorithms enable simultaneous detection of REM/NREM OSA and AF-associated OSA, improving diagnostics and addressing the link between sleep apnea and cardiovascular health.

2.5.4.2 Lightweight Models for Resource-Constrained Devices

Model compression techniques like pruning and quantization create lightweight architectures that ensure high accuracy while optimizing HSAT devices for real-time analysis in resource-limited settings.

2.5.4.3 Addressing Data Imbalance with SMOTE and MWMOTE

SMOTE and MWMOTE improve model reliability by augmenting rare apnea events in training, enhancing robustness, and ensuring accurate diagnostics even with imbalanced datasets.

2.5.4.4 Enhanced Diagnostic Precision Through Object Detection

Object detection algorithms focus on critical ECG signal regions, improving efficiency, reducing computational demands, and enhancing automated diagnostic accuracy.

2.5.5 Implementation of the Algorithmic Approaches

2.5.5.1 Integration with Wearable Single-Lead ECG Devices

Advanced algorithms enhance wearable single-lead ECG devices, enabling real-time detection of arrhythmias and apnea events without complex setups. These portable, user-friendly devices provide accurate diagnostics for clinical and home use, even in resource-limited settings.

2.5.5.2 Potential in Remote Healthcare

HSAT systems combined with cloud-based infrastructure enable real-time monitoring, secure data uploads, and timely alerts for severe apneas or arrhythmias. This approach improves patient safety, supports remote healthcare, and allows scalable data storage for population-level analysis.

2.6 Conclusion and Future Directions

2.6.1 Conclusion

HSAT devices provide a cost-effective, accessible alternative to in-lab PSG, benefiting underserved populations. They enable early OSA detection, reducing risks like cardiovascular disease and cognitive impairment, and effectively screen severe AF patients for PSG evaluation.

ECG-based HSAT advances OSA diagnosis by accurately identifying subtypes like REM/NREM-related and AF-associated OSA. With advanced algorithms, it differentiates OSA from CSA, detects sleep-stage events, and links AF to OSA. Lightweight deep learning models further improve real-time processing and detection efficiency.

2.6.2 Future Directions

Integrating multimodal data, including ECG, SpO_2, and sound signals, is essential for advancing HSAT systems, allowing precise monitoring of breathing, cardiac activity, and oxygen levels during sleep. Advanced algorithms like transformer models automate ECG analysis, enhancing diagnostic accuracy and reducing manual interpretation. Leveraging diverse datasets improves reliability, particularly for patients with comorbidities.

Improving wearable device hardware and software for non-laboratory use and conducting multicenter clinical trials are crucial for validating ECG-based HSAT. These advancements enable better diagnostics, treatment monitoring, and guidance of severe AF patients to sleep labs, solidifying ECG-based HSAT as a transformative tool for OSA management.

References

1. Punjabi NM. The epidemiology of adult obstructive sleep apnea. Proc Am Thorac Soc. 2008;5(2):136–43.
2. Senaratna CV, Perret JL, Lodge CJ, Lowe AJ, Campbell BE, Matheson MC, et al. Prevalence of obstructive sleep apnea in the general population: a systematic review. Sleep Med Rev. 2017;34:70–81.
3. Sullivan F. Hidden health crisis costing america billions: underdiagnosing and undertreating obstructive sleep apnea draining healthcare system. Am Acad Sleep Med. 2016;
4. Collop NA, Anderson WM, Boehlecke B, Claman D, Goldberg R, Gottlieb DJ, et al. Clinical guidelines for the use of unattended portable monitors in the diagnosis of obstructive sleep apnea in adult patients. J Clin Sleep Med-JCSM. 2007;3(7):737–47.
5. Mendonça F, Mostafa SS, Ravelo-García AG, Morgado-Dias F, Penzel T. Devices for home detection of obstructive sleep apnea: a review. Sleep Med Rev. 2018;41:149–60.
6. Teixeira RCP, Cahali MB. In-laboratory polysomnography worsens obstructive sleep apnea by changing body position compared to home testing. Sensors. 2024;24(9):2803.
7. Strollo PJ Jr, Rogers RM. Obstructive sleep apnea. N Engl J Med. 1996;334(2):99–104.
8. Sankri-Tarbichi AG. Obstructive sleep apnea-hypopnea syndrome: etiology and diagnosis. Avicenna J Med. 2012;2(01):3–8.
9. Kapur VK, Auckley DH, Chowdhuri S, Kuhlmann DC, Mehra R, Ramar K, et al. Clinical practice guideline for diagnostic testing for adult obstructive sleep apnea: an American Academy of Sleep Medicine clinical practice guideline. J Clin Sleep Med. 2017;13(3):479–504.
10. Suri JC, Suri JC, Suri TM. A review of the current status of home sleep apnea testing Vis-à-Vis in-lab polysomnography: is old still gold? Indian J Sleep Med. 2023;17(4):99–102.
11. Espinosa MA, Ponce P, Molina A, Borja V, Torres MG, Rojas M. Advancements in home-based devices for detecting obstructive sleep apnea: a comprehensive study. Sensors. 2023;23(23):9512.
12. Gomes DS, Seixas C, Cravo J. Performance of the Alice PDx device with the Somnolyzer automated scoring algorithm for the diagnosis of obstructive sleep apnea. Cureus. 2024;16(1)
13. ResMed. ApneaLink Air [Online]. Available from: https://bit.ly/4a5lpAz. Accessed 2 Jan 2025.
14. Kenny PP, Christine GG, David JT. A comparison of polysomnography and the WatchPAT in the diagnosis of obstructive sleep apnea. Otolaryngol Head Neck Surg. 2007;137(4):665–8.
15. Yuceege M, Firat H, Demir A, Ardic S. Reliability of the watch-PAT 200 in detecting sleep apnea in highway bus drivers. J Clin Sleep Med. 2013;9(4):339–44.
16. Gan YJ, Lim L, Chong YK. Validation study of WatchPat 200 for diagnosis of OSA in an Asian cohort. Eur Arch Otorrinolaringol. 2017;274:1741–5.
17. Zhang J, Wang C, Wang N, Liu J, Du J, Zhou G, et al. Evaluation of a portable device based on peripheral arterial tone in the detection of obstructive sleep apnea. Zhonghua er bi yan hou tou Jing Wai ke za zhi= Chinese J Otorhinolaryngol Head Neck Surg. 2012;47(2):112–6.
18. Driver HS, Pereira EJ, Bjerring K, Toop F, Stewart SC, Munt PW, et al. Validation of the MediByte® type 3 portable monitor compared with polysomnography for screening of obstructive sleep apnea. Can Respir J. 2011;18(3):137–43.
19. Lesser DJ, Haddad GG, Bush RA, Pian MS. The utility of a portable recording device for screening of obstructive sleep apnea in obese adolescents. J Clin Sleep Med. 2012;8(3):271–7.

20. Xu L, Han F, Keenan BT, Kneeland-Szanto E, Yan H, Dong X, et al. Validation of the Nox-T3 portable monitor for diagnosis of obstructive sleep apnea in Chinese adults. J Clin Sleep Med. 2017;13(5):675–83.
21. Labs R. Rooti Rx [Online]. Available from: https://bit.ly/40ijlBX. Accessed 23 Dec 2024.
22. Wings S. BEATINFO ECG Sensor [Online]. Available from: https://bit.ly/3C4hOGb. Accessed 23 Dec 2024.
23. Singh A, Al-Reefy H, Hewitt R, Kotecha B. Evaluation of ApneaGraph in the diagnosis of sleep-related breathing disorders. Eur Arch Otorrinolaringol. 2008;265:1489–94.
24. Xu H, Yan W, Lan K, Ma C, Wu D, Wu A, et al. Assessing electrocardiogram and respiratory signal quality of a wearable device (sensecho): semisupervised machine learning-based validation study. JMIR Mhealth Uhealth. 2021;9(8):e25415.
25. Xu H, Li P, Yang Z, Liu X, Wang Z, Yan W, et al. Construction and application of a medical-grade wireless monitoring system for physiological signals at general wards. J Med Syst. 2020;44:1–15.
26. Li P, Yang Z, Yan W, Yan M, He M, Yuan Q, et al., editors. Mobicardio: a clinical-grade mobile health system for cardiovascular disease management. 2019 IEEE International Conference on Healthcare Informatics (ICHI). IEEE; 2019.
27. Lan K, Liu X, Xu H, Li P, Yang Z, Yuan Q, et al., editors. Poster: DeePTOP: personalized tachycardia onset prediction using bi-directional LSTM in wearable embedded systems. EWSN; 2019.
28. Yoon DW, Hong I-H, Baik I, Shin H-W. Evaluation of the feasibility and preference of Nox-A1 type 2 ambulatory device for unattended home sleep test: a randomized crossover study. Sleep Biol Rhythms. 2019;17:297–304.
29. Park JH, Wang C, Shin H. FDA-cleared home sleep apnea testing devices. NPJ Digital Med. 2024;7(1):123.
30. Collop NA, Tracy SL, Kapur V, Mehra R, Kuhlmann D, Fleishman SA, et al. Obstructive sleep apnea devices for out-of-center (OOC) testing: technology evaluation. J Clin Sleep Med. 2011;7(5):531–48.
31. Abraham WT, Trupp RJ, Phillips B, Bourge RC, Bailey B, Harding SM, et al. Validation and clinical utility of a simple in-home testing tool for sleep-disordered breathing and arrhythmias in heart failure: results of the Sleep Events, Arrhythmias, and Respiratory Analysis in Congestive Heart Failure (SEARCH) STUDY. Congest Heart Fail. 2006;12(5):241–7.
32. Bar A, Pillar G, Dvir I, Sheffy J, Schnall RP, Lavie P. Evaluation of a portable device based on peripheral arterial tone for unattended home sleep studies. Chest. 2003;123(3):695–703.
33. Garg N, Rolle AJ, Lee TA, Prasad B. Home-based diagnosis of obstructive sleep apnea in an urban population. J Clin Sleep Med. 2014;10(8):879–85.
34. Pittman SD, Ayas NT, MacDonald MM, Malhotra A, Fogel RB, White DP. Using a wrist-worn device based on peripheral arterial tonometry to diagnose obstructive sleep apnea: in-laboratory and ambulatory validation. Sleep. 2004;27(5):923–33.
35. Zou D, Grote L, Peker Y, Lindblad U, Hedner J. Validation a portable monitoring device for sleep apnea diagnosis in a population based cohort using synchronized home polysomnography. Sleep. 2006;29(3):367–74.
36. Pépin J-L, Defaye P, Vincent E, Christophle-Boulard S, Tamisier R, Lévy P. Sleep apnea diagnosis using an ECG Holter device including a nasal pressure (NP) recording: validation of visual and automatic analysis of nasal pressure versus full polysomnography. Sleep Med. 2009;10(6):651–6.
37. Bouzid Z, Al-Zaiti SS, Bond R, Sejdić E. Remote and wearable ECG devices with diagnostic abilities in adults: a state-of-the-science scoping review. Heart Rhythm. 2022;19(7):1192–201.
38. Neri L, Oberdier MT, van Abeelen KC, Menghini L, Tumarkin E, Tripathi H, et al. Electrocardiogram monitoring wearable devices and artificial-intelligence-enabled diagnostic capabilities: a review. Sensors. 2023;23(10):4805.
39. Marulanda-Londono E, Chaturvedi S. The interplay between obstructive sleep apnea and atrial fibrillation. Front Neurol. 2017;8:668.

40. Moula AI, Parrini I, Tetta C, Lucà F, Parise G, Rao CM, et al. Obstructive sleep apnea and atrial fibrillation. J Clin Med. 2022;11(5):1242.
41. Rossi VA, Stradling JR, Kohler M. Effects of obstructive sleep apnoea on heart rhythm. Eur Respir J. 2013;41(6):1439–51.
42. Aurora RN, Crainiceanu C, Gottlieb DJ, Kim JS, Punjabi NM. Obstructive sleep apnea during REM sleep and cardiovascular disease. Am J Respir Crit Care Med. 2018;197(5):653–60.
43. Mokhlesi B, Punjabi NM. "REM-related" obstructive sleep apnea: an epiphenomenon or a clinically important entity? Sleep. 2012;35(1):5–7.
44. Alzoubaidi M, Mokhlesi B. Obstructive sleep apnea during REM sleep: clinical relevance and therapeutic implications. Curr Opin Pulm Med. 2016;22(6):545.
45. Lin C-Y, Wang Y-W, Setiawan F, Trang NTH, Lin C-W. Sleep apnea classification algorithm development using a machine-learning framework and bag-of-features derived from electrocardiogram spectrograms. J Clin Med. 2022;11(1):192.
46. Setiawan F, Lin C-W. A deep learning framework for automatic sleep apnea classification based on empirical mode decomposition derived from single-Lead electrocardiogram. Life. 2022;12(10):1509.
47. Setiawan F. Development of lightweight deep learning single-lead ECG algorithms: verification and accurate identification of atrial fibrillation patients and sleep apnea events in REM/NREM sleep. Tainan: National Cheng Kung University; 2024.

Interventional Drug-Induced Sleep Endoscopy

3

Ying-Shuo Hsu, Li-Jen Hsin, Li-Ang Lee, Wen-Te Liu, and Hsueh-Yu Li

Contents

Abstract

This chapter introduces one of the most rapidly advancing fields in the diagnosis and treatment of obstructive sleep apnea (OSA) in recent years: drug-induced sleep endoscopy (DISE). This diagnostic procedure not only provides valuable insights into the etiology and pathophysiology of OSA but also, when combined with various interventional techniques during the examination, offers critical guidance for personalized and precision-based treatment strategies tailored to each patient in the future.

Y.-S. Hsu (✉)
Otolaryngology, Shin Kong Wu Ho-Su Memorial Hospital, Taipei, Taiwan

Institute of brain science, National Yang Ming Chiao Tung University, Taipei, Taiwan

L.-J. Hsin · L.-A. Lee · W.-T. Liu · H.-Y. Li
Department of Otolaryngology, Chang Gung Memorial Hospital, Linkou, Taoyuan City, Taiwan

H.-Y. Li (ed.), *Snoring and Obstructive Sleep Apnea*,
https://doi.org/10.1007/978-981-95-2648-2_3

Keywords

Drug-induced sleep endoscopy · DISE · TCI-DISE · Interventional TCI-DISE · Sleep-disordered breathing · Obstructive sleep apnea · Positional therapy · Head rotation · Oral appliance · Intraoral negative airway pressure · iNAP

3.1 Introduction

In 1993, the estimated prevalence of obstructive sleep apnea (OSA) among women and men aged 30–60 years was 2% and 4%, respectively [1]. However, 20 years later, Peppard et al. reported that the prevalence of moderate-to-severe OSA was 10% among men aged 30–49 years and 17% among men aged 50–70 years. For women, the prevalence was 3% in those aged 30–49 years and 9% in those aged 50–70 years [2]. These figures are significantly higher than earlier estimates. Benjafield et al. estimated that approximately 425 million adults worldwide have moderate-to-severe OSA [3]. Consequently, OSA has become increasingly prevalent in modern society.

OSA is characterized by recurrent episodes of hypopnea or apnea caused by the collapse of the pharyngeal walls [4]. Approximately 1 billion adults worldwide are affected by OSA [3], which may contribute to hypertension [5], type 2 diabetes [6], excessive daytime sleepiness [7], and even life-threatening cerebrovascular events [8], and sudden cardiac arrest [9]. Although continuous positive airway pressure (CPAP) remains the first-line treatment and gold standard for OSA [10], many individuals are unable to tolerate long-term CPAP use [11].

Oral appliances (OA), including mandibular advancement devices (MAD), are a conservative treatment option for mild-to-moderate sleep-disordered breathing [12]. However, long-term use of OA carries a risk of temporomandibular joint (TMJ) disorders [13] and may not be effective for all patients. When conservative therapies fail or are poorly tolerated, surgical intervention becomes a viable alternative for treating OSA [14]. However, due to individual variations in the pattern of upper airway anatomical collapse, surgical outcomes can vary considerably. Even after surgery, persistent mouth-breathing habits may significantly diminish therapeutic effectiveness [15].

Positional therapy, which employs techniques to prevent the supine position during sleep, has demonstrated significant efficacy in reducing the apnea–hypopnea index (AHI) in many patients, emphasizing its therapeutic potential in sleep medicine [16]. However, it is not recommended for individuals with shoulder disorders or other physical disabilities that preclude sleeping in a lateral position, thereby preventing avoidance of the supine posture [17]. Moreover, determining which patients may benefit most from incorporating positional therapy into multimodal treatment strategies remains a significant challenge [18].

Drug-induced sleep endoscopy (DISE) is a pivotal tool in the surgical decision-making process within the field of sleep medicine [19]. Not only does it facilitate the investigation of the pathophysiological underpinnings of specific upper airway collapse patterns, but it also plays a crucial role in devising personalized, precision treatment approaches for patients [20]. In patients with OSA who are noncompliant with conservative therapy, upper airway surgery serves as an alternative treatment option [21]. However, surgical decision-making can be challenging due to the complex anatomy of the upper airway. DISE is a readily accessible method that enables surgeons to identify specific regions of airway collapse during sedation. Nevertheless, surgical planning based on DISE findings does not always result in superior postoperative outcomes. For example, Hsu and Jacobowitz [22] observed that among OSA patients who exhibited either multiple or single obstruction sites on DISE, a single surgical intervention at one site yielded similar postoperative improvements. One explanation for this unexpected finding is that manually administering propofol during DISE may lead to unstable sedation and exaggerated upper airway collapse, particularly at the tongue base.

3.2 Evolution of DISE

To accurately identify the anatomical sites and structures involved in upper airway obstruction before surgery, sleep endoscopy was first advocated by Croft and Pringle in the 1990s [19]. Subsequently, Kezirian refined the technique, renaming it "DISE" [23] and establishing the VOTE classification system, which categorizes obstruction at the velum, oropharyngeal lateral walls, tongue base, and epiglottis [24]. Given that this procedure provides real-time, dynamic visualization of upper airway collapse during simulated sleep, it has been widely adopted as a valuable tool to assist surgeons in personalizing surgical plans for patients with OSA [25].

Furthermore, multilevel upper airway obstruction is common among patients with OSA [14]. However, evaluations performed while the patient is awake often fail to accurately predict tongue base collapse during sleep [26]. Consequently, DISE has been proposed as a potentially useful modality to guide surgical decision-making in these patients. Despite its theoretical advantages, polysomnographic outcomes following DISE-guided upper airway surgery have not always been satisfactory [22, 25, 27–30]. Certain patterns of obstruction observed in preoperative DISE, including concentric velum collapse, complete anteroposterior tongue base obstruction, retrolingual lateral pharyngeal wall collapse, and supraglottic laryngeal collapse, have been associated with surgical failure [22, 28, 29].

Additionally, preoperative DISE findings do not always correlate with surgical success. For instance, Meraj et al [30] revealed that DISE findings were not significantly associated with surgical success following multilevel surgery combined with transoral robotic tongue base resection. These findings raise critical questions: Is multilevel surgery truly effective in treating multilevel airway obstruction identified through DISE? Furthermore, is multilevel surgery necessary for all patients in whom DISE reveals multilevel obstruction?

In a previous study [22], we found that among patients who underwent palatopharyngoplasty alone, surgical outcomes were comparable between those with multilevel obstruction and those with unilevel obstruction, as identified by

DISE. However, in the present study, DISE was performed using a manual administration method for propofol. This approach frequently induces drug-related tongue base collapse, potentially leading to DISE findings that do not accurately reflect the patient's natural sleep state. Consequently, the correlation between DISE findings and postoperative outcomes may be limited. This highlights the need for improved anesthetic administration techniques to enhance the accuracy and clinical applicability of DISE in guiding surgical interventions.

3.3 Target-Controlled Infusion (TCI) DISE

To address the aforementioned challenges, a more precise method of anesthetic administration is essential to better simulate the patient's natural sleep state. Furthermore, an improved approach would enhance both the safety and stability of DISE, thereby ensuring more reliable and clinically relevant findings. One potential strategy to further investigate these issues is the postoperative analysis of DISE findings, which may provide valuable insights into surgical outcomes and therapeutic effectiveness. A more reliable DISE technique is also necessary for accurately identifying collapse patterns during stable sleep. To achieve this, target-controlled infusion (TCI) has emerged as a promising approach [31, 32] (Fig. 3.1).

In our TCI-DISE study in 2021 [33], we analyzed preoperative and postoperative DISE staging patterns in patients with moderate to severe OSA following advanced palatopharyngoplasty. Our findings demonstrated that patients with postoperative velar collapse exhibited higher follow-up AHI values than those without velar collapse. Similarly, patients with preoperative complete base collapse had significantly higher follow-up AHI values compared to those with no or partial tongue base collapse. Additionally, patients with postoperative complete tongue base collapse had the highest follow-up AHI values, significantly exceeding those without complete collapse [33].

Our findings indicated that patients with complete tongue base collapse in TCI-DISE preoperatively experienced an AHI reduction of only 20.8%, whereas those with no or partial tongue base collapse preoperatively exhibited a significantly

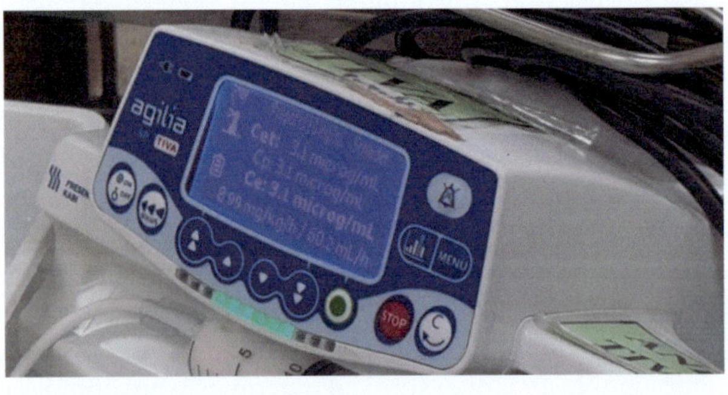

Fig. 3.1 Target-controlled infusion

greater AHI reduction of 45.8%. This indicates that the tongue base collapse observed in the TCI-DISE examination needs to be addressed. Failure to manage this condition may affect the surgical outcome, reducing the extent of improvement after the procedure.

In addition, DISE performed with TCI-DISE has demonstrated greater accuracy in predicting surgical outcomes, particularly regarding tongue base involvement [33]. The identification of tongue base collapse during TCI-DISE serves as a predictor of poorer surgical outcomes, highlighting the critical role of TCI-DISE in guiding personalized treatment strategies. Moreover, the TCI-DISE protocol has been further refined by incorporating adjunctive techniques, such as intraoral negative airway pressure (iNAP), head rotation, and the use of non-custom oral appliances, to enhance the assessment of therapeutic efficacy in individual patients [34, 35]. This approach is referred to as *interventional TCI-DISE*.

The following sections will provide a detailed discussion of the research findings on interventional TCI-DISE and its clinical benefits for patients with OSA.

3.4 Interventional TCI-DISE and iNAP

iNAP therapy (Fig. 3.2) has been associated with lower rates of oral tissue discomfort and potentially improved patient compliance compared to conventional treatments [36–38]. The iNAP device generates intermittent negative pressure within the oral cavity, which moves the soft palate and tongue anteriorly. This action increases both the retropalatal and retroglossal airway spaces, thereby reducing upper airway obstruction during sleep [38]. The device is battery-powered, and designed to automatically regulate airflow, stopping once the target negative pressure is reached and resuming only when necessary to maintain the desired pressure level [37]. The iNAP therapy may be a suitable alternative for patients with OSA who are intolerant to continuous positive airway pressure (CPAP) therapy.

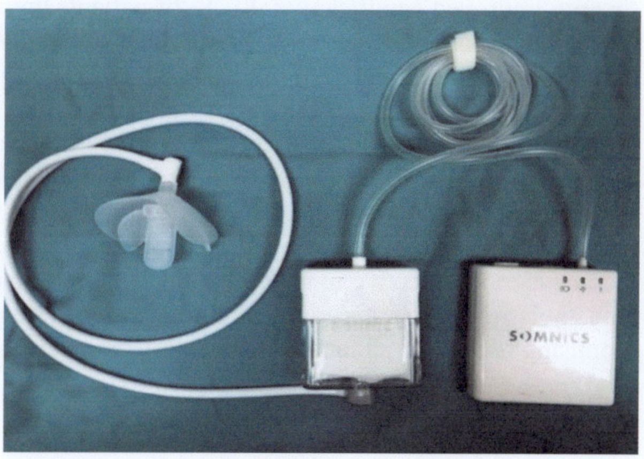

Fig. 3.2 Intraoral negative airway pressure (iNAP)

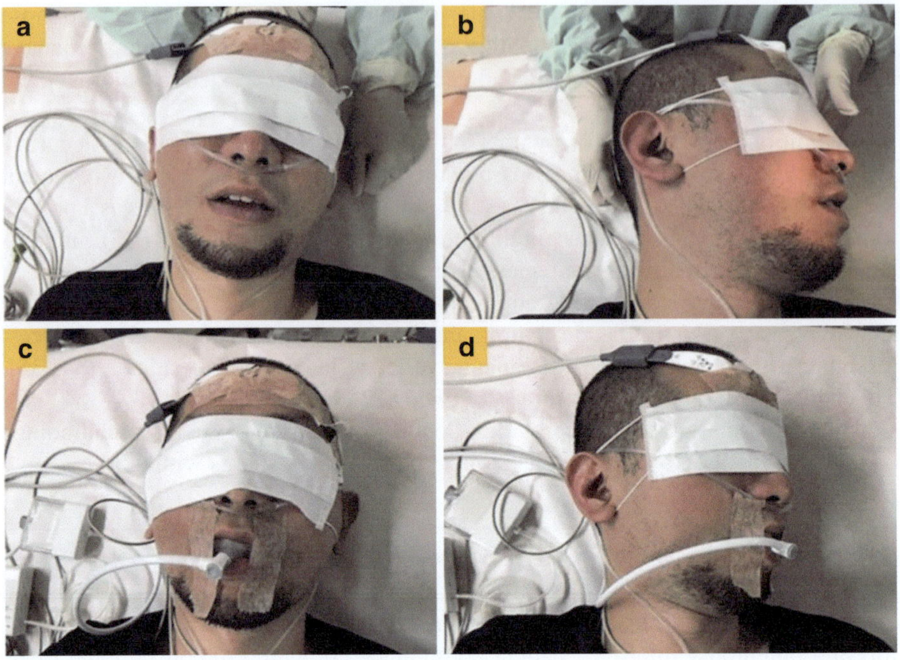

Fig. 3.3 A patient undergoing TCI-DISE in various positions with or without iNAP therapy. Notes: DISE was conducted multiple times for each patient in the following order: (**a**) supine position, (**b**) head rotation, (**c**) supine position with iNAP therapy, and (**d**) head rotation with iNAP therapy

In a previous study evaluating iNAP therapy using TCI-DISE [35], 92 patients were assessed for airway obstruction severity in the supine position, with and without lateral head rotation (Fig. 3.3).

Following the application of iNAP therapy in the supine position, obstruction severity improved significantly: 40.2% of patients with velar obstruction and 39% with tongue base obstruction experienced a reduction from complete or partial obstruction to partial or no obstruction, while 13% of patients with oropharyngeal obstruction showed similar improvement. When iNAP therapy was combined with head rotation, obstruction severity decreased further in 51% of patients with velar obstruction, 47% with tongue base obstruction, and 21% with epiglottic obstruction.

Therefore, iNAP therapy can be effectively assessed within the TCI-DISE framework to evaluate its therapeutic impact. Additionally, the findings suggest that iNAP therapy is more effective in lean individuals and those over the age of 50 years. Its efficacy is further enhanced when combined with positional therapy, making it a promising option for patients who benefit from lateral sleeping positions.

3.5 Interventional TCI-DISE with Oral Appliance and Body Weight Control

Obesity is a significant risk factor for OSA, and weight loss combined with lifestyle modifications is strongly recommended for all overweight and obese patients with OSA [39]. Study demonstrated that a 10% increase in body weight is associated with a 32% rise in the AHI, whereas a 10% weight loss leads to a 26% reduction in AHI [15, 40].

TCI-DISE can serve as a useful test model for evaluating conservative treatment strategies, such as positional therapy and OA therapy [20, 35, 41, 42]. Both treatments are effective non-invasive options for managing OSA. Previous studies have primarily focused on sleep study phenotypes and OSA severity when determining treatment indications. For example, positional therapy is most effective in patients with position-dependent sleep apnea (POSA) [43, 44], while OA therapy is generally recommended for those with mild-to-moderate OSA. Despite its benefits, some patients may be reluctant to change their sleep position in laboratory settings, potentially leading to an underestimation of the true prevalence of POSA [45, 46]. Additionally, OA therapy may be effective in some cases of severe OSA if TCI-DISE demonstrates a reduction in upper airway obstruction when the appliance is used. The combination of OA therapy and positional therapy represents a promising treatment approach and its effectiveness has been investigated using TCI-DISE in our previous study [34].

As shown in Fig. 3.4, 83 OSA patients underwent interventional TCI-DISE with OA. Among these patients, 23 exhibited persistent complete concentric velopharyngeal collapse (CCC) in the supine position despite concurrent head rotation and OA use. Their AHI was significantly higher than 60 patients without such collapse. Additionally, these 23 patients had a significantly higher body mass index (BMI). Those patients with complete tongue base obstructions in head rotation and OA use did not have a higher BMI, which means that anatomic obstruction of the tongue base may still need tailor-made surgeries according to their specific conditions. Higher BMI was an important factor in CCC. Therefore, patients do not succeed with OA and positional therapy may need weight control.

Patients who had a higher BMI presented with persistent velar obstruction even with head rotation and OA use in interventional TCI-DISE. This finding may be related to the larger pharyngeal wall fat pad in obese patients with OSA [47, 48], and they may need weight loss therapy [49–51]. More than 40% of patients still have OSA symptoms after bariatric surgery [52], which means that anatomic factors may still play an important role in these patients. Patients who showed complete tongue base collapse did not have a higher BMI. Therefore, anatomic obstruction-oriented upper airway surgery should also be considered in OSA patients with obesity [53, 54].

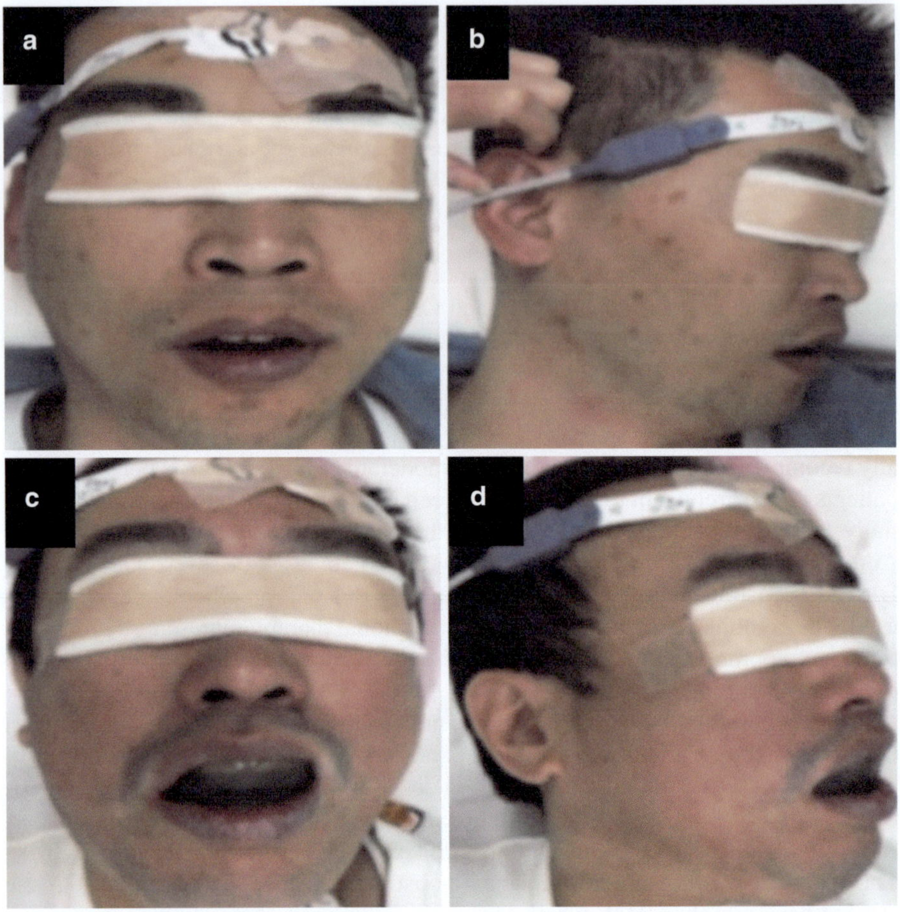

Fig. 3.4 The four positions utilized during interventional TCI-DISE with oral appliance (OA): (**a**) supine position, (**b**) head rotation, (**c**) mandibular advancement using an OA, and (**d**) head rotation and mandibular advancement using an OA

We can use these interventional TCI-DISE findings to adjust our future surgery and treatment protocol (Fig. 3.5). The integration of various conservative treatments, such as OA, positional therapy, intermittent negative airway pressure, weight control, and several kinds of surgical treatments, may effectively treat OSA, especially CCC [35, 43, 55].

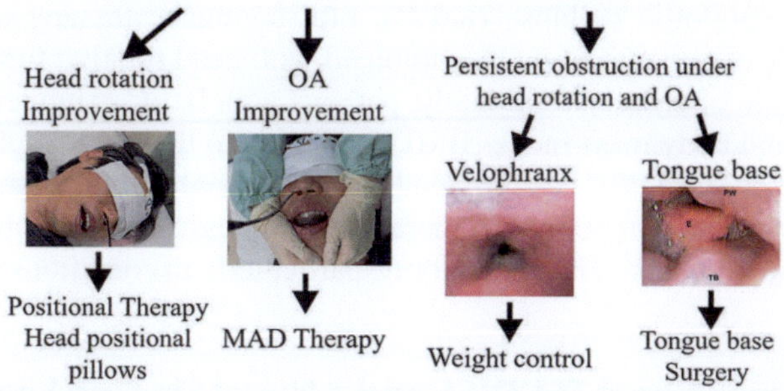

Fig. 3.5 Diagnosis and treatment recommendations after interventional TCI-DISE with oral appliance (OA). If upper airway obstruction is alleviated by head rotation, the patient may be advised to undergo positional therapy. If the obstruction improves with the use of an oral appliance (OA), mandibular advancement devices (MAD) therapy may be recommended. However, if upper airway obstruction persists despite the combined application of head rotation and OA, further management depends on the site of obstruction. If the obstruction is localized at the velopharyngeal level, weight management may be necessary. Conversely, if the obstruction occurs at the tongue base, surgical intervention targeting the tongue base may be required

3.6 Interventional TCI-DISE and Epiglottic Collapse

The identification of epiglottic collapse (EC) in patients with OSA is crucial, as EC can contribute to the failure of both uvulopalatopharyngoplasty (UPPP) and CPAP therapy [56]. Studies have explored the physiological mechanisms of EC and potential treatment strategies. According to Kuo et al., epiglottic length is a key factor influencing the risk of EC [57]. A 2021 case–control study by Kim et al. demonstrated that the jaw-thrust maneuver and head rotation are effective interventions for patients with EC [58]. Additionally, a study by Kent et al. found that four out of eleven patients with EC experienced symptom improvement with the use of OA [59]. Laser partial epiglottectomy, as reported by Golz et al. in 2000, resulted in significant improvement in 85% of patients based on postoperative polysomnography (PSG) findings [60]. Furthermore, Salamanca et al. described positive outcomes following epiglottis stiffening procedures in patients with EC [61].

In our previous study [45] evaluating 54 OSA patients with EC using interventional TCI-DISE, head rotation led to a significant reduction in anteroposterior epiglottic collapse (apEC) in 60% of patients, while mouth closure reduced apEC in 53%, oral appliance therapy in 47%, and iNAP therapy in 28%. Combination therapy showed even greater effectiveness in treating apEC, with head rotation plus mouth closure achieving a 77% success rate, head rotation plus OA 70%, and head rotation plus iNAP 68%. These findings suggest that head rotation combined with mouth closure is the most effective intervention for apEC as assessed by TCI-DISE.

For lateral epiglottic collapse (latEC), oral appliance therapy alone reduced severity in 54% of patients, and OA combined with head rotation further improved outcomes in 61% of cases. Additionally, patients with latEC exhibited higher body weight, AHI, and body mass index (BMI) compared to those with apEC.

Interventional TCI-DISE has provided valuable insights into the management of EC by not only identifying effective therapeutic strategies but also highlighting the impact of obesity on upper airway anatomy, particularly in conditions such as latEC.

3.7 Interventional TCI-DISE and a Novel Finding Leading to Oral Breathing: Expiratory Velopharyngeal Obstruction

In recent years, a significant correlation has been observed between oral breathing and the severity of sleep-disordered breathing in patients with OSA [51]. However, attempts to reduce mouth breathing through various interventions have led to the identification of a phenomenon termed "mouth puffing" in a substantial subset of patients [62, 63]. This term describes a condition in which patients, when instructed to maintain oral closure during sleep, experience difficulty exhaling through the nasal passages. This suggests an obstruction in the oropharyngeal–nasal airway during expiration.

In our recent study [64], we used interventional TCI-DISE to explore the anatomical etiology of oral breathing. During interventional TCI-DISE in this study, some patients exhibited the mouth puffing phenomenon when their mouths were taped or forcibly closed. This observation correlated with an endoscopic finding identified as expiratory velopharyngeal obstruction (EVO) (Figs. 3.6 and 3.7). EVO

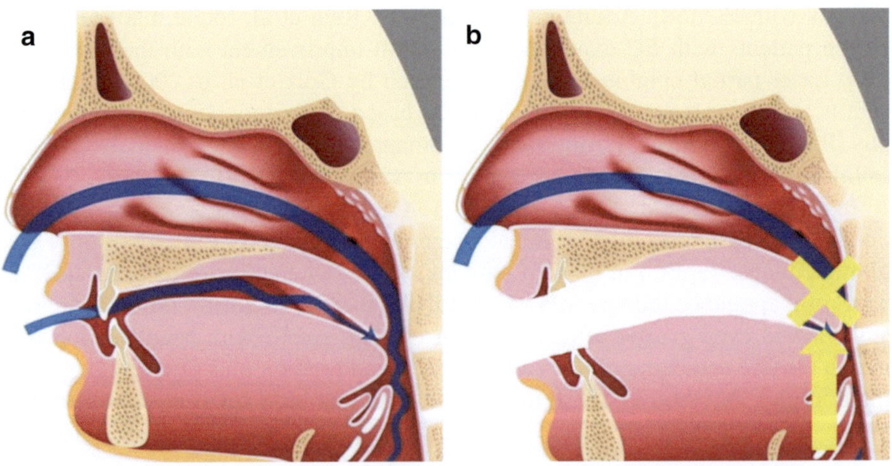

Fig. 3.6 Expiratory velopharyngeal obstruction (EVO). (**a**) Depiction of the inspiratory phase. (**b**) Demonstration of paradoxical velopharyngeal collapse during expiration. EVO is characterized by a dynamic collapse of the velopharyngeal soft tissues during exhalation, leading to partial or complete obstruction of the nasal airway. This obstruction impairs nasal exhalation, necessitating compensatory oral breathing to maintain adequate airflow

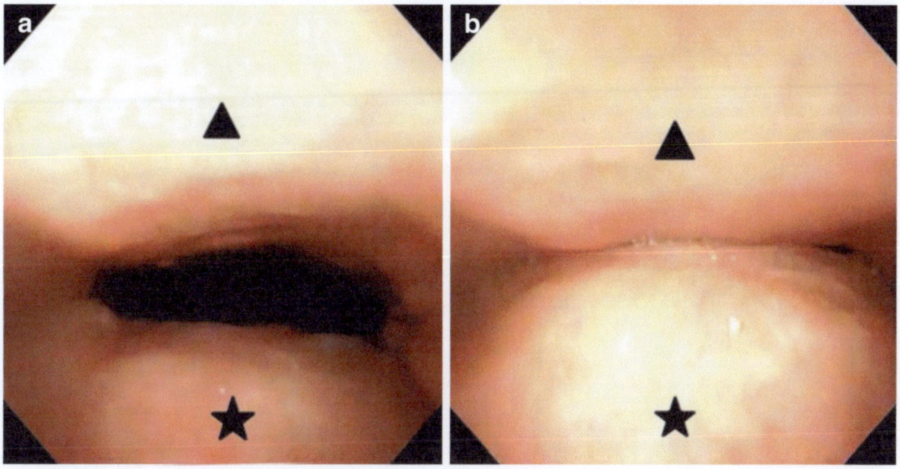

Fig. 3.7 Expiratory velopharyngeal obstruction (EVO) observed during interventional drug-induced sleep endoscopy (DISE) with target-controlled infusion (TCI). (**a**) Representation of the inspiratory phase. (**b**) Demonstration of EVO during exhalation in interventional TCI-DISE. EVO is characterized by dynamic collapse of the velopharyngeal structures during expiration, resulting in partial or complete airway obstruction at the velopharyngeal level. Anatomical landmarks include the posterior pharyngeal wall (▲) and the soft palate (★)

occurs due to the inversion of soft palate tissue during expiration, leading to constriction of the passage between the pharynx and the nasal cavity. The presence or absence of EVO may serve as an important therapeutic indicator in OSA management.

In 72 OSA patients with particular attention given to the presence of EVO in interventional TCI-DISE, mouth closure and iNAP were associated with an increased occurrence of EVO, observed in 43.1% and 34.7% of cases, respectively, compared to OA therapy (20.1%). However, head rotation was associated with a reduction in EVO during DISE compared to the supine position (26% vs. 35.8%). Notably, increasing age was identified as a risk factor for EVO, with each additional year of age increasing the odds of EVO.

Therefore, head rotation and OA therapy can reduce EVO and improve mouth breathing in patients with OSA. These interventional TCI-DISE findings provide valuable insights for future treatment strategies, particularly in patients who experience oral breathing, mouth puffing or expiratory airway obstruction during sleep.

3.8 Interventional TCI-DISE with Palatal Suture

Temporary relief of palatal obstruction can be achieved through anterior displacement of the velum (Vad) using a suture during interventional TCI-DISE. The Vad test is performed by placing a No. 2 Dexon® stay suture at the central base of the soft palate (Fig. 3.8).

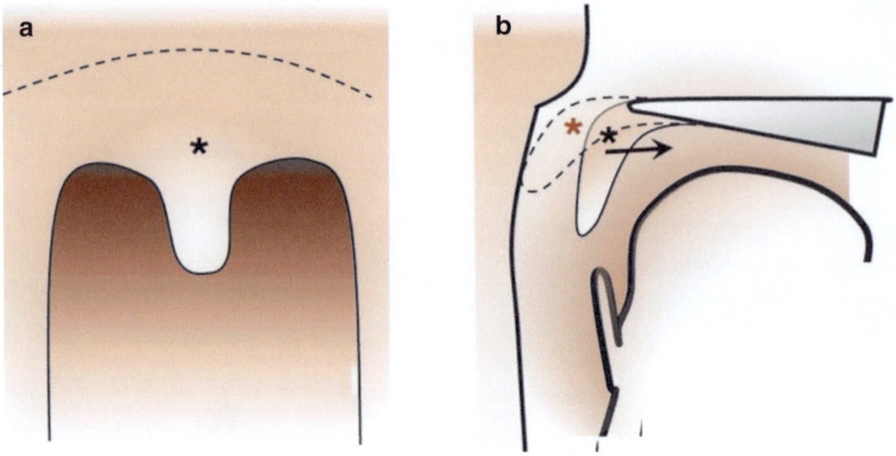

Fig. 3.8 Anterior displacement of the velum (Vad). (**a**) A stay suture is placed at the central base of the patient's soft palate (*), with the dashed line indicating the margin of the hard palate. (**b**) Anterior displacement of the velum (Vad) results in the opening of the velopharynx (arrow)

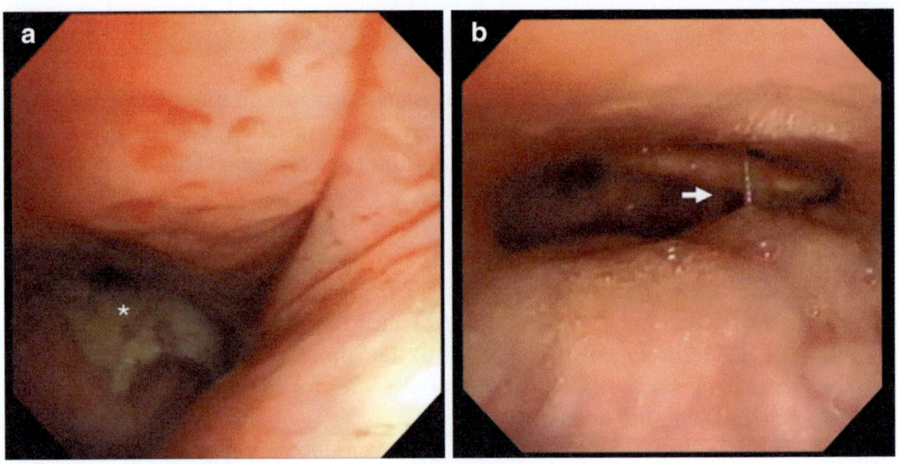

Fig. 3.9 Secondary airway collapse following anterior displacement of the velum (Vad) (**a**) Complete collapse of the tongue base (*) following anterior displacement of the velum (Vad). (**b**) Subsequent complete collapse of the epiglottis (arrow) after Vad

Following Vad, snoring was often eliminated; however, in some cases, it shifted from the velopharynx to the hypopharynx. The vibratory components subsequently relocated to the pharyngeal mucosa, the tongue base, or the epiglottis (Fig. 3.9).

Incorporating Vad into interventional TCI-DISE allows for the identification of patients at higher risk for complex airway collapse, facilitating the development of more individualized surgical strategies. In cases where complete collapse occurs at the tongue base or epiglottis, UPPP alone may be inadequate, necessitating a more comprehensive surgical approach.

3.9 Summary and Future Directions

Interventional TCI-DISE allows clinicians to directly visualize the true sleep-related airway obstruction in patients with OSA. Additionally, it provides an accurate simulation of therapeutic interventions such as iNAP, positional therapy, and OA, making it an invaluable tool for sleep surgeons.

Proficiency in the operation, documentation, and interpretation of interventional TCI-DISE enables sleep surgeons to precisely determine the optimal therapeutic approach for each patient. Furthermore, in cases where surgical outcomes do not meet expectations, this technique allows for the formulation of additional treatment strategies tailored to the patient's specific airway collapse patterns.

Future research directions may integrate artificial intelligence (AI) and big data analytics to facilitate more rapid and precise diagnosis while enabling automated generation of personalized treatment recommendations. Additionally, the accumulation of postoperative interventional TCI-DISE data will enhance our understanding of the clinical significance of findings observed during the procedure. This, in turn, will contribute to the development of more effective therapeutic solutions, particularly for the management of EVO and mouth breathing.

References

1. Young T, Palta M, Dempsey J, Skatrud J, Weber S, Badr S. The occurrence of sleep-disordered breathing among middle-aged adults. N Engl J Med. 1993;328(17):1230–5. https://doi.org/10.1056/NEJM199304293281704.
2. Peppard PE, Young T, Barnet JH, Palta M, Hagen EW, Hla KM. Increased prevalence of sleep-disordered breathing in adults. Am J Epidemiol. 2013;177(9):1006–14. https://doi.org/10.1093/aje/kws342.
3. Benjafield AV, Ayas NT, Eastwood PR, et al. Estimation of the global prevalence and burden of obstructive sleep apnoea: a literature-based analysis. Lancet Respir Med. 2019;7(8):687–98. https://doi.org/10.1016/S2213-2600(19)30198-5.
4. Sundaram S, Bridgman SA, Lim J, Lasserson TJ. Surgery for obstructive sleep apnoea. Cochrane Database Syst Rev. 2005;4:CD001004. https://doi.org/10.1002/14651858.CD001004.pub2.
5. Peppard PE, Young T, Palta M, Skatrud J. Prospective study of the association between sleep-disordered breathing and hypertension. N Engl J Med. 2000;342(19):1378–84.
6. Reichmuth KJ, Austin D, Skatrud JB, Young T. Association of sleep apnea and type II diabetes: a population-based study. Am J Respir Crit Care Med. 2005;172(12):1590–5. https://doi.org/10.1164/rccm.200504-637OC.
7. Engleman HM, Douglas NJ. Sleep. 4: Sleepiness, cognitive function, and quality of life in obstructive sleep apnoea/hypopnoea syndrome. Thorax. 2004;59(7):618–22.

8. Yaggi HK, Concato J, Kernan WN, Lichtman JH, Brass LM, Mohsenin V. Obstructive sleep apnea as a risk factor for stroke and death. N Engl J Med. 2005;353(19):2034–41. https://doi.org/10.1056/NEJMoa043104.
9. Gami AS, Olson EJ, Shen WK, et al. Obstructive sleep apnea and the risk of sudden cardiac death: a longitudinal study of 10,701 adults. J Am Coll Cardiol. 2013;62(7):610–6. https://doi.org/10.1016/j.jacc.2013.04.080.
10. Kushida CA, Littner MR, Hirshkowitz M, et al. Practice parameters for the use of continuous and bilevel positive airway pressure devices to treat adult patients with sleep-related breathing disorders. Sleep. 2006;29(3):375–80. https://doi.org/10.1093/sleep/29.3.375.
11. Kohler M, Smith D, Tippett V, Stradling JR. Predictors of long-term compliance with continuous positive airway pressure. Thorax. 2010;65(9):829–32. https://doi.org/10.1136/thx.2010.135848.
12. Doff MHJ, Hoekema A, Wijkstra PJ, et al. Oral appliance versus continuous positive airway pressure in obstructive sleep apnea syndrome: a 2-year follow-up. Sleep. 2013;36(9):1289–96. https://doi.org/10.5665/sleep.2948.
13. Van den Bossche K, Op de Beeck S, Dieltjens M, et al. Multimodal phenotypic labelling using drug-induced sleep endoscopy, awake nasendoscopy and computational fluid dynamics for the prediction of mandibular advancement device treatment outcome: a prospective study. J Sleep Res. Published online June 22, 2022:e13673. https://doi.org/10.1111/jsr.13673.
14. Lin HC, Weaver EM, Lin HS, Friedman M. Multilevel obstructive sleep apnea surgery. Adv Otorhinolaryngol. 2017;80:109–15. https://doi.org/10.1159/000470879.
15. Ramaswamy AT, Li C, Suurna MV. A case of hypoglossal nerve stimulator-resistant obstructive sleep apnea cured with the addition of a chin strap. Laryngoscope. 2018;128(7):1727–9. https://doi.org/10.1002/lary.27010.
16. Joosten SA, Edwards BA, Wellman A, et al. The effect of body position on physiological factors that contribute to obstructive sleep apnea. Sleep. Published online September 1, 2015. https://doi.org/10.5665/sleep.4992.
17. Verbraecken J, Dieltjens M, Op de Beeck S, et al. Non-CPAP therapy for obstructive sleep apnoea. Breathe Sheff Engl. 2022;18(3):220164. https://doi.org/10.1183/20734735.0164-2022.
18. Dieltjens M, Vroegop AV, Verbruggen AE, et al. A promising concept of combination therapy for positional obstructive sleep apnea. Sleep Breath. 2015;19(2):637–44. https://doi.org/10.1007/s11325-014-1068-8.
19. Croft CB, Pringle M. Sleep nasendoscopy: a technique of assessment in snoring and obstructive sleep apnoea. Clin Otolaryngol Allied Sci. 1991;16(5):504–9. https://doi.org/10.1111/j.1365-2273.1991.tb01050.x.
20. Safiruddin F, Koutsourelakis I, de Vries N. Upper airway collapse during drug induced sleep endoscopy: head rotation in supine position compared with lateral head and trunk position. Eur Arch Otorrinolaringol. 2015;272(2):485–8. https://doi.org/10.1007/s00405-014-3215-z.
21. Aurora RN, Casey KR, Kristo D, et al. Practice parameters for the surgical modifications of the upper airway for obstructive sleep apnea in adults. Sleep. 2010;33(10):1408–13. https://doi.org/10.1093/sleep/33.10.1408.
22. Hsu YS, Jacobowitz O. Does sleep endoscopy staging pattern correlate with outcome of advanced palatopharyngoplasty for moderate to severe obstructive sleep apnea? J Clin Sleep Med. 2017;13(10):1137–44.
23. Kezirian EJ. Drug-induced sleep endoscopy. Oper Tech Otolaryngol-Head Neck Surg. 2006;17(4):230–2. https://doi.org/10.1016/j.otot.2006.10.005.
24. Kezirian EJ, Hohenhorst W, de Vries N. Drug-induced sleep endoscopy: the VOTE classification. Eur Arch Otorrinolaringol. 2011;268(8):1233–6. https://doi.org/10.1007/s00405-011-1633-8.
25. Blumen M, Bequignon E, Chabolle F. Drug-induced sleep endoscopy: a new gold standard for evaluating OSAS? Part II: results. Eur Ann Otorhinolaryngol Head Neck Dis. 2017;134(2):109–15. https://doi.org/10.1016/j.anorl.2016.12.004.

26. Lin HY, Lin YC, Hsu YS, et al. Comparison of findings between clinical examinations and drug-induced sleep endoscopy in patients with obstructive sleep apnea syndrome. Int J Environ Res Public Health. 2020;17(17) https://doi.org/10.3390/ijerph17176041.
27. Blumen MB, Latournerie V, Bequignon E, Guillere L, Chabolle F. Are the obstruction sites visualized on drug-induced sleep endoscopy reliable? Sleep Breath. 2015;19(3):1021–6. https://doi.org/10.1007/s11325-014-1107-5.
28. Koutsourelakis I, Safiruddin F, Ravesloot M, Zakynthinos S, de Vries N. Surgery for obstructive sleep apnea: sleep endoscopy determinants of outcome. Laryngoscope. 2012;122(11):2587–91. https://doi.org/10.1002/lary.23462.
29. Soares D, Sinawe H, Folbe AJ, et al. Lateral oropharyngeal wall and supraglottic airway collapse associated with failure in sleep apnea surgery. Laryngoscope. 2012;122(2):473–9. https://doi.org/10.1002/lary.22474.
30. Meraj TS, Muenz DG, Glazer TA, Harvey RS, Spector ME, Hoff PT. Does drug-induced sleep endoscopy predict surgical success in transoral robotic multilevel surgery in obstructive sleep apnea? Laryngoscope. 2017;127(4):971–6. https://doi.org/10.1002/lary.26255.
31. De Vito A, Agnoletti V, Berrettini S, et al. Drug-induced sleep endoscopy: conventional versus target controlled infusion techniques—a randomized controlled study. Eur Arch Otorrinolaringol. 2011;268(3):457–62. https://doi.org/10.1007/s00405-010-1376-y.
32. De Vito A, Agnoletti V, Zani G, et al. The importance of drug-induced sedation endoscopy (D.I.S.E.) techniques in surgical decision making: conventional versus target controlled infusion techniques-a prospective randomized controlled study and a retrospective surgical outcomes analysis. Eur Arch Otorrinolaringol. 2017;274(5):2307–17. https://doi.org/10.1007/s00405-016-4447-x.
33. Chiu FH, Chang Y, Liao WW, et al. Post-operative sleep endoscopy with target-controlled infusion after palatopharyngoplasty for obstructive sleep apnea: anatomical and polysomnographic outcomes. Nat Sci Sleep. 2021;13:1181–93. https://doi.org/10.2147/NSS.S311702.
34. Lin HC, Wang CH, Kuo TBJ, et al. Upper airway surgery or weight control? Modified drug-induced sleep endoscopy for obstructive sleep apnea. Otolaryngol--Head Neck Surg. 2023;169(5):1345–55. https://doi.org/10.1002/ohn.364.
35. Kuo YH, Liu TJ, Chiu FH, et al. Novel intraoral negative airway pressure in drug-induced sleep endoscopy with target-controlled infusion. Nat Sci Sleep. 2021;13:2087–99. https://doi.org/10.2147/NSS.S327770.
36. Yamaguchi Y, Kato M. Pilot study of oral negative pressure therapy for obstructive sleep apnea-hypopnea syndrome. J Sleep Disord Ther. 2017;06(03) https://doi.org/10.4172/2167-0277.1000271.
37. Hung TC, Liu TJ, Hsieh WY, et al. A novel intermittent negative air pressure device ameliorates obstructive sleep apnea syndrome in adults. Sleep Breath. 2019;23(3):849–56. https://doi.org/10.1007/s11325-018-01778-z.
38. Hung TC, Liu TJ, Lu TM, et al. Building a model to precisely target the responders of a novel intermittent negative air pressure device-with mechanism definition. Sleep Med. 2020;72:20–7. https://doi.org/10.1016/j.sleep.2020.03.014.
39. Araghi MH, Chen YF, Jagielski A, et al. Effectiveness of lifestyle interventions on obstructive sleep apnea (OSA): systematic review and meta-analysis. Sleep. 2013;36(10):1553–62. 1562A–1562E. https://doi.org/10.5665/sleep.3056.
40. Peppard PE, Young T, Palta M, Dempsey J, Skatrud J. Longitudinal study of moderate weight change and sleep-disordered breathing. JAMA. 2000;284(23):3015–21. https://doi.org/10.1001/jama.284.23.3015.
41. Safiruddin F, Koutsourelakis I, de Vries N. Analysis of the influence of head rotation during drug-induced sleep endoscopy in obstructive sleep apnea. Laryngoscope. 2014;124(9):2195–9. https://doi.org/10.1002/lary.24598.
42. Shih HW, Chen YL, Lin HC, et al. Potential treatments for Epiglottic collapse in obstructive sleep apnea: how modified drug-induced sleep endoscopy help? Otolaryngol—Head Neck Surg. Published online November 23, 2023. https://doi.org/10.1002/ohn.572.

43. Yalamanchili R, Mack WJ, Kezirian EJ. Drug-induced sleep endoscopy findings in supine vs nonsupine body positions in positional and nonpositional obstructive sleep apnea. JAMA Otolaryngol- Head Neck Surg. 2019;145(2):159–65. https://doi.org/10.1001/jamaoto.2018.3692.
44. Edwards BA, Joosten SA, Sands SA. A detailed mechanistic understanding of positional and nonpositional OSA: what side are you on? Chest. 2024;166(1):7–9. https://doi.org/10.1016/j.chest.2024.03.025.
45. Hsu YS, Chen TY, Wu D, Lin CM, Juang JN, Liu WT. Screening of obstructive sleep apnea in patients who snore using a patch-type device with electrocardiogram and 3-axis accelerometer. J Clin Sleep Med. 2020;16(7):1149–60. https://doi.org/10.5664/jcsm.8462.
46. Liu WT, Lin SY, Tsai CY, et al. Comparison of hospital-based and home-based obstructive sleep apnoea severity measurements with a single-lead electrocardiogram patch. Sensors. 2021;21(23):8097. https://doi.org/10.3390/s21238097.
47. Huon LK, Liu SYC, Shih TTF, Chen YJ, Lo MT, Wang PC. Dynamic upper airway collapse observed from sleep MRI: BMI-matched severe and mild OSA patients. Eur Arch Oto-Rhino-Laryngol. 2016;273(11):4021–6. https://doi.org/10.1007/s00405-016-4131-1.
48. Shelton KE, Woodson H, Gay S, Suratt PM. Pharyngeal fat in obstructive sleep apnea. Am Rev Respir Dis. 1993;148(2):462–6. https://doi.org/10.1164/ajrccm/148.2.462.
49. Hedjoudje A, Abu Dayyeh BK, Cheskin LJ, et al. Efficacy and safety of endoscopic sleeve Gastroplasty: a systematic review and meta-analysis. Clin Gastroenterol Hepatol. 2020;18(5):1043–1053.e4. https://doi.org/10.1016/j.cgh.2019.08.022.
50. Chen TY, Liu CT, Chung CH, Hung SL, Chien WC, Chen JH. Bariatric surgery may provide better protection than uvulopalatopharyngoplasty against major adverse cardiovascular events in obese patients with obstructive sleep apnea. Surg Obes Relat Dis. 2021;17(4):780–91. https://doi.org/10.1016/j.soard.2020.11.018.
51. Wilding JPH, Batterham RL, Calanna S, et al. Once-weekly semaglutide in adults with overweight or obesity. N Engl J Med. 2021;384(11):989–1002. https://doi.org/10.1056/NEJMoa2032183.
52. Currie AC, Kaur V, Carey I, et al. Obstructive sleep apnea remission following bariatric surgery: a national registry cohort study. Surg Obes Relat Dis. 2021;17(9):1576–82. https://doi.org/10.1016/j.soard.2021.05.021.
53. Li HY, Lee LA, Tsai MS, et al. How to manage continuous positive airway pressure (CPAP) failure –hybrid surgery and integrated treatment. Auris Nasus Larynx. Published online May 2020:S0385814620300742. https://doi.org/10.1016/j.anl.2020.03.007.
54. Lin CC, Liu KH, Lee LA, et al. Combined airway and bariatric surgery (CABS) for obstructive sleep apnea patients with morbid obesity: a comprehensive alternative preliminary study. J Clin Med. 2022;11(23):7078. https://doi.org/10.3390/jcm11237078.
55. Cavaliere M, De Luca P, De Santis C, et al. Drug-Induced Sleep Endoscopy (DISE) with simulation bite to predict the success of Oral appliance therapy in treating obstructive sleep apnea/hypopnea syndrome (OSAHS). Transl Med UniSa. 2020;23:58–62. https://doi.org/10.37825/2239-9747.1011.
56. Catalfumo FJ, Golz A, Westerman ST, Gilbert LM, Joachims HZ, Goldenberg D. The epiglottis and obstructive sleep apnoea syndrome. J Laryngol Otol. 1998;112(10):940–3. https://doi.org/10.1017/s0022215100142136.
57. Kuo IC, Hsin LJ, Lee LA, et al. Prediction of Epiglottic collapse in obstructive sleep apnea patients: epiglottic length. Nat Sci Sleep. 2021;13:1985–92. https://doi.org/10.2147/NSS.S336019.
58. Kim HY, Sung CM, Jang HB, Kim HC, Lim SC, Yang HC. Patients with epiglottic collapse showed less severe obstructive sleep apnea and good response to treatment other than continuous positive airway pressure: a case-control study of 224 patients. J Clin Sleep Med JCSM Off Publ Am Acad Sleep Med. 2021;17(3):413–9. https://doi.org/10.5664/jcsm.8904.
59. Kent DT, Rogers R, Soose RJ. Drug-induced sedation endoscopy in the evaluation of OSA patients with incomplete oral appliance therapy response. Otolaryngol Neck Surg. 2015;153(2):302–7. https://doi.org/10.1177/0194599815586978.

60. Golz A, Goldenberg D, Westerman ST, et al. Laser partial epiglottidectomy as a treatment for obstructive sleep apnea and laryngomalacia. Ann Otol Rhinol Laryngol. 2000;109(12 Pt 1):1140–5. https://doi.org/10.1177/000348940010901211.
61. Salamanca F, Leone F, Bianchi A, Bellotto RGS, Costantini F, Salvatori P. Surgical treatment of epiglottis collapse in obstructive sleep apnoea syndrome: epiglottis stiffening operation. Acta Otorhinolaryngol Ital Organo Uff Della Soc Ital Otorinolaringol E Chir Cerv-facc. 2019;39(6):404–8. https://doi.org/10.14639/0392-100X-N0287.
62. Jau JY, Kuo TBJ, Li LPH, et al. Mouth puffing phenomena of patients with obstructive sleep apnea when mouth-taped: device's efficacy confirmed with physical video observation. Sleep Breath. Published online March 11, 2022. https://doi.org/10.1007/s11325-022-02588-0.
63. Jau JY, Kuo TBJ, Li LPH, et al. Mouth puffing phenomenon and upper airway features may be used to predict the severity of obstructive sleep apnea. Nat Sci Sleep. 2023;15:165–74. https://doi.org/10.2147/NSS.S384387.
64. Hsu YS, Ke YK, Kuo TBJ, et al. Expiratory velopharyngeal obstruction: sleep endoscopy-guided treatment strategies to prevent Oral breathing during sleep. Otolaryngol--Head Neck Surg. 2024;171(5):1591–600. https://doi.org/10.1002/ohn.902.

Cardiopulmonary Coupling

4

Liang-Wen Hang and Wan-Ju Cheng

Contents

Abstract

Cardiopulmonary coupling (CPC) refers to the relationship between the cardiovascular and respiratory systems. Cardiopulmonary synchronization is a physiological phenomenon observed during sleep. The synchronization intensity of heart rate variability (HRV) and respiration allows the measurement of sleep characteristics such as sleep stability and fragmentation. In this chapter, the application of CPC in identifying obstructive sleep apnea will be reviewed.

Using Fourier transformation, cross-spectral power and coherence of HRV and respiration signals are calculated, and further used to calculate the ratio of

L.-W. Hang (✉)
College of Medicine, China Medical University, Taichung, Taiwan
Sleep Medicine Center, Department of Pulmonary and Critical Care Medicine, China Medical University Hospital, Taichung, Taiwan
e-mail: lungwen.hang@gmail.com

W.-J. Cheng
National Center for Geriatrics and Welfare Research, National Health Research Institutes, Zhunan, Miaoli County, Taiwan

Department of Public Health, China Medical University, Taichung, Taiwan
e-mail: s871056@nhri.edu.tw

H.-Y. Li (ed.), *Snoring and Obstructive Sleep Apnea*,
https://doi.org/10.1007/978-981-95-2648-2_4

coherent cross-power in different frequencies. The low-frequency coupling (LFC) reflects cyclic pattern across multiple breathing cycles and is associated with cyclic alternating pattern in electroencephalogram. LFC is often linked to disturbed sleep or fragmented sleep. Elevated LFC coupling (e-LFC) is identified apnea/hypopneas has been defined as a minimum low-frequency power > 0.05 normalized units with a low-to-high frequency ratio >30 (Seo et al. Clin Exp Otorhinolaryngol 14:424–6, 2021). Third, very-low-frequency coupling occurs in rapid eye movement (REM) sleep and wakefulness. The logarithm of high-to-low frequency CPC ratio is often used for measuring continuous varying CPC.

Patients with obstructive sleep apnea (OSA) have increased LFC, and continuous positive airway pressure (CPAP) therapy decreases LFC. Simple CPC showed a high correlation with AHI from polysomnography (PSG). CPC data combined with oxygen desaturation to calculate AHI has been approved by the USA Food and Drug Administration for diagnosis of OSA. Notably, AHI from CPC was lower than PSG in pediatric patients with non-severe OSA.

Keywords

Cardiopulmonary coupling (CPC) · Obstructive sleep apnea

4.1 Introduction

Cardiopulmonary coupling (CPC) refers to the relationship between the cardiovascular and respiratory systems, particularly the synchronization of heart rate with respiratory patterns. Cardiopulmonary synchronization is a physiological phenomenon observed during sleep, where the heart's electrical activity and respiratory activity interact in specific patterns. Continuous electrocardiogram (ECG) or photoplethysmography (PPG) signals are used to retrieve pulse rate variability between RR intervals and respiration using R-wave amplitude fluctuations, called the ECG-derived respiration (EDR) signal. The synchronization intensity of heart rate variability (HRV) and respiration allows the measurement of sleep characteristics [1] (Fig. 4.1).

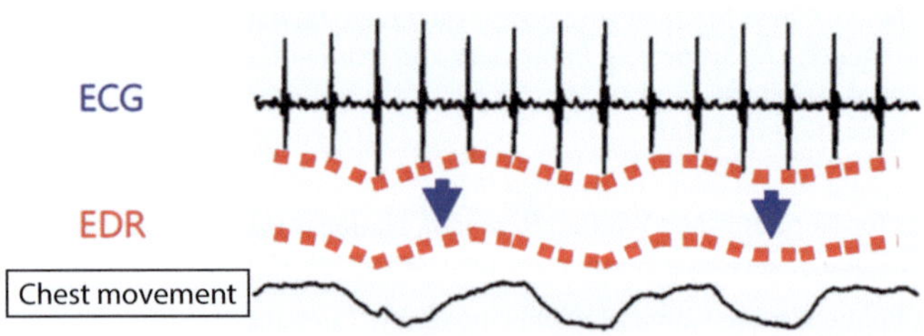

Fig. 4.1 ECG-derived respiration from R-wave amplitude coincides with chest movement

Its main applications include sleep quality assessment, e.g., sleep stability and fragmentation, and cardiac and respiratory health, which reflects autonomic dysregulation in chronic heart failure or chronic obstructive pulmonary disease. For sleep disorders, CPC has been used to evaluate periodic limb movements and obstructive sleep apnea. In this chapter, the application of CPC in identifying obstructive sleep apnea will be reviewed.

4.2 Key Features of CPC

Using Fourier transformation, cross-spectral power, and coherence of HRV and respiration signals are calculated, and further used to calculate the ratio of coherent cross-power in different frequencies: very low (0.004–0.01 Hz), low (0.01–0.1 Hz), high (0.1–0.4 Hz) (Table 4.1). The coupling signal strengthens when heart rate variations are synchronized with breathing signals and weakens when they are not synchronized. High-frequency coupling is associated with stable sleep, particularly during deep non-REM sleep including part of stage 2 and all stage 3 NREM sleep (Fig. 4.2a). It reflects physiological sinus arrhythmia due to breath-to-breath fluctuation, representing an efficient interaction between the autonomic nervous system and respiratory control. Whereas low-frequency coupling (LFC) reflects cyclic pattern across multiple breathing cycles and is associated with cyclic alternating pattern in electroencephalogram. LFC is often linked to disturbed sleep or fragmented sleep, including all stage 1 and part of stage 2 sleep. It is also seen in conditions like sleep apnea or insomnia (Fig. 4.2b). Elevated power in LFC coupling (e-LFC) to identify apnea/hypopneas has been defined as a minimum low-frequency power > 0.05 normalized units with a low-to-high frequency ratio > 30 [2]. Third, very-low-frequency coupling occurs in REM sleep and wakefulness (Fig. 4.2b). The logarithm of high-to-low frequency CPC ratio is often used for measuring continuous varying CPC.

Table 4.1 Characteristics of CPC spectrum

CPC characteristics	Frequency	Sleep stage
High frequency coupling	0.1–0.4 Hz	Deep NREM sleep
Low frequency coupling	0.01–0.1 Hz	Fragmented sleep and light NREM sleep
Elevated power in low frequency coupling	0.01–0.1 Hz with high power	Respiratory events
Very low frequency coupling	<0.01 Hz	REM/awake

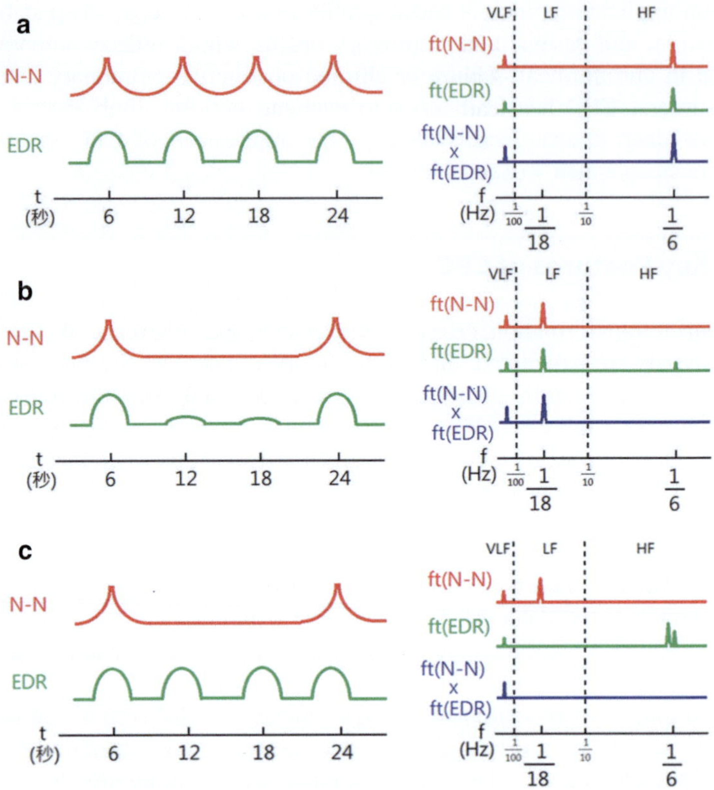

Fig. 4.2 (**a**) Left panel shows signals in time domain and right panel for frequency domain. The blue line in the right panel shows the coupling signal, with the strongest energy at 1/6 Hz, indicating a 6-second cycle of heart rate variation synchronized with breathing. This cycle falls within the normal breathing frequency range (HF). (**b**) The coupling signal (blue curve) shows the strongest energy at 1/18 Hz, indicating an 18-second cycle of heart rate variation synchronized with breathing. This cycle falls within the frequency range of respiratory events (LF) and is associated with elevated low-frequency coupling (e-LFC), indicating the occurrence of a sleep apnea (sleep-disordered breathing) event during this period. (**c**) The coupling signal (blue curve) shows the strongest energy at below 1/100 Hz, indicating heart rate variations synchronized with breathing occurring at intervals longer than 100 seconds. This cycle falls within the REM and wake frequency range (VLF) and can be identified as either REM sleep or wakefulness

4.3 CPC and OSA

Patients with OSA have increased LFC, and CPAP therapy decreases LFC. Simple CPC showed a high correlation with apnea-hypoponea index (AHI) from PSG. [3] CPC data combined with oxygen desaturation to calculate AHI has been approved by the USA FDA for diagnosis of OSA. Notably, AHI from CPC was lower than PSG in pediatric patients with non-severe OSA [4].

Case Example

In a patient with OSA, Fig. 4.3a shows the CPC during deep sleep. Breathing rate is stable at approximately 12–24 breaths per minute. The synchronization between heart rate and EDR signals is clearly reflected in the CPC coupling spectrum in the 0.1–0.4 Hz range (high frequency). Figure 4.3b shows the CPC during awake or REM sleep, showing very low frequency coupling (<0.01 Hz). During light sleep, the heart

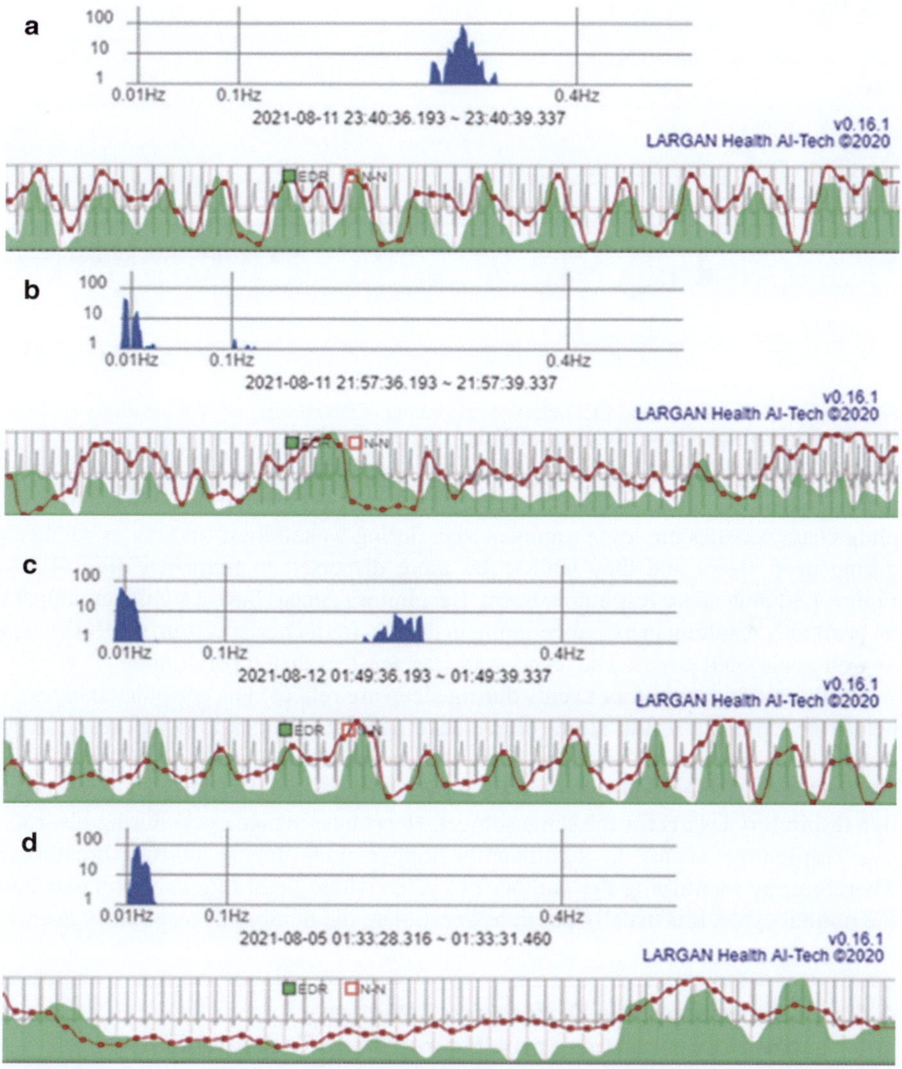

Fig. 4.3 (**a**) CPC in high frequency spectrum during deep sleep. (**b**) CPC in very low frequency spectrum during awake or REM sleep. (**c**) CPC in low frequency spectrum during light NREM sleep. (**d**) CPC during a respiratory event. Red line indicates R–R intervals, and green shade indicates ECG-derived respiration (EDR) signal

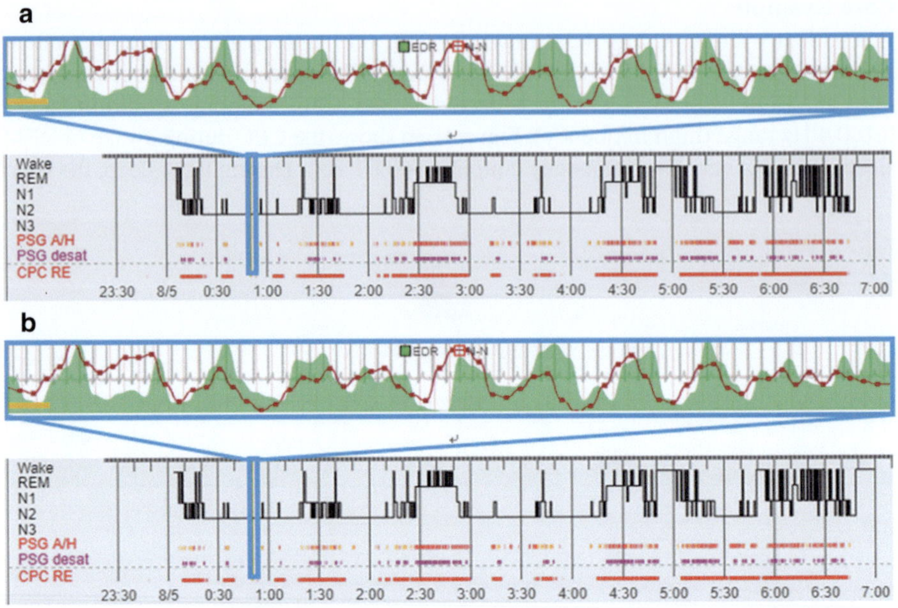

Fig. 4.4 (**a**) R–R interval and ECG-derived respiration (EDR) during normal breathing at sleep. (**b**) R–R interval and ECG-derived respiration (EDR) during obstructive sleep apneas

rate is sometimes influenced by breathing and sometimes by other factors. The coupling characteristics are less prominent than during wakefulness and not as stable as during deep sleep, and they tend to be more dispersed in frequency (Fig. 4.3c). Figure 4.3d indicates a respiratory event. Respiratory apneas have a significant impact on heart rate, resulting in a clear coupling in the low frequency spectrum (0.01–0.1 Hz) as well as elevated power. The respiratory rate was less than 6 per minute.

Furthermore, respiratory events during sleep are reflected as periodic changes in heart rate [5]. As a result, studies have developed techniques to detect respiratory events by analyzing HRV, such as cyclic variation of heart rate (CVHR) [6]. Figure 4.4 shows heart rate changes (red line) during normal breathing and obstructive respiratory events for the same subject. Heart rate change cycle during obstructive respiratory events is significantly longer than during normal breathing. Therefore, by identifying the number of cycles where heart rate variation exceeds the normal cycle, it is usually possible to estimate the number of respiratory events.

4.4 Comparing the Difference Between the Polysomnography, Home Sleep Test and CPC

Compared to PSG, which determines sleep states by observing overt physiological phenomena (e.g., EEG, airflow, chest and abdominal movements), CPC uses heart rate variability characteristics that are less affected by individual differences to assess sleep states. Therefore, CPC is well suited for automated computer analysis, providing rapid and objective detection results. Table 4.2 compares the signals collected by CPC, home sleep tests, and PSG, along with their tolerability and disease detection target in clinical use. Using CPC to identify sleep apneas and hypopneas hold the advantage of the ease to integrate into wearable devices. Due to its single signal nature, it is easy to use by non-health care professionals. Additionally, the cost is low and can be used to track changes over time and intervention by repeat measurement.

Although CPC can extract valid cardiopulmonary coupling features from complex heart rate variations, it has limitations. Some conditions may result in unclear coupling, such as severe arrhythmias or patients with pacemakers. Additionally, cardiopulmonary systems of children under 6 years old is under development, and due to incomplete clinical data, CPC is not suitable for sleep monitoring in this age group.

Table 4.2 Comparison between CPC, home sleep test, and PSG in clinical use

	CPC	Home sleep test	PSG
Signals	ECG, respiration, SpO_2, body position and movements	Nasal air flow, respiration, pulse, SpO_2	EEG, EOG, nasal flow, oronasal thermistor, sound recordings, piezoelectric sensor for snore detection, videotaping, SpO_2, chest and abdominal movements, ECG, position, EMG
Tolerability	High	Medium	Low
Complexity of usage	Self-administered	Self-administered	Sleep technician needed
Consumables	ECG patches	Nasal flow sensor	ECG patches and nasal flow sensor
Abnormality detection	OSA, insomnia	OSA	OSA, limb movements, parasomnia
Parameters	Total sleep time, sleep quality, position, desaturation, leave bed, HRV	Desaturation	Total sleep time, sleep quality, position, desaturation, leave bed

*The characteristics for CPC were exemplified using the commercial instrument Largan DOMII®

4.5 CPC and Loop Gain in OSA

Identifying endotypic traits that contribute to sleep apnea, i.e., upper airway collapsibility, arousal threshold, upper airway dilator muscle function, and loop gain, is gaining importance in personalized treatment. CPC is able to distinguish a respiratory pattern of periodic breathing, complex sleep apnea, and central sleep apnea, using narrow spectral band (0.006–0.1 Hz) e-LFC [2]. This finding suggest that high loop gain may be identifiable using CPC; nevertheless, no studies have examined the association between CPC and endotypic traits directly.

References

1. Lu M, Penzel T, Thomas RJ. Cardiopulmonary coupling. Adv Exp Med Biol. 2022;1384:185–204.
2. Seo MY, Yoo J, Hwang SJ, Lee SH. Diagnosis of obstructive sleep apnea in adults using the cardiopulmonary coupling-derived software-generated apnea-hypopnea index. Clin Exp Otorhinolaryngol. 2021;14(4):424–6.
3. Zhai F, Li Y, Chen J. Comparison of polysomnography, sleep apnea screening test and cardiopulmonary coupling in the diagnosis of pediatric obstructive sleep apnea syndrome. Int J Pediatr Otorhinolaryngol. 2021;149:110867.

4. Mietus JE, Peng CK, Ivanov PC, Goldberger AL. Detection of obstructive sleep apnea from cardiac interbeat interval time series. Comput Cardiol. 2000;27:753–6.
5. Hayano J, Watanabe E, Saito Y, Sasaki F, Fujimoto K, Nomiyama T, et al. Screening for obstructive sleep apnea by cyclic variation of heart rate. Circ Arrhythm Electrophysiol. 2011;4(1):64–72.
6. Thomas RJ, Mietus JE, Peng CK, Gilmartin G, Daly RW, Goldberger AL, et al. Differentiating obstructive from central and complex sleep apnea using an automated electrocardiogram-based method. Sleep. 2007;30(12):1756–69.

Application of Ultrasound in Obstructive Sleep Apnea Management

5

Jeng-Wen Chen, Wan-Ni Lin, and Chih-Chung Huang

Contents

Abstract

This chapter investigates the promising role of ultrasound imaging in managing obstructive sleep apnea (OSA). OSA affects millions globally and is associated with severe health risks, yet conventional diagnostic techniques like polysomnography often fall short in identifying specific sites of airway obstruction.

J.-W. Chen (✉)
Department of Otorhinolaryngology–Head and Neck Surgery, Cardinal Tien Hospital and Fu Jen Catholic University, New Taipei City, Taiwan
e-mail: 086365@mail.fju.edu.tw

W.-N. Lin
Department of Otorhinolaryngology–Head and Neck Surgery, Sleep Center, Linkou Chang Gung Memorial Hospital, Taoyuan, Taiwan

C.-C. Huang
Department of Biomedical Engineering, National Cheng Kung University, Tainan, Taiwan

H.-Y. Li (ed.), *Snoring and Obstructive Sleep Apnea*,
https://doi.org/10.1007/978-981-95-2648-2_5

Ultrasound is a non-invasive, radiation-free, and cost-effective alternative that allows for real-time evaluation of upper airway structures such as the tongue and pharyngeal walls. Various ultrasound modalities, including amplitude mode (A-mode), brightness mode (B-mode), Doppler mode, backscatter ultrasound imaging, and elastography, enable detailed assessments of tissue morphology, vascularity, and stiffness, offering valuable insights into OSA's pathophysiology. While advantages exist, challenges remain regarding user dependency and the need for standardized protocols. This chapter advocates for future research to enhance the diagnostic capabilities of ultrasound and its integration with other diagnostic modalities, paving the way for improved patient outcomes in OSA management.

Keywords

Obstructive sleep apnea · Ultrasound · Elastography · Upper airway · Tongue · Pharynx

Abbreviations

A-mode	amplitude mode
AHI	apnea-hypopnea index
BUI	backscatter ultrasound imaging
BMI	body mass index
B-mode	brightness mode
CT	computed tomography
DLA	distance between the lingual arteries
ICA	internal carotid artery
LPWT	lateral pharyngeal wall thickness
MRI	magnetic resonance imaging
MAD	mandibular advancement device
MM	Müller maneuver
OSA	obstructive sleep apnea
PSG	polysomnography
SWE	shear wave elastography
TBT	tongue base thickness
TORS	transoral robotic surgery
UA	upper airway
UAS	upper airway stimulation

5.1 Overview

Obstructive sleep apnea (OSA) is a prevalent sleep disorder causing recurrent partial or complete upper airway (UA) blockages during sleep, leading to intermittent hypoxemia, frequent arousals, and disrupted sleep patterns [1]. Globally, OSA impacts millions and is linked to significant health issues, such as cardiovascular and cerebrovascular conditions, metabolic disorders, and cognitive decline [2]. While polysomnography (PSG) remains the diagnostic gold standard, it has limitations in pinpointing the sites of UA obstruction and underlying mechanisms of OSA [3].

Ultrasound imaging has emerged as a promising, non-invasive, cost-effective, and readily available diagnostic tool for evaluating the UA in patients with OSA [4]. Unlike other imaging modalities such as magnetic resonance imaging (MRI) or computed tomography (CT), ultrasound does not involve radiation, is portable, and can be performed at the point of care [5]. It allows for real-time, dynamic assessment of soft tissue structures, including the tongue and pharyngeal walls [6, 7].

Ultrasound imaging encompasses various modalities that provide distinct types of information about the UA in the context of OSA [8]. For instance, brightness mode (B-mode) ultrasound, the most commonly used modality, visualizes tissue morphology by generating a grayscale image based on the intensity of reflected sound waves [3, 4, 6, 9, 10]. Doppler mode evaluates tissue vascularization by detecting blood flow within vessels [11, 12]. Elastography, which includes compression sonoelastography and shear wave elastography (SWE), measures the mechanical properties of tissues, such as stiffness [13, 14]. Additionally, other quantitative ultrasound techniques for tissue characterization [8], such as backscatter ultrasound imaging (BUI) [15, 16] and amplitude mode (A-mode) imaging [17], analyze ultrasound waves to examine tissue microstructure.

These ultrasound modalities offer a comprehensive toolkit for assessing the UA in OSA. B-mode provides fundamental morphological data, while Doppler and elastography supply information on tissue vascularization and stiffness. Quantitative techniques, such as BUI and A-mode, offer detailed insights into tissue composition and interfaces, supporting the potential of machine learning algorithms applied to ultrasound data for OSA severity classification [18]. Dynamic measurements, including the Müller maneuver (MM) during wakefulness [4, 6, 13, 19], drug-induced sleep [20, 21], and whole-night monitoring during natural sleep [17, 22, 23], combined with methods such as the modified optical flow-based method [24], facilitate the analysis of physiological changes during airway obstruction events. These modalities are invaluable for both research and clinical applications in the management of OSA. This chapter will further explore the various applications of ultrasound in the management of OSA and its integration with other diagnostic and treatment modalities.

5.2 Anatomical Assessment

5.2.1 Tongue Base Thickness

The ultrasound measurement of tongue base thickness (TBT) is a significant area of focus in OSA research, with studies suggesting its potential as both a diagnostic indicator and marker of the disease's underlying mechanism [21]. The rationale for this focus stems from the fact that OSA is characterized by repetitive collapse of the pharyngeal airway during sleep, with the tongue base being a critical anatomical structure that can contribute significantly to this obstruction [6, 25]. Research indicates that an increase in tongue size [16], altered tongue mechanics [19], or posterior displacement of the tongue base [4, 6] measured during wakefulness can be associated with the narrowing or collapse of the UA during sleep. Assessing TBT, therefore, offers a way to identify individuals at risk of OSA and to better understand the pathophysiology of the disease, with the potential to aid in treatment planning.

Typically, a submental approach is used, with the transducer placed under the chin to visualize the tongue base in either a sagittal [6] or coronal view [4]. In the sagittal view (Fig. 5.1), the distance from the skin surface to the dorsal surface of the tongue base, namely air-mucosa interface, is measured, while the coronal view provides a cross-sectional assessment (Fig. 5.2). Measurements can be taken during quiet breathing as well as during the MM [4, 6, 10, 13, 18, 19, 24], a forced inspiratory effort against an obstructed airway that simulates the airway collapse seen in OSA. While ultrasound is the most common, other imaging modalities, such as MRI and CT, can also be used to assess tongue base dimensions, though they are less available for dynamic clinical assessment [26, 27].

Current literature suggests a strong association between TBT and OSA, with several studies highlighting that patients with OSA tend to have thicker tongues compared to controls [11, 16]. Dynamic TBT measurements during the MM, or the difference in TBT with and without the MM, are considered a better predictor of OSA than static measurements [4, 6]. Some studies also report a correlation between TBT and the severity of OSA, where patients with more severe OSA show greater tongue base thickness [19]. Furthermore, TBT has been shown to increase during drug-induced sleep [20, 21], which is linked to tongue base collapse and OSA, as well as during snoring, hypopnea, and apnea events during natural sleep [17, 22, 23], suggesting a posterior displacement of the tongue base during obstructive

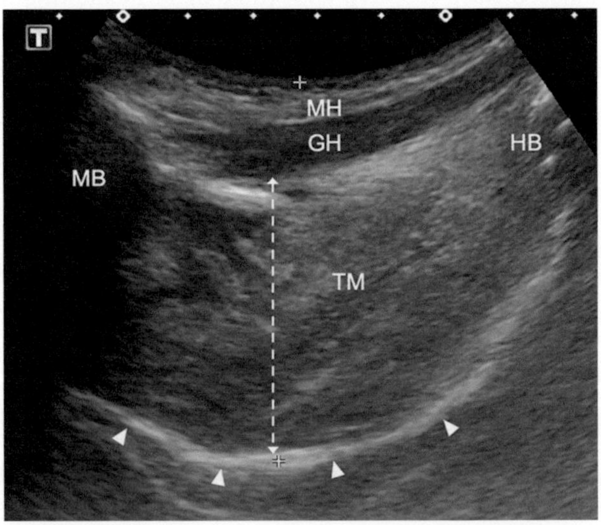

Fig. 5.1 Ultrasound imaging of the tongue in the midline sagittal plane. The arrowheads indicate the dorsal surface of the tongue or air–mucosa interface. The length of the white-dashed line indicates the thickness of the tongue. TM tongue muscle, MH mylohyoid muscle, GH geniohyoid muscle, MB mandible, and HB hyoid bone

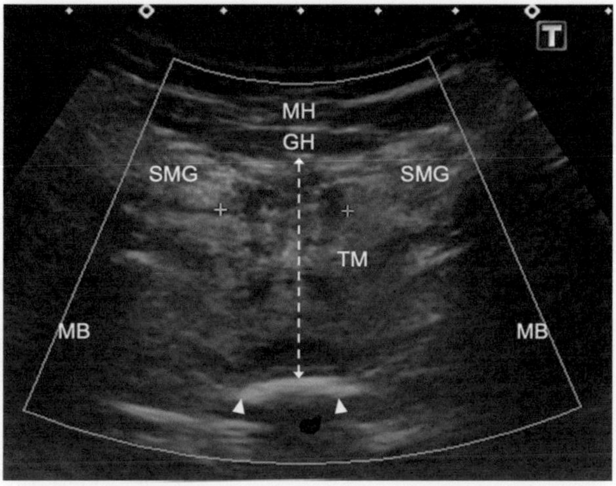

Fig. 5.2 Ultrasound imaging of the tongue in the coronal plane. The arrowheads indicate the dorsal surface of the tongue or air–mucosa interface. The length of the white-dashed line indicates the thickness of the tongue. TM tongue muscle, MH mylohyoid muscle, GH geniohyoid muscle, SMG submandibular gland, MB mandible, and HB hyoid bone

respiratory events. Recent studies are emphasizing the use of standardized ultrasound protocols and automated algorithms to minimize bias and improve measurement reliability [19]. However, the current literature is not without limitations, including variations in measurement methods, a lack of standardized protocols, small sample sizes, and underrepresentation of certain populations [28].

5.2.2 Distance Between Lingual Arteries

The research on the measurement of the distance between the lingual arteries (DLA) indicate its potential as a marker of disease severity and a means of understanding UA anatomy [4]. The rationale behind measuring DLA stems from the fact that the lingual arteries run along the lower lateral borders of the tongue base, and the distance between them reflects the width of the tongue base. Changes in this distance may indicate alterations in tongue size or position, which are known to play a critical role in the pathophysiology of OSA [3, 10, 20, 29, 30].

Through ultrasound imaging, a submental approach is used, with the transducer positioned in the transverse plane at a midpoint between the inferior border of the mandible and the hyoid bone [20, 30]. Color Doppler mode is often employed to visualize the lingual arteries as they enter the tongue base [11, 12]. The distance between the two arteries is then measured. Some studies take measurements during wakefulness, [4] while others take measurements during drug-induced sleep to assess changes in DLA [20, 21].

Current literature reveals that DLA is significantly wider in patients with OSA compared to controls. Moreover, studies have found a positive correlation between DLA and both the apnea–hypopnea index (AHI) and body mass index (BMI) [3, 20, 30]. Research suggests that an increased BMI might lead to fat accumulation in the tongue, pushing the lingual arteries laterally, thus increasing the DLA. Furthermore, it has been observed that during drug-induced sleep, the DLA elongates significantly, which is hypothesized to be due to the flattening of the tongue muscle [20]. Studies also suggest that a DLA greater than 30 mm is associated with an increased risk of OSA [3]. Some researchers have used DLA as a parameter to define safety zones during tongue base surgeries [11].

5.2.3 Lateral Pharyngeal Wall Thickness

The rationale for measuring lateral pharyngeal wall thickness (LPWT) is based on the understanding that the lateral pharyngeal walls, composed of muscles, lymphoid tissue, and fat, can contribute to airway narrowing and collapse in OSA [9]. An increase in LPWT can reduce the volume of the UA, leading to increased collapsibility and an increased risk of OSA [7].

The typical approach involves placing a linear transducer on the lateral side of the neck, just below the lateral border of the occipital bone, in an oblique coronal plane [7]. Using color Doppler, the internal carotid artery (ICA) is identified, and the LPW is visualized as an echogenic air–mucosa interface. The LPWT is measured as the distance between the medial wall of the ICA and the echogenic surface of the pharynx [9, 31]. Measurements are taken during quiet breathing, and sometimes during the MM, to assess dynamic changes [9]. Some studies also measure the distance from the echogenic surface of the pharynx to the skin, which represents the parapharyngeal neck thickness [9, 28].

The current literature indicates that LPWT is significantly increased in patients with OSA compared to non-apneic individuals [8, 29, 30]. Moreover, a positive correlation exists between LPWT and AHI, indicating that a greater LPWT is associated with more severe OSA [8, 9, 30]. This correlation persists even after adjusting for confounding factors such as age, sex, body mass index (BMI), and neck circumference [7, 10, 30]. Studies have shown that the LPWT during MM and the change in LPWT between rest and the MM are also associated with OSA severity [9]. Some studies suggest that LPWT is a heritable trait that may be genetically linked to OSA severity [7]. While ultrasound measurements are considered reliable, some research indicates that they may overestimate LPWT compared to MRI measurements [7, 30].

5.3 Dynamic Evaluation

Dynamic evaluation of the UA, especially the tongue base, is crucial for understanding OSA because it captures changes in airway structures during different states and correlates these changes with respiratory events. This approach is essential because OSA is characterized by a dynamic collapse of the UA rather than a static anatomical abnormality [21].

During wakefulness, the MM is often employed to simulate airway collapse [19, 24]. Ultrasound imaging is used during the MM to assess changes in TBT, the DLA, and LPWT in response to negative pressure. Modified optical flow-based methods are used to track the complex dynamic movements of the tongue during the MM [19, 24]. During natural sleep, PSG is used to monitor respiratory events concurrently with wearable ultrasound device to capture changes in TBTs. This allows for the correlation of real-time changes in tongue dimensions with apneas, hypopneas, and snoring [17, 22, 23]. Additionally, drug-induced sleep ultrasonography is used to evaluate TBT and other parameters during induced sleep, which allows for more accurate assessment of tongue base collapse [20, 21].

Current literature suggests that patients with moderate-to-severe OSA demonstrate minimal tongue movement compared to controls or those with mild OSA during wakefulness, which indicates a compromised neuromuscular response [19, 27]. Moreover, during natural sleep, TBT increases significantly during obstructive respiratory events, with the greatest increase when apnea occurred [17, 22]. Different patterns of tongue deformation during sleep help in understanding the specific mechanisms of obstruction [23]. Overall, dynamic evaluation provides a more comprehensive understanding of the biomechanical and physiological factors contributing to OSA.

5.4 Tissue Characterization

Tissue characterization of the UA, particularly the tongue, is an emerging area of focus in understanding the pathophysiology of OSA. The rationale behind tissue characterization lies in the fact that the composition and mechanical properties of soft tissues in the UA, such as muscle, fat, and connective tissue, can significantly affect its collapsibility and contribute to the development of OSA [1, 13–15]. Characterizing the tissue microstructure and mechanical properties can offer insights into the underlying causes of UA obstruction, beyond what can be ascertained from anatomical measurements alone [13–15].

Several methods are utilized for tissue characterization, with ultrasound-based techniques being prominent due to their non-invasiveness and accessibility. SWE is a quantitative ultrasound technique that assesses tissue stiffness by measuring the propagation speed of shear waves through the tissue (Fig. 5.3). This technique allows for the objective evaluation of the mechanical properties of tongue muscles and other soft tissues [13, 14]. Studies using SWE have quantified the stiffness of tongue tissue during wakefulness and during the MM. Additionally, BUI is another quantitative ultrasound technique that analyzes the backscatter intensity of ultrasound waves, which is sensitive to changes in tissue microstructure [8, 15, 16]. BUI can provide detailed information about tissue composition, such as the presence of fat or fibrosis, that are not readily apparent in conventional B-mode ultrasound.

Current literature indicates that tongue tissue stiffness, as measured by SWE, is altered in patients with OSA. Studies have found that tongue tissues, especially in the posterior third of the tongue, are significantly stiffer in awake patients with OSA than in controls during the MM [13]. Other studies have found that tongue hardness is lower in patients with OSA compared to controls, possibly due to increased fatty infiltration at the base of the tongue [27]. BUI analysis has shown that the tissue

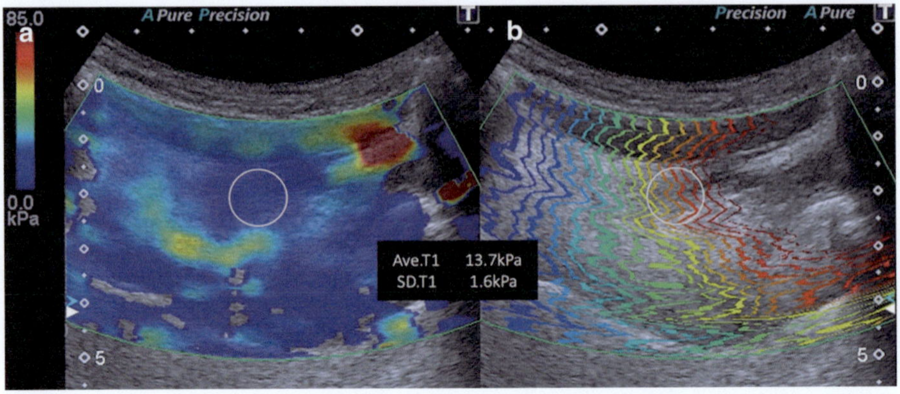

Fig. 5.3 The color diagram (**a**) and propagation modes (**b**) of shear wave ultrasound elastography imaging for the tongue in the midline sagittal plane. The average shear modulus of the tongue in the selected region of interest is 13.7 kPa with a standard deviation of 1.6 kPa

microstructure of the tongue correlates with OSA severity, with the predictive model of OSA severity being more robust when BUI parameters are included with age and BMI [15]. These findings suggest that the mechanical properties and composition of tongue tissue play a significant role in the development of OSA. Furthermore, some studies have shown that the stiffness of tongue muscles does not vary according to the severity of OSA, indicating that other factors such as muscle thickness might be more significant in the pathology of OSA [32]. These studies are still relatively new, and further investigation is needed to determine the clinical implications of these findings and how they can be applied to OSA diagnosis and treatment planning.

5.5 Treatment Guidance

Ultrasound treatment guidance in OSA is evolving toward a more personalized approach, driven by the increasing recognition of the heterogeneity in OSA pathophysiology [16]. The rationale for individualized treatment stems from the understanding that OSA is not a single disease entity but rather a syndrome with various contributing factors, including anatomical, neuromuscular, and physiological traits [16]. Therefore, effective treatment strategies should be tailored to address the specific underlying mechanisms contributing to airway obstruction in each patient.

Methods for treatment guidance involve a comprehensive assessment of the UA and its function. Specifically, dynamic ultrasound assessments during wakefulness and sleep, including the MM and drug-induced sleep ultrasonography, are used to evaluate the dynamic changes in TBT and other parameters. These techniques help to identify the specific sites and patterns of airway collapse, guiding the selection of appropriate interventions [21]. Furthermore, tissue characterization using SWE and BUI can provide additional information about the mechanical properties and composition of the tongue tissue, which can be relevant for treatment planning. For example, patients with increased tongue stiffness or fat infiltration may be less likely to respond to certain therapies, such as mandibular advancement devices (MADs) [15].

Current literature highlights the potential of ultrasound to guide treatment decisions. Studies have shown that TBT during drug-induced sleep ultrasonography is correlated with OSA severity, suggesting that this measure can be used to evaluate the efficacy of treatments targeting the tongue base [21]. In addition, intraoperative ultrasound is used to guide transoral robotic surgery (TORS) for tongue base reduction, allowing surgeons to safely remove a larger volume of tissue while avoiding damage to the lingual artery [11]. These studies demonstrate that intraoperative ultrasound-assisted surgery leads to a greater resected base of tongue (BOT) volume, and is associated with better improvement in AHI and reduced bleeding risk. Moreover, the use of ultrasound to assess UA muscle collapsibility can help determine the suitability of different treatment options, such as upper airway stimulation (UAS), by evaluating the response to electrical stimulation [16]. Additionally, artificial intelligence models using ultrasound images to classify OSA severity may potentially aid clinicians in categorizing OSA severity and tailoring treatment plans [18]. Overall, ultrasound and its related techniques are shown to be promising in bridging the gap between anatomy and physiology, thus leading to more informed treatment choices for patients with OSA.

5.6 Pediatric OSA Applications

Pediatric OSA applications of ultrasound are increasingly being explored due to the unique challenges in diagnosing and managing OSA in children [28]. The rationale for using ultrasound in pediatric OSA stems from the fact that traditional methods like PSG are often difficult to perform, resource-intensive, and not easily accessible. Furthermore, questionnaires used for screening may not be as effective in children as in adults. Ultrasound provides a non-invasive, radiation-free, and portable alternative that can be used to evaluate UA structures, which is particularly advantageous in the pediatric population. Additionally, ultrasound can be performed without sedation, unlike CT or MRI, which minimizes effects on airway tone [28, 33].

Methods for ultrasound assessment in pediatric OSA primarily focus on measuring various anatomical parameters of the UA. These include LPWT, total neck thickness at the retropharyngeal level, tonsil volume, and adenoid thickness. LPWT is measured as the distance between the ICA and the echogenic surface of the pharynx [9]. Tonsil volume can be measured using submental ultrasound and compared with surgical specimens, although it does not appear to correlate with OSA diagnosis or severity [9]. Adenoid thickness is also measured and has been shown to correlate with radiological and endoscopic measures of size and occlusion [34]. Some studies also use the MM to assess the collapsibility of the UA during forced deep inspiration [9].

Current literature indicates that several ultrasound parameters correlate with OSA diagnosis or other reference measurements in children. LPWT and total neck thickness have been shown to correlate with moderate-to-severe OSA [9]. Some studies have found that LPWT is associated with an AHI >1, and more strongly with moderate-to-severe OSA (AHI ≥ 5) [28]. Increased LPWT is suggested to compress the airway, increasing airway collapsibility. A study that compared ultrasound and MRI measurements of LPWT in children found moderate agreement between the two [7]. While tonsil volume by ultrasound correlates well with the volume of surgical specimens, it does not correlate with OSA diagnosis [28]. However, some studies suggest that the space occupied by the tonsils in the airway is more predictive of OSA severity than the size itself [34]. Adenoid thickness has been shown to correlate with endoscopic and radiological measures, but its direct correlation with OSA has not been fully explored [28]. Overall, while ultrasound is a promising tool, there is a need for more research to standardize the definitions of pediatric OSA, validate diagnostic value, and report findings in relation to established ranges.

5.7 Advantages of Ultrasound in OSA

Ultrasound offers several advantages in the evaluation of OSA, primarily due to its non-invasive, radiation-free, and portable nature. Unlike other imaging modalities such as CT or MRI, ultrasound is inexpensive and readily available in various clinical settings. It allows for dynamic assessments of the UA and is not limited to static images. Ultrasound enables real-time visualization of soft tissues, including the tongue, pharyngeal walls, and surrounding structures. Additionally, it can be used to assess muscle function with techniques like elastography and evaluate blood flow with Doppler mode. These features make ultrasound a versatile tool for OSA assessment, suitable for repeated measurements and monitoring treatment responses.

5.8 Limitations and Future Directions

Ultrasound in OSA has limitations, including its user-dependent nature requiring specialized training to acquire and interpret images. Image quality can be affected by patient factors, such as neck circumference and tissue composition, as well as probe positioning and pressure. Standardization of ultrasound protocols is necessary to overcome operator bias and ensure reproducibility. Ultrasound is limited in its ability to visualize all areas of the pharynx due to acoustic shadows from bony structures, and it may not fully evaluate circumferential or lateral wall obstructions. Furthermore, measurements are often taken during wakefulness, not during sleep, which may not fully represent the dynamic changes that occur during sleep in patients with OSA.

Future research should focus on standardizing ultrasound protocols and exploring dynamic ultrasound assessments during sleep. Combining ultrasound with other techniques such as PSG and drug-induced sleep endoscopy may provide more comprehensive insights into OSA pathophysiology. Further investigation into tissue characterization techniques, like SWE and BUI, may enhance the diagnostic and prognostic value of ultrasound. Additionally, integrating artificial intelligence models with ultrasound imaging may help automate image analysis and improve the accuracy of OSA classification. Future studies should focus on larger populations, balanced representation of sex and race, and examination of other UA muscles involved in OSA.

Take Home Messages

- Ultrasound provides a non-invasive, radiation-free method for real-time evaluation of the UA structures, allowing for dynamic assessments that can capture changes occurring during sleep.
- Although ultrasound is a versatile and cost-effective tool for diagnosing OSA, it has limitations related to operator dependency and the need for standardized protocols. Future research is essential to enhance the reproducibility and accuracy of ultrasound assessments in clinical practice.
- Combining ultrasound with other diagnostic modalities, such as PSG or drug-induced sleep endoscopy, holds promise for providing a more comprehensive understanding of OSA pathophysiology and improving clinical outcomes for patients.

References

1. Strollo PJ Jr, Rogers RM. Obstructive sleep apnea. N Engl J Med. 1996;334(2):99–104.
2. Young T, Palta M, Dempsey J, Skatrud J, Weber S, Badr S. The occurrence of sleep-disordered breathing among middle-aged adults. N Engl J Med. 1993;328(17):1230–5.
3. Lahav Y, Rosenzweig E, Heyman Z, Doljansky J, Green A, Dagan Y. Tongue base ultrasound: a diagnostic tool for predicting obstructive sleep apnea. Ann Otol Rhinol Laryngol. 2009;118(3):179–84.
4. Shu CC, Lee P, Lin JW, Huang CT, Chang YC, Yu CJ, et al. The use of sub-mental ultrasonography for identifying patients with severe obstructive sleep apnea. PLoS One. 2013;8(5):e62848.
5. Kwan BCH, Juge L, Gandevia SC, Bilston LE. Sagittal measurement of tongue movement during respiration: comparison between ultrasonography and magnetic resonance imaging. Ultrasound Med Biol. 2019;45(4):921–34.
6. Chen JW, Chang CH, Wang SJ, Chang YT, Huang CC. Submental ultrasound measurement of dynamic tongue base thickness in patients with obstructive sleep apnea. Ultrasound Med Biol. 2014;40(11):2590–8.
7. Yuen HM, Lai ACY, Liu EKH, Lee MC, Chu WCW, Chan JWY, et al. Validation of the sonographic measurement of lateral parapharyngeal wall thickness in childhood obstructive sleep apnea. Nat Sci Sleep. 2022;14:2013–21.
8. de Angelo LA, Pereira FL, Duarte BB, Cahali MB. Use of ultrasonography in the evaluation of patients with sleep apnea: a systematic review. Braz J Otorhinolaryngol. 2024;90(6):101468.
9. Lin CY, Chen CN, Kang KT, Hsiao TY, Lee PL, Hsu WC. Ultrasonographic evaluation of upper airway structures in children with obstructive sleep apnea. JAMA Otolaryngol Head Neck Surg. 2018;144(10):897–905.
10. Singh M, Tuteja A, Wong DT, Goel A, Trivedi A, Tomlinson G, et al. Point-of-care ultrasound for obstructive sleep apnea screening: are we there yet? A systematic review and meta-analysis. Anesth Analg. 2019;129(6):1673–91.

11. Chang CC, Wu JL, Hsiao JR, Lin CY. Real-time, intraoperative, ultrasound-assisted transoral robotic surgery for obstructive sleep apnea. Laryngoscope. 2021;131(4):E1383–E90.
12. Hsu HJ, Wu JL, Hsiao JR, Lin CY. Quantification of the impact of intraoperative ultrasound in transoral robotic tongue base reduction. Laryngoscope. 2022;132(5):1125–31.
13. Chang CH, Huang CC, Wang YH, Chou FJ, Chen JW. Ultrasound shear-wave elastography of the tongue in adults with obstructive sleep apnea. Ultrasound Med Biol. 2020;46(7):1658–69.
14. Arens P, Fischer T, Dommerich S, Olze H, Lerchbaumer MH. Ultrasound shear wave elastography of the tongue during selective hypoglossal nerve stimulation in patients with obstructive sleep apnea syndrome. Ultrasound Med Biol. 2021;47(10):2869–79.
15. Liu SYC, Bosschieter PFN, Abdelwahab M, Chao PY, Chen A, Kushida C. Association of backscattered ultrasonographic imaging of the tongue with severity of obstructive sleep apnea in adults. JAMA Otolaryngol Head Neck Surg. 2023;149(7):580–6.
16. Bosschieter PFN, Liu SYC, Chao PY, Chen A, Kushida CA. Using standardized ultrasound imaging to correlate OSA severity with tongue morphology. Sleep Med. 2024;120:15–21.
17. Weng CK, Chen JW, Lee PY, Huang CC. Implementation of a wearable ultrasound device for the overnight monitoring of tongue base deformation during obstructive sleep apnea events. Ultrasound Med Biol. 2017;43(8):1639–50.
18. Ontimare Manlises C, Chen JW, Huang CC. A gated recurrent unit model based on ultrasound images of dynamic tongue movement for determining the severity of obstructive sleep apnea. Ultrasonics. 2024;141:107320.
19. Manlises CO, Chen JW, Huang CC. Dynamic tongue area measurements in ultrasound images for adults with obstructive sleep apnea. J Sleep Res. 2020;29(4):e13032.
20. Abuan MRA, Lin WN, Hsin LJ, Lee LA, Fang TJ, Chen NH, et al. Tongue imaging during drug-induced sleep ultrasound in obstructive sleep apnea patients. Auris Nasus Larynx. 2020;47(5):828–36.
21. Huang YC, Hsu YB, Lan MY, Yang MC, Kao MC, Huang TT, et al. Dynamic tongue base thickness measured by drug-induced sleep ultrasonography in patients with obstructive sleep apnea. J Formos Med Assoc. 2021;120(1 Pt 2):354–60.
22. Chen JW, Huang CC, Weng CK, Chang CH, Wang SJ. Simultaneous recording of ultrasound and polysomnography during natural sleep in patients with obstructive sleep apnea: a pilot study. J Sleep Res. 2017;26(4):481–6.
23. Yu RB, Huang CC, Chang CH, Wang YH, Chen JW. Prevalence and patterns of tongue deformation in obstructive sleep apnea: a whole-night simultaneous ultrasonographic and polysomnographic study. J Sleep Res. 2021;30(3):e13131.
24. Chien CY, Chen JW, Chang CH, Huang CC. Tracking dynamic tongue motion in ultrasound images for obstructive sleep apnea. Ultrasound Med Biol. 2017;43(12):2791–805.
25. Lun HM, Liu RC, Hu Q, Liu YL, Wei LS, Wu D, et al. Potential ultrasonic anatomical markers of obstructive sleep apnoea-hypopnoea syndrome. Clin Radiol. 2023;78(2):e137–e42.
26. Kapur VK, Auckley DH, Chowdhuri S, Kuhlmann DC, Mehra R, Ramar K, et al. Clinical practice guideline for diagnostic testing for adult obstructive sleep apnea: an American Academy of Sleep Medicine Clinical Practice Guideline. J Clin Sleep Med. 2017;13(3):479–504.
27. Brown EC, Cheng S, McKenzie DK, Butler JE, Gandevia SC, Bilston LE. Tongue stiffness is lower in patients with obstructive sleep apnea during wakefulness compared with matched control subjects. Sleep. 2015;38(4):537–44.
28. Burns DW, Chan VWS, Trivedi A, Englesakis M, Munshey F, Singh M. Ready to scan? A systematic review of point of care ultrasound (PoCUS) for screening of obstructive sleep apnea (OSA) in the pediatric population. J Clin Anesth. 2022;83:110973.
29. Hussein SA, Kamel KM, Kaddah SZ, Abd El-Hamid EE, Shaban MM. Role of ultrasonography in assessment of anatomic upper airway changes in patients with obstructive sleep apnea. Adv Respir Med. 2020;88(6):548–57.
30. Kalkanis A, Testelmans D, Papadopoulos D, Van den Driessche A, Buyse B. Insights into the use of point-of-care ultrasound for diagnosing obstructive sleep apnea. Diagnostics (Basel). 2023;13(13)

31. Mohan Lal B, Vyas S, Malhotra A, Ray A, Gupta G, Pandey S, et al. Ultrasonography of the neck in patients with obstructive sleep apnea. Sleep Breath. 2023;27(3):903–12.
32. Chu CA, Chen YJ, Chang KV, Wu WT, Ozcakar L. Reliability of sonoelastography measurement of tongue muscles and its application on obstructive sleep apnea. Front Physiol. 2021;12:654667.
33. Deniz I, Eyigor H, Yalcin M, Ozturk Yilmaz G, Yilmaz MD, Osma U, et al. The usefulness of transcervical ultrasonography for the prediction of obstructive sleep apnea in children. Cranio. 2024;43:1–8.
34. Sagtas E, Mengi E, Kara CO, Senol H. A novel assessment method with ultrasound for obstructive tonsillar hypertrophy in children. J Ultrasound Med. 2021;40(9):1795–801.

Snoring Sound Analysis

6

Li-Ang Lee, Guo-She Lee, Hai-Hua Chuang, Yi-Ping Chao, and Hsueh-Yu Li

Contents

L.-A. Lee (✉)
Department of Otorhinolaryngology, Head and Neck Surgery, Sleep Center, Linkou Medical Center, Chang Gung Memorial Hospital, Chang Gung University, Taoyuan, Taiwan

School of Medicine, College of Medicine, Chang Gung University, Taoyuan, Taiwan

School of Medicine, College of Life Science and Medicine, National Tsing Hua University, Hsinchu, Taiwan
e-mail: 5738@cgmh.org.tw

G.-S. Lee
Faculty of Medicine, National Yang Ming Chiao Tung University, Taipei, Taiwan

Department of Otolaryngology, Taipei City Hospital, Ren-Ai Branch, Taipei, Taiwan

H.-H. Chuang
School of Medicine, College of Life Science and Medicine, National Tsing Hua University, Hsinchu, Taiwan

Department of Community Medicine, Cathay General Hospital, Taipei, Taiwan

Department of Industrial Engineering and Management, National Taipei University of Technology, Taipei, Taiwan

Y.-P. Chao
Department of Otorhinolaryngology, Head and Neck Surgery, Sleep Center, Linkou Medical Center, Chang Gung Memorial Hospital, Chang Gung University, Taoyuan, Taiwan

Department of Computer Science and Information Engineering, Chang Gung University, Taoyuan, Taiwan

H.-Y. Li
Department of Otorhinolaryngology, Head and Neck Surgery, Sleep Center, Linkou Medical Center, Chang Gung Memorial Hospital, Chang Gung University, Taoyuan, Taiwan

School of Medicine, College of Medicine, Chang Gung University, Taoyuan, Taiwan

H.-Y. Li (ed.), *Snoring and Obstructive Sleep Apnea*,
https://doi.org/10.1007/978-981-95-2648-2_6

Abstract

This chapter delves into the sophisticated realm of snoring sound analysis, emphasizing its critical role in understanding and managing sleep-disordered breathing (SDB). It covers a range of topics from the basic acoustic properties of snoring to advanced analytical techniques using Snore Map technology. Discussions include the identification of primary snoring generators such as the soft palate and tongue base, and the relationship between snoring sounds and carotid artery health. The section evaluates various therapeutic interventions and their efficacy in reducing snoring severity. Noteworthy is the exploration of objective snoring sound characteristics that aid in assessing disease burden and therapeutic outcomes. Despite the relative accuracy of current acoustic analysis methods, the text highlights the need for more rigorous research to refine diagnostic criteria and personalize treatment options for SDB. This section underscores the importance of snoring sound analysis in the broader context of sleep medicine and patient care.

Keywords

Snoring · Sleep apnea syndrome · Microphone · Neck piezoelectric sensor Smartphone · Vibration site · Carotid Artery Atherosclerosis · Anti-snoring therapy

6.1 Introduction

Sleep-disordered breathing (SDB) encompasses a range of conditions including habitual snoring (HS), sleep apnea syndrome (SAS), obesity hypoventilation, and upper airway resistance syndrome, and presents a significant public health issue due to its prevalence among adult populations [1, 2]. Beyond being a disturbance, snoring is a vital clinical indicator of SAS and is associated with numerous comorbidities such as hypertension, cardiovascular disease, diabetes mellitus, arthritis, and backache [3, 4]. The primary mechanics of snoring involve the vibration of structures within the pharyngeal airway, particularly the fluttering of the soft palate [5].

Although traditional polysomnography (PSG) remains the gold standard for diagnosing SAS, its accessibility is limited due to the need for specialized facilities and equipment, posing barriers to widespread screening and early diagnosis [6]. While conventional PSG techniques using piezoelectric sensors effectively minimize environmental noise and quantify snoring events and indices, they do not correlate well with subjective snoring scores or accurately pinpoint the anatomical sources of snoring [7, 8].

In light of the critical role of snoring in evaluating SDB, there has been an increasing focus on the precise, breath-by-breath identification and quantification of snoring. This has spurred technological advancements aimed at enhancing the accuracy of snoring detection and the analysis of snoring patterns, contributing to better diagnostic strategies and monitoring of sleep disorders [3, 9].

6.2 Techniques for Snoring Detection

The detection of snoring sounds is pivotal in diagnosing and managing sleep disorders such as SAS. This involves a variety of methods that cater to different settings, from clinical to home environments. The technologies deployed range from acoustic analysis using microphones [10], smartphone applications [11], wearable devices that capture vibrations [12], and PSG [7] in clinical scenarios. These technologies aim to differentiate snoring from other ambient sounds, evaluate sleep quality, and monitor the effectiveness of therapeutic interventions (Table 6.1).

6.2.1 Technology Overview

6.2.1.1 Microphone Systems

Both standalone microphones and integrated smartphone microphones fall under this category. Positioned near the sleeping area, they capture audio which includes snoring sounds, as well as background noises. Advanced artificial intelligence (AI) algorithms are crucial for these systems to filter out non-snoring sounds, achieving high classification accuracies.

6.2.1.2 Smartphone Applications

Leveraging the widespread availability of smartphones, these applications use built-in microphones to record snoring sounds. They are highly convenient and user-friendly, although they may encounter privacy concerns from app permissions and potential interference from ambient noise.

6.2.1.3 Neck-Wearable Piezoelectric Sensors

The neck-wearable piezoelectric sensor (NPS) is attached directly to the skin. They detect physical vibrations from snoring, offering high accuracy due to minimal ambient noise interference. Such devices provide an additional benefit of privacy since they do not record audio.

Table 6.1 Comparison of snoring detection technologies

Feature	Microphone-based Systems	Smartphone-based Systems	Neck-wearable Piezoelectric Sensor	Polysomnography
Data acquisition	Captures snoring sounds audibly	Records snoring using device microphones	Senses neck vibrations for snoring detection	Detects multiple physiological signals through multiple sensors, snoring commonly acquired using piezoelectric sensors positioned at the submental or tracheal areas
Confidentiality	Reduced, prone to capturing background noise	Reduced, captures background noise and requires permissions	Enhanced, does not record audio	Very high, performed in secure environments
User convenience	Contactless, generally more comfortable for extended use	Contactless, leverages user's own device for convenience	Requires skin contact, possibly uncomfortable	Less comfortable, involves multiple
Precision	Susceptible to background noise interference	Subject to variability based on device position and ambient noise	High accuracy, even in environments with noise	Benchmark for sleep disorder diagnosis
Expense	Relatively low, utilizes readily available devices	Minimal, uses user's smartphone	Generally higher, requiring specialized equipment	High cost, requiring advanced equipment and skilled operators
User-friendliness	Easy to set up, place microphone near the sleeper	Usually user-friendly, integrated within personal devices	Setup needed for proper sensor placement	Requires professional installation and setup
Handiness	Moderately portable, depending on recording devices	Exceptionally portable, an integral part of smartphones	Very portable, designed for individual use	Limited, confined to clinic or specialized setting
Data safety	Depends on how recordings and data are stored	Tied to app and smartphone security settings	Based on the security protocols of the device and software	Highly secure, complies with stringent medical regulations

6.2.1.4 PSG

As the clinical gold standard for diagnosing SAS, PSG is comprehensive but often cumbersome and expensive, limiting its accessibility. Recent advancements have, however, made home-based tests more practical and less intrusive.

The choice between non-contact microphone systems and contact-based NPS systems for snoring detection depends on factors such as accuracy, comfort, privacy, and specific data requirements [13]. Microphone systems, which include smartphone integrations, are placed near the sleeper and capture both snoring and ambient sounds [14]. These systems can use advanced artificial intelligence algorithms to filter out background noise, achieving classification accuracies exceeding 90% [15]. Their non-contact nature and user-friendliness make them particularly suitable for consumer and home use, offering simplicity and ease of operation.

In contrast, contact-based NPS systems, positioned around the neck or directly attached to the skin, specifically detect vibrations from snoring and carotid artery pulsations, thereby significantly reducing the interference of ambient noise and enhancing privacy since they do not record actual sounds [16]. Our recent studies utilizing deep learning models have demonstrated that these systems can achieve diagnostic accuracies of up to 92% [17]. Despite the potential discomfort associated with direct skin contact, recent technological advancements have led to the development of less intrusive and more comfortable models, making them more suitable for home use [18].

6.3 Energy Characteristics of Snoring Sounds

To conduct a comprehensive analysis of full-night snoring sounds, we recommend using a portable digital recorder, such as the PCM-D100 (Sony Electronics Inc., Tokyo, Japan). This device is equipped with two high-performance unidirectional dynamic microphones arranged in a 90-degree X-Y pattern. For optimal capture of snoring and breathing sounds, recommended settings are low-cut filter, record level to “5”, and turning off of all limiters [15]. The recorder should be positioned 45–100 cm from the patient’s head [10, 19]. Recording background noise for 1–10 min before sleep is crucial for controlling ambient sounds, ensuring the bedroom remains quiet (intensity of background noise <40 dBA) and the subject sleeps alone.

Sound recordings are captured at a sampling rate of 44,100 Hz with a 16-bit resolution and typically run continuously for at least 6 hours. Figure 6.1 displays the snoring sound signals, illustrating both the sound waves in time domain (A) and in frequency domain (B) over time. In prior studies, we found that background noise levels consistently ranged between 3.15 and 40 Hz [20]. We employ high-pass filtering to minimize the impact of these low-frequency noises and low-pass filtering to reduce ceiling effects, focusing our analysis on snoring sound frequencies between 40 and 2000 Hz [15].

For frequency-domain analysis, we use fast Fourier transformation to obtain the power spectra of the snoring sounds. Table 6.2 outlines the key metrics used to characterize snoring sounds, which are divided into three primary domains: time, intensity,

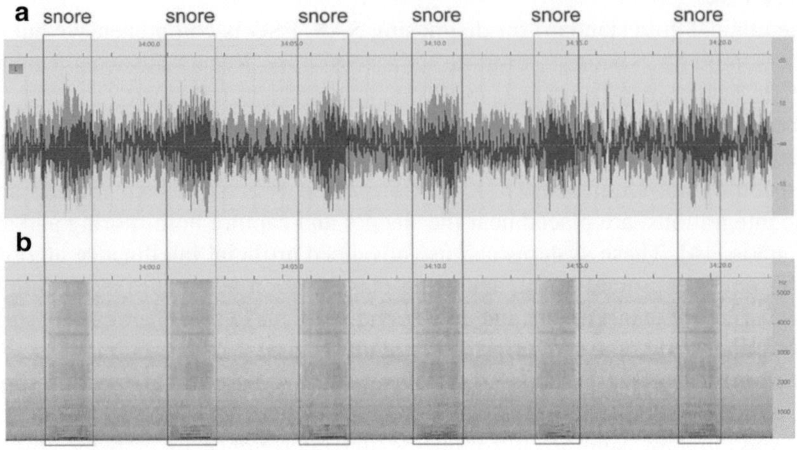

Fig. 6.1 Demonstration of original snoring sound signals in a home environment. (**a**) Time domain: Visualization of snoring sound signals over time. (**b**) Frequency domain: Visualization of spectrogram of snoring sounds over time. Red boxes highlight specific snoring events

Table 6.2 Snoring sound characteristic metrics

Domain	Definition
Time-domain metrics	
Snoring index (events/hour)	The number of snoring events per hour of sleep
Snoring time ratio	The proportion of total sleep time spent snoring
Snoring duration (second)	Total duration of all snoring events
Intensity-domain metrics	
Mean intensity (dB)	Average intensity of snoring events
Maximal intensity (dB)	Peak intensity observed in snoring events
Frequency-domain metrics	
Fundamental frequency (Hz)	The lowest/base frequency in the sound spectrum of a snore
Peak frequency (Hz)	The frequency at which the maximum power occurs within the spectrum
Mean frequency (Hz)	The average frequency calculated across the spectrum
Power ratio (Hz)	The ratio of power above a specific frequency to the power below it
Mel-scale frequency cepstral coefficient	Measures of the energy spectrum of a sound based on the Mel scale, focusing on human auditory perception
Maximum power frequency (HZ)	Frequency at which the highest power of a single snoring event occurs
Mean power frequency (Hz)	Frequency at which the mean power of a single snoring event occurs

and frequency [10, 21–23]. Acoustic sound analysis with these metrics helps assess snoring severity, pinpoint the origin of snoring sounds, and estimate the severity of SAS.

We acquire the power spectra of all snores during sleep, averaging them to calculate the overall mean energy across frequency units throughout the, classifying them into three frequency bands: low-frequency band (B1, 40–299 Hz), mid-frequency band (B2, 300–849 Hz), and high-frequency band (B3, 850–2000 Hz). Snoring sounds are categorized into five types based on these bands, ranging from minimal energy snoring to complex multi-frequency snoring patterns (Fig. 6.2) [10]. Analysis of the characteristics of each band help identifying the origin of snoring sounds and estimating SAS severity.

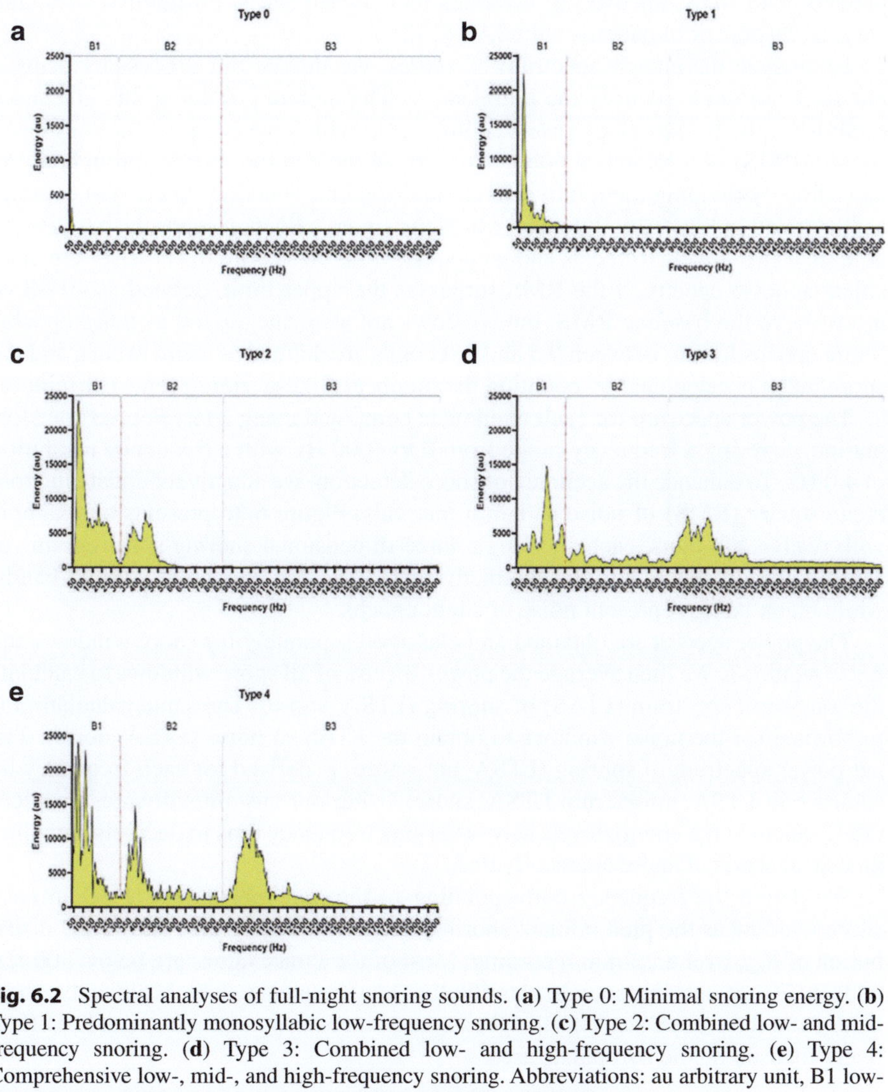

Fig. 6.2 Spectral analyses of full-night snoring sounds. (**a**) Type 0: Minimal snoring energy. (**b**) Type 1: Predominantly monosyllabic low-frequency snoring. (**c**) Type 2: Combined low- and mid-frequency snoring. (**d**) Type 3: Combined low- and high-frequency snoring. (**e**) Type 4: Comprehensive low-, mid-, and high-frequency snoring. Abbreviations: au arbitrary unit, B1 low-frequency band, B2 mid-frequency band, B3 high-frequency band

6.4 Mapping Snoring Patterns: The Snore Map

Since 2012, our team at Linkou Medical Center, Chang Gung Memorial Hospital in Taoyuan City, Taiwan, has developed a specialized software named Snore Map to analyze full-night snoring sound signals. This method has demonstrated high sensitivity and positive predictive values for detecting snoring events [10, 15, 24–28]. Our findings have illuminated significant associations between snoring sound energy (SSE) and various health parameters, including intima-media thickness and profiles of the common carotid artery [15, 25], as well as auditory dysfunction in patients with SAS [29]. Additionally, we utilized the Snore Map system to evaluate the effects of positional therapy [24] and surgical treatments [27, 30] on SSE in patients with SDB, uncovering associations between SSEs, obstructive sites, and surgical responses in patients with SAS [28].

To analyze the energy spectrum of snores, we initiate our process by using a playback system to identify the initial snore. The minute preceding this first snore establishes the background noise baseline, from which we compute the root mean square (RMS) as a reference point. We analyze the snoring sounds throughout the recording by dividing them into discrete 0.25-second windows that do not overlap, as illustrated in Fig. 6.3a. Windows where the signal's RMS exceeds the baseline by at least 10 dB are classified as snore epochs, while those with lower RMS are classified as noise epochs. If the RMS surpasses the upper limit, defined as 40 dB or more above the baseline RMS, the windows are also categorized as noise epochs. Snore epochs lasting between 0.5 and 3 seconds are defined as snore events, and the snore index is calculated by counting the number of these snore events per minute.

The power spectrum for each window is computed using a fast Fourier transformation, covering a frequency range from 0 to 4000 Hz with a frequency resolution of 4.0 Hz. To enhance the accuracy of snore detection, we adaptively update the root mean square (RMS) of noise at 1-min intervals. Figure 6.3b presents an example with twelve sub-plots, each showing a three-dimensional snoring spectrograms of 1-min duration. Each sub-plot distinctly marks snoring epochs in colored bands, while blank bands represent noise or silent epochs.

The power spectra are obtained and classified separately for snore windows and noise windows. We then average the power spectra of all snore windows to calculate the long-term spectrum (LTAS) of snoring (LTSA_snore). The same calculation is performed for the noise windows to obtain the LTAS of noise (LTSA_noise). The net power spectrum of snoring (LTSA_net_snore) is derived for each frequency by subtracting LTSA_noise from LTSA_snore. Using our custom software, we accurately quantify the energy levels across various frequency bins in decibels, enabling precise analysis of snore characteristics.

We define the frequency corresponding to the maximal power (P_{max}) in each snore window as the predominant snoring frequency. Figure 6.3c shows the distribution of P_{max} over a 22-min recording. Most of the Pmax values are below 300 Hz, with 99.7% of snoring sounds classified as low frequency, which typically originates from the soft palate. This insight suggests that sleep treatments could be effectively focused on the soft palate area.

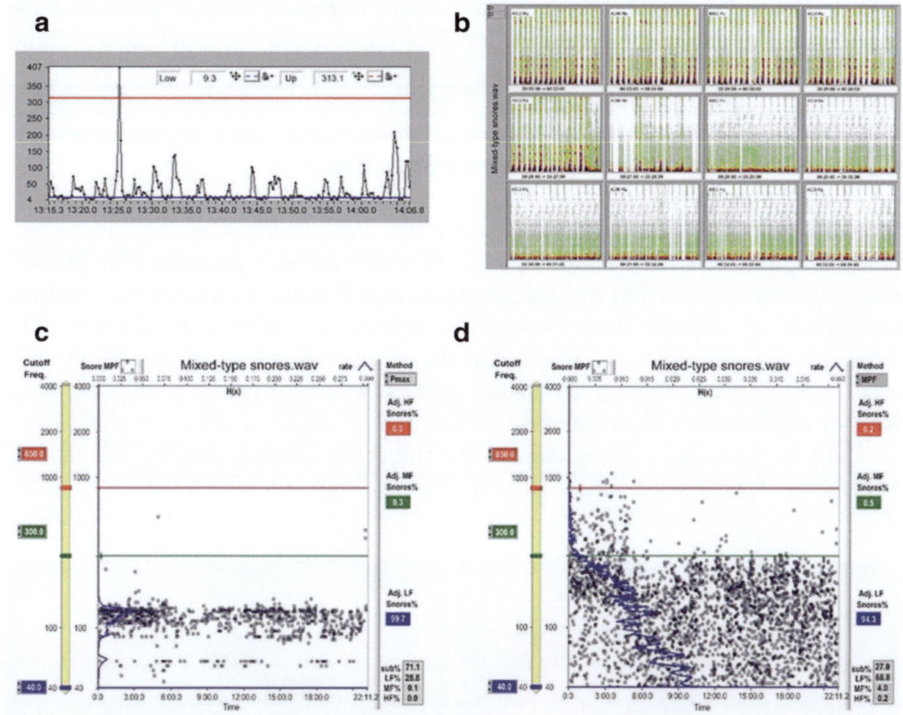

Fig. 6.3 Demonstration of Snore Map. (**a**) Root mean square (RMS) plot of a 10-second recording. Snore and noise epoch classification: Displays the discrete 0.25-second windows used in the analysis of snoring sounds. Each window is classified as a snore epoch if its RMS is 10 dB above the baseline RMS (blue line) and below 40 dB of the baseline RMS (red line); otherwise, it is classified as a noise epoch. (**b**) Three-dimensional snoring spectrum: Visualizes each epoch distinctly. Colored bands signify snoring epochs, while blank bands denote noise or silent epochs. (**c**) Maximum power frequency (P_{max}) spectrum: Illustrates the distribution of P_{max} values across the recording time, emphasizing that most snoring frequencies fall below 300 Hz. The distribution includes low-frequency (LF) snoring epochs at 99.7%, mid-frequency (MF) at 0.3%, and high-frequency (HF) at 0%. (**d**) Mean power frequency (MPF) spectrum: Displays the MPF values over the recording time, reflecting the complexity of snoring dynamics. The distribution covers LF snoring epochs at 94.3%, MF at 5.5%, and HF at 0.2%

Additionally, we calculate the mean power frequency (MPF) of each signal window using the formula:

$$\mathrm{MPF} = \frac{\sum_{f=40}^{2000} f \times P_f}{\sum_{f=40}^{2000} P_f}$$

where f represents the frequency and P_f represents the power at each frequency. Figure 6.3d illustrates the distribution of MPF over time for another 22-min recording. Snoring sounds within the mid-frequency range often indicate multiple snoring sources and suggest multiple levels of upper airway obstruction, highlighting the complexity of snoring dynamics in some patients.

6.5 Loco-Sonic Recognition of Snoring

To identify snoring generators and obstructive sites, some researchers have combined drug-induced sleep endoscopy (DISE) with acoustic analysis [28, 31, 32]. Initial findings reveal that the spectrograms and formant frequencies of snoring sounds during induced sleep closely resemble those during natural sleep [33]. Typically, the soft palate is identified as the primary generator of snoring in most cases, while isolated snoring generators such as the tongue base, lateral pharyngeal wall, or epiglottis are less common.

The soft palate primarily produces low-frequency snoring sounds, classified as Snore Map Type 1 (Fig. 6.4a). Snoring sounds originating from the tongue base

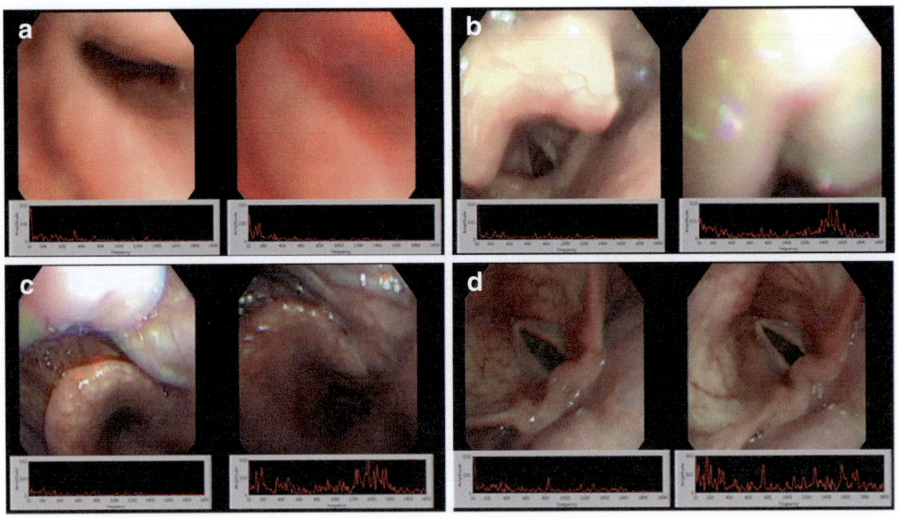

Fig. 6.4 Power spectra of snoring sounds and their generators. (**a**) Soft palate snore demonstrates how vibrations of the soft palate produce low-frequency snoring sounds. The left panel shows the soft palate in a static state, while the right panel displays it in a vibrating state. (**b**) Tongue base snore illustrates how vibrations from the tongue base contribute to high-frequency snoring sounds. The left panel depicts the tongue base at rest, and the right panel shows it vibrating. (**c**) Lateral pharyngeal wall snore shows how vibrations of the lateral pharyngeal wall enhance mid-frequency snoring sounds. The left panel portrays the lateral pharyngeal wall as static, with the right panel showing it vibrating. (**d**) Epiglottis snore explains how vibrations of the epiglottis intensify mid-frequency snoring sounds. The left panel features the epiglottis at rest, while the right panel depicts it vibrating

contribute high-frequency components, categorized under Snore Map Type 3 (Fig. 6.4b). Vibrations from the lateral pharyngeal wall enhance mid-frequency snoring sounds, noted as Snore Map Type 4 (Fig. 6.4c). Similarly, epiglottis-induced snoring intensifies mid-frequency sounds, also identified as Snore Map Type 4 (Fig. 6.4d).

Additionally, patients with at least two complete obstruction sites as identified by DISE show a significant positive correlation with maximal low-frequency intensity and body mass index, according to logistic regression analysis. An inverse correlation exists between tonsil obstruction and mean mid-frequency intensity [28]. There is also a positive relationship between snoring event duration and complete concentric collapse of the soft palate [32].

6.6 Acoustic Impact on the Common Carotid Artery

Atherosclerotic cardiovascular diseases (CVD), encompassing coronary and cerebrovascular conditions, rank among the top causes of morbidity and premature mortality worldwide [34]. Factors like increased carotid intima-media thickness (CIMT) and carotid atherosclerosis are well-known predictors of CVD, alongside hypertension, diabetes, smoking, and hyperlipidemia [35].

Emerging research suggests that snoring transmits vibrational energy from the upper airway to surrounding tissues and into the carotid artery wall [36]. Studies in animal models have demonstrated that such vibrations induce endothelial dysfunction [37] and diminish baroreflex sensitivity [38], processes known to accelerate carotid atherosclerosis in humans [39]. Notably, Kim et al. confirmed an independent link between habitual snoring and increased CIMT [40]. Further studies by Lee et al. have shown that specific frequencies of SSE are correlated with variations in CIMT [15].

Our recent investigations have further detailed these associations. We discovered that normalized mid-frequency SSE holds an independent relationship with both CIMT and the types of carotid artery profiles [25]. Moreover, findings indicate that normalized low-frequency SSE and being female are separately linked to the occurrence of carotid stenosis. A comprehensive regression model considering the type of carotid artery profile incorporated variables such as normalized mid-frequency SSE, snoring duration, female gender, and age (Fig. 6.5). This model adjusted for additional factors including scores from the Snore Outcomes Survey, the 3% oxygen desaturation index, total SSE, normalized high-frequency SSE, smoking status, and lipid profiles.

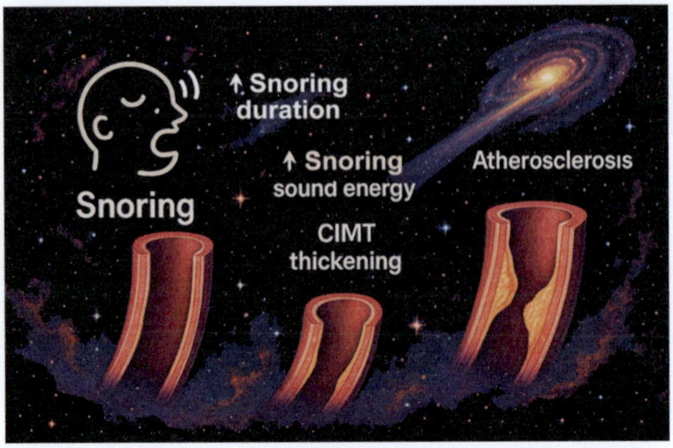

Fig. 6.5 The relationship between snoring sound characteristics and carotid artery profile. This figure illustrates the associations between different characteristics of snoring sound energy (SSE) and carotid artery health. Increased normalized mid-frequency SSE correlates with greater carotid intima-media thickness (CIMT) and a higher incidence of unhealthy carotid profiles, such as carotid stenosis. Conversely, decreased normalized low-frequency SSE is associated with an elevated risk of carotid stenosis. Additionally, longer snoring durations are linked to increased CIMT, highlighting the potential impact of snoring characteristics on vascular health [25]. (Source: Figure created using Microsoft Copilot)

6.7 Evaluating Post-Therapeutic Changes in Snoring

Numerous anti-snoring treatments are available, including positional therapy, nasal interventions (sprays and surgeries), oral appliances, minimally invasive surgeries, laser-assisted uvuloplatoplasty, and uvulopalatopharyngoplasty [41–43]. Most studies evaluate the effectiveness of these treatments by assessing changes in subjective snoring severity or using objective measures such as the snoring index and snoring duration in patients with SDB.

Our Snore Map technology has been instrumental in quantifying these changes. For instance, positional therapy using a head positional pillow predominantly reduces the snoring index [24]. Palatal implants primarily decrease maximum low-frequency intensity, while radiofrequency-assisted tissue volume reduction of the palate lowers the snoring index [30]. Relocation pharyngoplasty significantly improves overall mean intensity, peak intensity, Snore Map type, mean intensity across all three frequency bands, mid- and high-frequency snoring indexes, and mean frequency in the low-frequency band [27]. In children with SAS, significant reductions in SSE within the 41–200, 201–400, and 801–1000 Hz frequency bands have been observed post-adenotonsillectomy [19].

Thus, analyzing snoring sound characteristics provides important objective outcome measures for evaluating the efficacy of treatments in patients with SDB.

6.8 Summary and Future Directions

In summary, analyzing objective snoring sound characteristics is crucial for assessing various aspects of SDB, including snoring severity, disease burden, identification of the primary snoring generator, multi-level airway obstructions, carotid artery health, and the effectiveness of therapeutic interventions. Over the past decades, acoustic analysis of snoring has proven relatively accurate and provides valuable insights, although it has not emerged as a definitive method for diagnosing SDB [44].

There remains a significant need for more comprehensive studies that utilize large sample sizes and focus on single snore event analysis. Such studies should aim to develop an efficacy criterion that specifically addresses the unique acoustic features of snoring for better diagnosis of SDB and to guide personalized treatment strategies for snoring and SDB. Given our experience and the preliminary successes in snoring sound analysis, this field holds considerable potential for future research that could lead to more effective diagnosis and treatment options for SDB.

References

1. Kreitinger K, Light M, Patel S, Malhotra A. Sleep-disordered breathing. In: Sleep medicine and mental health. Springer; 2020. p. 131–50.
2. Cumpston E, Chen P. Sleep apnea syndrome. In: StatPearls. Treasure Island (FL): StatPearls Publishing; 2024. Disclosure: Philip Chen declares no relevant financial relationships with ineligible companies.
3. Sowho M, Sgambati F, Guzman M, Schneider H, Schwartz A. Snoring: a source of noise pollution and sleep apnea predictor. Sleep. 2020;43(6)
4. Chuang LP, Hsu SC, Lin SW, Ko WS, Chen NH, Tsai YH. Prevalence of snoring and witnessed apnea in Taiwanese adults. Chang Gung Med J. 2008;31(2):175–81.
5. Pevernagie D, Aarts RM, De Meyer M. The acoustics of snoring. Sleep Med Rev. 2010;14(2):131–44.
6. Markun LC, Sampat A. Clinician-focused overview and developments in polysomnography. Curr Sleep Med Rep. 2020;6(4):309–21.
7. Chen NH, Li HY, Gliklich RE, Chu CC, Liang SC, Wang PC. Validation assessment of the Chinese version of the snore outcomes survey. Qual Life Res. 2002;11(6):601–7.
8. Quinn SJ, Huang L, Ellis PD, Williams JE. The differentiation of snoring mechanisms using sound analysis. Clin Otolaryngol Allied Sci. 1996;21(2):119–23.
9. Hsu YC, Wang JD, Huang PH, Chien YW, Chiu CJ, Lin CY. Integrating domain knowledge with machine learning to detect obstructive sleep apnea: snore as a significant bio-feature. J Sleep Res. 2022;31(2):e13487.
10. Lee LA, Yu JF, Lo YL, Chen YS, Wang DL, Cho CM, Ni YL, Chen NH, Fang TJ, Huang CG, et al. Energy types of snoring sounds in patients with obstructive sleep apnea syndrome: a preliminary observation. PLoS One. 2012;7(12):e53481.
11. Cho SW, Jung SJ, Shin JH, Won TB, Rhee CS, Kim JW. Evaluating prediction models of sleep apnea from smartphone-recorded sleep breathing sounds. JAMA Otolaryngol Head Neck Surg. 2022;148(6):515–21.
12. Chao Y-P, Chuang H-H, Lo Y-L, Huang S-Y, Zhan W-T, Lee G-S, Li H-Y, Shyu L-Y, Lee L-A. Automated sleep apnea detection from snoring and carotid pulse signals using an innovative neck wearable piezoelectric sensor. Measurement. 2025;242:116102.

13. Arnardottir ES, Isleifsson B, Agustsson JS, Sigurdsson GA, Sigurgunnarsdottir MO, Sigurdarson GT, Saevarsson G, Sveinbjarnarson AT, Hoskuldsson S, Gislason T. How to measure snoring? A comparison of the microphone, cannula and piezoelectric sensor. J Sleep Res. 2016;25(2):158–68.
14. Xie J, Aubert X, Long X, van Dijk J, Arsenali B, Fonseca P, Overeem S. Audio-based snore detection using deep neural networks. Comput Methods Prog Biomed. 2021;200:105917.
15. Lee GS, Lee LA, Wang CY, Chen NH, Fang TJ, Huang CG, Cheng WN, Li HY. The frequency and energy of snoring sounds are associated with common carotid artery intima-media thickness in obstructive sleep apnea patients. Sci Rep. 2016;6:30559.
16. Xu C, Yang Y, Gao W. Skin-interfaced sensors in digital medicine: from materials to applications. Matter. 2020;2(6):1414–45.
17. Chao YP, Chuang LP, Lee ZH, Huang SY, Zhan WT, Shyu LY, Lo YL, Lee GS, Li HY, Lee LA. Distinguishing severe sleep apnea from habitual snoring using a neck-wearable piezoelectric sensor and deep learning: a pilot study. Comput Med Biol. 2025;187:110070.
18. Kwon S, Kim H, Yeo WH. Recent advances in wearable sensors and portable electronics for sleep monitoring. iScience. 2021;24(5):102461.
19. Lu CT, Li HY, Lee GS, Huang YS, Huang CG, Chen NH, Lee LA. Snoring sound energy as a potential biomarker for disease severity and surgical response in childhood obstructive sleep apnoea: a pilot study. Clin Otolaryngol. 2019;44(1):47–52.
20. Yu J-F, Chen Y-s, Li H-Y, Li L-A. Sleeping position can reduce the effect of snoring on sleeping partners. J Acoust Soc Am. 2014;135(4_Supplement):2165.
21. Fiz JA, Jane R, Sola-Soler J, Abad J, Garcia MA, Morera J. Continuous analysis and monitoring of snores and their relationship to the apnea-hypopnea index. Laryngoscope. 2010;120(4):854–62.
22. Qian K, Janott C, Pandit V, Zhang Z, Heiser C, Hohenhorst W, Herzog M, Hemmert W, Schuller B. Classification of the excitation location of snore sounds in the upper airway by acoustic multifeature analysis. IEEE Trans Biomed Eng. 2017;64(8):1731–41.
23. Huang Z, Lobbezoo F, Vanhommerig JW, Volgenant CMC, de Vries N, Aarab G, Hilgevoord AAJ. Effects of demographic and sleep-related factors on snoring sound parameters. Sleep Med. 2023;104:3–10.
24. Chen WC, Lee LA, Chen NH, Fang TJ, Huang CG, Cheng WN, Li HY. Treatment of snoring with positional therapy in patients with positional obstructive sleep apnea syndrome. Sci Rep. 2015;5:18188.
25. Chuang HH, Liu CH, Wang CY, Lo YL, Lee GS, Chao YP, Li HY, Kuo TBJ, Yang CCH, Shyu LY, et al. Snoring sound characteristics are associated with common carotid artery profiles in patients with obstructive sleep apnea. Nat Sci Sleep. 2021;13:1243–55.
26. Yu JF, Chen YS, Li HY. The characteristics of snoring at pharyngeal anatomy in natural sleep: snoring duration. J Mech. 2012;28(1):91–5.
27. Li HY, Lee LA, Yu JF, Lo YL, Chen NH, Fang TJ, Hsin LJ, Lin WN, Huang CG, Cheng WN. Changes of snoring sound after relocation pharyngoplasty for obstructive sleep apnoea: the surgery reduces mean intensity in snoring which correlates well with apnoea-hypopnoea index. Clin Otolaryngol. 2015;40(2):98–105.
28. Lee LA, Lo YL, Yu JF, Lee GS, Ni YL, Chen NH, Fang TJ, Huang CG, Cheng WN, Li HY. Snoring sounds predict obstruction sites and surgical response in patients with obstructive sleep apnea hypopnea syndrome. Sci Rep. 2016;6:30629.
29. Lu CT, Lee LA, Lee GS, Li HY. Obstructive sleep apnea and auditory dysfunction-does snoring sound play a role? Diagnostics (Basel). 2022;12(10)
30. Lee LA, Yu JF, Lo YL, Chen NH, Fang TJ, Huang CG, Cheng WN, Li HY. Comparative effects of snoring sound between two minimally invasive surgeries in the treatment of snoring: a randomized controlled trial. PLoS One. 2014;9(5):e97186.
31. Peng H, Xu H, Xu Z, Huang W, Jia R, Yu H, Zhao Z, Wang J, Gao Z, Zhang Q, et al. Acoustic analysis of snoring sounds originating from different sources determined by drug-induced sleep endoscopy. Acta Otolaryngol. 2017;137(8):872–6.

32. Huang Z, Bosschieter PFN, Aarab G, van Selms MKA, Vanhommerig JW, Hilgevoord AAJ, Lobbezoo F, de Vries N. Predicting upper airway collapse sites found in drug-induced sleep endoscopy from clinical data and snoring sounds in patients with obstructive sleep apnea: a prospective clinical study. J Clin Sleep Med. 2022;18(9):2119–31.
33. Koo SK, Kwon SB, Moon JS, Lee SH, Lee HB, Lee SJ. Comparison of snoring sounds between natural and drug-induced sleep recorded using a smartphone. Auris Nasus Larynx. 2018;45(4):777–82.
34. Nedkoff L, Briffa T, Zemedikun D, Herrington S, Wright FL. Global trends in atherosclerotic cardiovascular disease. Clin Ther. 2023;45(11):1087–91.
35. Kablak-Ziembicka A, Przewlocki T. Clinical significance of carotid intima-media complex and carotid plaque assessment by ultrasound for the prediction of adverse cardiovascular events in primary and secondary care patients. J Clin Med. 2021;10(20)
36. Howitt L, Kairaitis K, Kirkness JP, Garlick SR, Wheatley JR, Byth K, Amis TC. Oscillatory pressure wave transmission from the upper airway to the carotid artery. J Appl Physiol (1985). 2007;103(5):1622–7.
37. Cho JG, Witting PK, Verma M, Wu BJ, Shanu A, Kairaitis K, Amis TC, Wheatley JR. Tissue vibration induces carotid artery endothelial dysfunction: a mechanism linking snoring and carotid atherosclerosis? Sleep. 2011;34(6):751–7.
38. Narayan J, Amatoury J, Cho JG, Verma M, Kairaitis K, Wheatley J, Amis T. Snoring effects on the baroreflex: an animal model. Respir Physiol Neurobiol. 2012;180(2–3):342–51.
39. Ulleryd MA, Prahl U, Borsbo J, Schmidt C, Nilsson S, Bergstrom G, Johansson ME. The association between autonomic dysfunction, inflammation and atherosclerosis in men under investigation for carotid plaques. PLoS One. 2017;12(4):e0174974.
40. Kim J, Pack A, Maislin G, Lee SK, Kim SH, Shin C. Prospective observation on the association of snoring with subclinical changes in carotid atherosclerosis over four years. Sleep Med. 2014;15(7):769–75.
41. Li HY, Lee LA, Wang PC, Chen NH, Lin Y, Fang TJ. Nasal surgery for snoring in patients with obstructive sleep apnea. Laryngoscope. 2008;118(2):354–9.
42. Li HY, Lee LA, Wang PC, Fang TJ, Chen NH. Can nasal surgery improve obstructive sleep apnea: subjective or objective? Am J Rhinol Allergy. 2009;23(6):e51–5.
43. Lee YC, Lee LA, Li HY. The palatal septal cartilage implantation for snoring and obstructive sleep apnea. Auris Nasus Larynx. 2018;45(6):1199–205.
44. Jin H, Lee LA, Song L, Li Y, Peng J, Zhong N, Li HY, Zhang X. Acoustic analysis of snoring in the diagnosis of obstructive sleep apnea syndrome: a call for more rigorous studies. J Clin Sleep Med. 2015;11(7):765–71.

Pharmacotherapy of Obstructive Sleep Apnea in Adults

7

Yu-Shu Huang, Wei-Chih Chin, and Tsai-Yu Wang

Contents

Abstract

Obstructive sleep apnea (OSA) is a common disorder involving repeated upper airway collapse during sleep, contributing to cardiovascular, metabolic, and neurocognitive complications. Standard treatments include weight loss, oral appliances, surgery, and continuous positive airway pressure (CPAP), the gold standard, though its use is often limited by poor adherence. Pharmacotherapy presents a promising alternative or adjunct by targeting key pathophysiological

Y.-S. Huang (✉) · W.-C. Chin
Department of Psychiatry and Sleep center, Chang Gung Memorial Hospital, Linkou, Taiwan

College of Medicine, Chang Gung University, Taoyuan, Taiwan

T.-Y. Wang
College of Medicine, Chang Gung University, Taoyuan, Taiwan

Department of Thoracic Medicine, Chang Gung Memorial Hospital, Linkou, Taiwan

H.-Y. Li (ed.), *Snoring and Obstructive Sleep Apnea*,
https://doi.org/10.1007/978-981-95-2648-2_7

mechanisms such as airway collapsibility, ventilatory instability, and low arousal thresholds. This article reviews recent studies on both established and emerging pharmacological treatments, including ventilatory stimulants, serotonergic agents, Rapid Eye Movement (REM)-suppressants, acetylcholinesterase inhibitors, and drugs that reduce loop gain or increase arousal thresholds. Medications aimed at weight control and lowering upper airway resistance are also discussed. Combination therapies are being explored for greater effectiveness. While pharmacotherapy remains controversial, ongoing research highlights its potential role within personalized, multidimensional treatment plans. The article concludes with challenges and future directions in optimizing drug-based OSA management.

Keywords

Obstructive sleep apnea · Pharmacotherapy · Pathophysiological mechanisms · Combination drugs

7.1 Introduction

Obstructive sleep apnea (OSA) is characterized by recurrent episodes of complete or partial obstruction of the upper airway during sleep [1]. It affects approximately 10–30% of adults globally [2–5], with prevalence varying by criteria and demographics. Its key features—intermittent hypoxia, hypercapnia, and sleep fragmentation—can lead to systemic issues like hypertension, diabetes, cardiovascular disease, and increased mortality, while also impairing cognition, alertness, and quality of life [6–8]. Research in recent years further reveals its correlation with neurodegeneration and dementia [9, 10]. Therefore, untreated sleep apnea and related is an important major public health issue.

Generally speaking, treatment strategies of OSA can be divided into three major categories according to their pathophysiological causes: (1) The standard treatment is the continuous positive airway pressure (CPAP) therapy; (2) non-CPAP treatment such as oral appliances, bariatric surgery, maxillofacial surgery, and upper airway surgery including hypoglossal nerve stimulation; (3) Behavioral therapy and lifestyle modification such as oromyofunctional therapy, positional therapy, avoiding aggravating factors including alcohol, sedatives, and muscle relaxants before bedtime, weight gain, lack of exercise, and smoking [11]. Although oral appliances can be alternative treatments for mild-to-moderate OSA and supine position-related sleep apnea, long-term effects remain uncertain [12, 13]. Surgical options including septoplasty, adenotonsillectomy, pharyngoplasty, maxillomandibular advancement (MMA) [14], and bariatric surgery are invasive and can be very expensive. Additionally, there is a lack of sufficient long-term follow-up research on their outcomes and complications [15]. CPAP therapy is effective but its adherence rates are typically below 50%, relating to discomfort, inconvenience, and dissatisfaction [16–18]. These results highlight critical gaps in OSA treatment and have driven interest in pharmacological therapies, which aim to mitigate the underlying mechanisms or alleviate symptoms of OSA. This chapter delves into the landscape of OSA pharmacotherapy, examining its current status, challenges, and future opportunities.

7.2 Structured Discussion

7.2.1 Mechanisms of Pharmacological Treatments

Previous studies have found that pharmacotherapy shows promises for patients with mild-to-moderate OSA, CPAP-intolerant individuals and those with specific phenotypes responsive to targeted drug treatments. The most important issue is that pharmacotherapy should target the underlying mechanisms of OSA. Considering that OSA is a heterogeneous disease and there may be more than one underlying causes and endotypes of OSA in adults or elderly. Therefore, pharmacotherapy may have potential roles as a monotherapy and as an adjunctive treatment.

Proposed mechanisms of medication to reduce OSA severity include (1) increasing upper airway(UA) dilator muscle tone, reducing REM sleep proportion, elevating cholinergic tone during sleep; (2) enhancing ventilatory drive and decreasing loop gain; (3) raising arousal thresholds; (4) decreasing collapsibility and lowering airway resistance; and (5) other medications that enhance daytime wakefulness [19–24]. We review and classify potential OSA medications, evaluate their effectiveness and side effects, and explore combination drug therapy.

7.2.1.1 Increasing Upper Airway Dilator Muscle Tone

Pharmacological agents including noradrenergic agonists, serotonergic agonists, muscarinic inhibitor, cannabinoid agonists, TASK1/3 channel inhibitors may increase upper airway dilator tone.

1. *Noradrenergic Agonists:* Noradrenergic activity plays a role in upper airway control. Studies have investigated the effects of tricyclic antidepressants (TCAs) and norepinephrine reuptake inhibitors (SNRIs) on reducing AHI and improving sleep-related outcomes.
   (a) *Protriptyline and desipramine* are TCA drugs that may reduce AHI but are associated with significant side effect [25–28]
   (b) *Atomoxetine and oxybutynin:* A randomized, double-blind, placebo-controlled crossover trial with combination drug of atomoxetine and oxybutynin (noradrenergic agonists + muscarinic antagonist) showed 50–75% AHI reduction [29]. These drugs reduced upper airway dilator muscle tone through decreasing noradrenergic activity in Non – Rapid Eye Movement (NREM) sleep and muscarinic cholinergic inhibition in REM sleep.
   (c) *Reboxetine combination with hyoscine butylbromide:* A double-blind, placebo-controlled trial showed improved tensor palatini muscle activity and reduced pharyngeal resistance during sleep. It led to a 35% reduction in AHI

and a significant improvement in nadir oxygen saturation in 12 OSA patients [30].

(d) *Atomoxetine and aroxybutynin*: It is another new combination drug of MARIPOSA Phase II Trial for OSA [31], and had clinically significant improvement in OSA. Further data are needed to determine its long-term efficacy and safety.

2. *Serotonergic Agonists:* Serotonin plays a crucial role in modulating upper airway motor neurons and respiratory control. Its activity diminishes during REM sleep and contributes to airway collapse. In the brainstem, endogenous serotonin release promotes drive to the upper airway dilators in the waking state, primarily via postsynaptic 5-HT2 receptors [32]. Peripherally, serotonin release at 5-HT3 receptors in the nodose ganglion promotes the expression of REM-related apnea [33]. Due to its involvement in central nervous system (CNS) and peripheral nervous system (PNS) respiratory regulation, serotonin reuptake inhibitors (SRIs) have been investigated as potential treatments for OSA.
   (a) *Buspirone:* Acts on 5-HT1A, a small study showed a 36% reduction in AHI [34].
   (b) *Paroxetine:* One double-blind, placebo-controlled crossover study with eight patients on paroxetine (40 mg) study showed no change in AHI [35].
   (c) *Fluoxetine:* A study comparing fluoxetine and protriptyline showed that fluoxetine reduced REM sleep percentage and found it was more effective than protriptyline in decreasing apnea or hypopnea, but neither fully resolved sleep-disordered breathing [26]
   (d) *Fluoxetine and ondansetron* (5-HT3 receptor antagonist) combination study: A randomized, double-blind, placebo-controlled trial with 44 patients showed a 40% reduction in AHI in the high-dose fluoxetine + ondansetron combination group (fluoxetine 10 mg daily and ondansetron 24 mg daily) [36].
   (e) *Venlafaxine, mirtazapine, and L-tryptophon:* There was no significant improvement in AHI with venlafaxine. Both *mirtazapine and L-tryptophan* were not recommended for the treatment of OSA because of side effects [37, 38].

3. *Potassium channel blockers:* Recent research has identified potassium channel blockers as a potential treatment for OSA by modulating motor neuron excitability in REM and NREM sleep.

   (a) *4-Aminopyridine:* A trial showed increased genioglossus activity in REM sleep but the effects are limited due to subtherapeutic dosing [39].
   (b) *Doxapram*: It showed some benefits but lack of a specific drug target limits its treatment potential [29].
   (c) *Dronabinol*: *Cannabinoid receptors (CBRs)* which are functionally coupled to K+ channels are targeted for OSA treatment. CBR2 is highly expressed in the hypoglossal motor nucleus. Dronabinol, a partial CBR1/2 agonist, was tested in a proof-of-concept study and a phase-2 trial. In 17 patients, 21-day

dronabinol use reduced AHI by 29%. A larger trial with 73 patients showed an 8.5 event/hour reduction at high doses [40]

7.2.1.2 Enhancing Ventilatory Drive and Decreasing Loop Gain

Loop gain quantifies the instability of ventilatory chemoreflex control. High loop gain leads to exaggerated ventilatory responses, causing airway collapse. This has driven interest in targeting ventilatory control mechanisms for OSA treatment.

(a) *Acetazolamide*: Carbonic anhydrase inhibitors (CAIs) like acetazolamide lower PCO_2 by inducing metabolic acidosis and relative hyperventilation. A study by Whyte et al. (250 mg/QID acetazolamide for 2 weeks) in ten moderate-to-severe OSA patients showed a 52% AHI reduction, metabolic acidosis (pH drop from 7.42 to 7.36, $p < 0.01$), and decreased PCO_2. However, there was no significant improvement in arousals or daytime sleepiness. Edwards et al. studied twelve CPAP users with AHI >10/h using 500 mg acetazolamide twice daily. The drug reduced loop gain and AHI by 41% and 52%, but had no significant symptom improvement [41].

(b) *Acetazolamide combined with eszopiclone:* A double-blind and crossover trial with twenty OSA patients showed the combination significantly reduced OSA severity and hypoxic burden, improving AHI during NREM supine sleep (45%) and sleep architecture without major side effects [42].

(c) *Draonabinol combined with acetazolamide*: A randomized dose finding showed the significant reductions in AHI and OSA severity, and demonstrated to be well tolerated and safe at low, medium, and high doses [43].

(d) *Sulthiame*: Another CAI which is a second-line drug for treatment of benign childhood epilepsy which reduced OSA and had a satisfactory safety profile in moderate and severe OSA [44].

(e) *Donepezil*: An acetylcholinesterase inhibitor acting on ventilator drive reduced AHI by 50% in patients with Alzheimer's disease [45]. However, other trials found no significant effect on AHI, oxygen saturation, or sleepiness [46]

(f) Infusion of *physostigmine* showed a 24% AHI reduction, mainly in REM sleep [47]

7.2.1.3 Raising Arousal Thresholds

Sedatives and hypnotics were discouraged in OSA due to concerns of delayed arousal, worsening blood gas derangement and airway obstruction from muscle relaxation. Nevertheless, low arousal thresholds can be found in 30–50% of OSA patients and cause frequent awakenings, sleep fragmentation, and breathing instability [21]. Sedatives and hypnotics that increase arousal thresholds while preserving upper airway muscle activity are under investigation [48].

1. *Benzodiazepines* act on $GABA_A$ receptors and produce sedative, hypnotic, and muscle-relaxing effects. It is contraindicated due to concerns of worsened symptoms, but may benefit low arousal threshold OSA patients.

(a) *Triazolam* (0.25 mg) increased arousal threshold (24–33%) in healthy individuals and severe OSA patients [49], but did not improve AHI.
(b) *Nitrazepam* (5–10 mg) increased arousal threshold but did not improve AHI. Temazepam trials also showed no effect on AHI [50]

2. *Z-Drugs (Non-benzodiazepine Hypnotics)* selectively act on *a*1 subunits of $GABA_A$ receptors and do not have muscle-relaxing effects, making them potentially safer for OSA patients.

    (a) *Zolpidem* (10 mg) and zopiclone (7.5 mg): Large trials showed no genioglossus activity impairment, but no significantly reduction in AHI or OSA severity was found despite increased arousal threshold, and sleep efficiency [51–53]
    (b) *Eszopiclone* (3 mg) increased arousal threshold by 30% and reduced 45% AHI in a small study [54].

3. *Alternative Hypnotics*

    (a) S*odium oxybate* can treat narcolepsy patients. Though non-significant, a single dose could reduce AHI by 17% but increase central apneas in OSA patients. A two-week trial showed a significant 32% AHI reduction and 24% more slow-wave sleep, suggesting improved sleep stability and higher arousal thresholds [55].
    (b) *Trazodone*, a serotonin antagonist and reuptake inhibitor, increases respiratory arousal threshold by 32% without affecting airway muscle activity. While some studies found no impact on AHI, one showed a reduction from 38.7 to 28.5 events/hour ($p = 0.041$) and may stabilize breathing and reduce N1 sleep [56]

7.2.1.4 Decreasing Collapsibility and Lowering Airway Resistance

Patients with OSA are predisposed to airway collapse due to narrowing skeletal structure, enlarged tonsils and adenoids, fat deposition, or nasal congestion. Most of these airway obstruction problems require surgery or CPAP to solve. However, there are medications that can improve airway collapse include weight loss drugs, diuretics, and nasal decongestants.

1. *Weight Loss Drugs*: Approximately 45% of patients with OSA are obese. Excess fat around the neck, tongue, upper airway, and abdomen increases the risk of pharyngeal collapse. A magnetic resonance imaging (MRI) study showed that weight loss reduces upper airway soft tissue volume, correlating with lower AHI [57]. The American Thoracic Society recommends weight loss medications as adjunctive therapy for OSA patients who have a body mass index (BMI) ≥ 27 kg/m^2 and struggle with lifestyle interventions [58]. The US Food and Drug Administration (FDA)-approved weight loss drugs such as orlistat, liraglutide,

tirzepatideand, phentermine/topiramate extended-release (ER) have been studied in OSA.

(a) *Phentermine 15 mg/topiramate 92 mg ER:* A randomized small sample size trial of moderate to severe OSA found a significant AHI reduction from 44 to 14 events/hour. The treatment group experienced a 10.2% weight loss which significantly correlated with AHI improvement. However, no significant differences were found in the arousal index, desaturation index, or glycemic/lipid values [59].

(b) *Liraglutide*: A glucagon-like peptide-1 (GLP-1) receptor agonist was evaluated in the SCALE Sleep Apnea trial, which had a large sample size and used polysomnography at specialized sleep centers. After 32 weeks, AHI decreased by 12.2 events/hour in the liraglutide group versus 6.1 in the placebo group, with greater weight loss (5.7% vs. 1.6%) [60, 61].

(c) *Tirzepatide:* A new molecule capable of controlling glucose blood levels by combining the dual agonism of glucose-dependent insulinotropic polypeptide (GIP) and GLP-1 receptors. It is the first FDA-approved drug to treat obstructive sleep apnea. A trial of moderate-to-severe OSA patients with obesity showed significantly reduced AHI (25 events/hour), body weight, hypoxic burden, hsCRP, and systolic blood pressure versus placebo at 52 weeks. Gastrointestinal side effects were common but mild [62]

2. *Nasal Decongestants and Controlling Allergy*: Nasal obstruction and allergy has been hypothesized to contribute to OSA, and thus nasal decongestants and anti-allergy may improve its symptoms, such as oxymetazoline, phenylephrine, pseudoephedrine (60 mg), domperidone (10 mg), montelukast, and fluticasone and other nasal corticosteroid [63]

7.2.2 Combination Drugs Therapy

While many drugs have been explored for OSA, current results are mixed, and most successful single-drug interventions provide only partial relief. Considering the complexity and heterogeneity of OSA, drugs targeting a single endotype or phenotype are unlikely to achieve satisfactory therapeutic effects. Combination of drugs with different mechanism for the three key traits, unstable ventilatory control (high loop gain), low arousal threshold, and poor upper airway dilator muscle compensation can be an important future direction. Besides combination drugs currently under investigation such as atomoxetine and oxybutynin, ronabinol and acetazolamide, GLIP-1RA and dual agonists, these traits can be targeted pharmacologically by acetazolamide to stabilize ventilation, eszopiclone to increase the arousal threshold, and noradrenergic drugs such as venlafaxine to increase upper airway dilator muscle compensation.

7.2.3 Challenges in Pharmacotherapy

1. Heterogeneity in phenotypes and endotypes: Due to OSA's multifactorial nature, treatment must be tailored to individual phenotypes and endotypes, as variations in anatomy, ventilatory control, and arousal thresholds affect drug response; identifying biomarkers for stratification is key to optimizing pharmacotherapy.
2. Safety and tolerability: Adverse effects like sedation, gastrointestinal (GI) discomfort, and cardiovascular risks may hinder adherence and long-term use of pharmacologic agents. Comprehensive safety evaluations are essential for expanding their clinical use.
3. Limited evidence base: Most studies had small sizes and followed patients only short-term. Additionally, the lack of comparisons between pharmacological agents and CPAP complicates the integration of drugs into standard care algorithms. High-quality, multicenter randomized controlled trials are needed.

7.2.4 Future Directions

1. Personalized medicine: Advancements in genomics, phenotyping, and digital health technologies could enable personalized pharmacological regimens tailored to individual OSA profiles.
2. Multidimensional approach and management: It is often necessary to combine drugs with different mechanisms to treat medical conditions such as diabetes or hypertension. Some recent studies have promising results by combining different OSA interventions. Furthermore, combining pharmacotherapy with behavioral, surgical, or device-based interventions may yield synergistic benefits, such as oxygen and eszopiclone, lemborexant with positional therapy.
3. Longitudinal research: Future studies should focus on the long-term efficacy, safety, and cost-effectiveness of pharmacological treatments for establishing their role in comprehensive OSA care.

7.3 Conclusion

Pharmacotherapy represents a promising frontier in the management of OSA, and can offer alternatives or adjuncts for selected patients. While current literature demonstrates varying degrees of efficacy and safety, the integration of personalized approaches, combination drug therapy and multimodal strategies has the potential to optimize outcomes. Further research and innovation are essential to addressing the unmet needs of patients with OSA.

References

1. ICSD-3TR: The AASM International Classification of Sleep Disorders—Third Edition, Text Revision (ICSD-3-TR).
2. Lee W, Nagubadi S, Kryger MH, Mokhlesi B. Epidemiology of obstructive sleep apnea: a population-based perspective. Expert Rev Respir Med. 2008;2(3):349–64. https://doi.org/10.1586/17476348.2.3.349. PMID: 19690624; PMCID: PMC2727690.
3. Young T, Peppard PE, Gottlieb DJ. Epidemiology of obstructive sleep apnea: a population health perspective. Am J Respir Crit Care Med. 2002;165(9):1217–39.
4. Peppard PE, Young T, Barnet JH, Palta M, Hagen EW, Hla KM. Increased prevalence of sleep-disordered breathing in adults. Am J Epidemiol. 2013;177:1006–14.
5. Benjafield AV, Ayas NT, Eastwood PR, Heinzer R, Ip MSM, Morrell MJ, Nunez CM, Patel SR, Penzel T, Pépin J-L, et al. Estimation of the global prevalence and burden of obstructive sleep apnoea: a literature-based analysis. Lancet Respir Med. 2019;7:687–98.
6. DiCaro MV, Lei K, Yee B, Tak T. The effects of obstructive sleep apnea on the cardiovascular system: a comprehensive review. J Clin Med. 2024;13:3223. https://doi.org/10.3390/jcm13113223.
7. Karkoulias K, Lykouras D, Sampsonas F, Karaivazoglou K, Sargianou M, Drakatos P, Spiropoulos K, Assimakopoulos K. The impact of obstructive sleep apnea syndrome severity on physical performance and mental health. The use of SF-36 questionnaire in sleep apnea. Eur Rev Med Pharmacol Sci. 2013;17(4):531–6.
8. Weihs A, Frenzel S, Grabe HJ. The link between obstructive sleep apnoea and neurodegeneration and cognition. Curr Sleep Med Rep. 2021;7:87–96. https://doi.org/10.1007/s40675-021-00210-5.
9. Gnoni V. et al. Distinct cognitive changes in male patients with obstructive sleep apnoea without co-morbidities. Front Sleep. https://doi.org/10.3389/frsle.2023.1097946.
10. Tian Q, Sun J, Li X, et al. Association between sleep apnoea and risk of cognitive impairment and Alzheimer's disease: a meta-analysis of cohort-based studies. Sleep Breath. 2024;28:585–95. https://doi.org/10.1007/s11325-023-02934-w.
11. Verbraecken J, Dieltjens M, Op de Beeck S, Vroegop A, Braem M, Vanderveken O, Randerath W. Non-CPAP therapy for obstructive sleep apnoea. Breathe (Sheff). 2022;18(3):220164. https://doi.org/10.1183/20734735.0164-2022. Epub 2022 Oct 11. PMID: 36340820; PMCID: PMC9584565.
12. Kushida CA, Morgenthaler TI, Littner MR, Alessi CA, Bailey D, Coleman J Jr, Friedman L, Hirshkowitz M, Kapen S, Kramer M, Lee-Chiong T, et al. Practice parameters for the treatment of snoring and obstructive sleep apnea with oral appliances: an update for 2005. Sleep. 2006;29(2):240–3.
13. Huang YS, Chuang LC, Hervy-Auboiron M, Paiva T, Lin CH, Guilleminault C. Neutral supporting mandibular advancement device with tongue bead for passive myofunctional therapy: a long term follow-up study. Sleep Med. 2018;pii: S1389-9457(18):30447. https://doi.org/10.1016/j.sleep.2018.09.013.
14. Li KK, Riley RW, Powell NB, Guilleminault C. Patient's perception of the facial appearance after maxillomandibular advancement for obstructive sleep apnea syndrome. J Oral Maxillofac Surg. 2001;59(4):377–80; discussion 380–1. https://doi.org/10.1053/joms.2001.21870.
15. Li K, Guilleminault C, Amat P. Traitement du SAOS par avancement maxillomandibulaire: complications graves et échecs [Treatment of OSA by maxillomandibular advancement: serious complications and failures]. Orthod Fr. 2022;93(2):155–68. French. https://doi.org/10.1684/orthodfr.2022.83.
16. Cistulli PA, Armitstead J, Pepin J-L, Woehrle H, Nunez CM, Benjafield A, Malhotra A. Short-term CPAP adherence in obstructive sleep apnea: a big data analysis using real world data. Sleep Med. 2019;59:114.

17. Lin H-S, Zuliani G, Amjad EH, Prasad AS, Badr MS, Pan C-JG, Rowley JA. Treatment compliance in patients lost to follow-up after polysomnography. Otolaryngol—Head Neck Surg. 2007;136:236–40.
18. Patel SR, Bakker JP, Stitt CJ, Hardy W, Aloia MS, Nouraie SM. Impact of an extended telemonitoring and coaching program on continuous positive airway pressure adherence. Ann Am Thorac Soc. 2022;19(12):2070–6. https://doi.org/10.1513/AnnalsATS.202204-310OC.
19. Mason M, Welsh EJ, Smith I. Drug therapy for obstructive sleep apnoea in adults. Cochrane Database Syst Rev. 2013;2013(5):CD003002. https://doi.org/10.1002/14651858.CD003002.pub3. PMID: 23728641; PMCID: PMC11623339.
20. Taranto-Montemurro L, Messineo L, Wellman A. Targeting endotypic traits with medications for the pharmacological treatment of obstructive sleep apnea. A review of the current literature. J Clin Med. 2019;8(11):1846. https://doi.org/10.3390/jcm8111846. PMID: 31684047; PMCID: PMC6912255.
21. Arredondo E, DeLeon M, Masozera I, Panahi L, Udeani G, Tran N, Nguyen CK, Atphaisit C, de la Sota B, Gonzalez G Jr, Liou E, Mayo Z, Nwosu J, Shiver TL. Overview of the role of pharmacological management of obstructive sleep apnea. Medicina (Kaunas). 2022;58(2):225. https://doi.org/10.3390/medicina58020225. PMID: 35208549; PMCID: PMC8874508.
22. Ip MSM, Chung KF. Pharmacotherapy for obstructive sleep apnoea: is the magic bullet in sight? Respirology. 2021;26(9):835–6. https://doi.org/10.1111/resp.14128. Epub 2021 Aug 3.
23. Landry SA, Edwards BA. Pharmacotherapy for sleep apnoea: a search for the right therapeutic key. Respirology. 2021;26(5):411–2. https://doi.org/10.1111/resp.14038. Epub 2021 Mar 17.
24. Schmickl CN, Edwards BA, Malhotra A. Drug therapy for obstructive sleep apnea: are we there yet? Am J Respir Crit Care Med. 2022;205(12):1379–81. https://doi.org/10.1164/rccm.202202-0301ED. PMID: 35320066; PMCID: PMC9875900.
25. Brownell LG, West P, Sweatman P, Acres JC, Kryger MH. Protriptyline in obstructive sleep apnea: a double-blind trial. N Engl J Med. 1982;307:1037–42.
26. Hanzel DA, Proia NG, Hudgel DW. Response of obstructive sleep apnea to fluoxetine and protriptyline. Chest. 1991;100:416–21.
27. Whyte KF, Gould GA, Airlie MAA, Shapiro CM, Douglas NJ. Role of protriptyline and acetazolamide in the sleep apnea/hypopnea syndrome. Sleep. 1988;11:463–72.
28. Taranto-Montemurro L, Sands SA, Edwards BA, Azarbarzin A, Marques M, De Melo C, Eckert DJ, White DP, Wellman A. Desipramine improves upper airway collapsibility and reduces OSA severity in patients with minimal muscle compensation. Eur Respir J. 2016;48:1340–50.
29. Taranto-Montemurro L, Messineo L, Sands SA, Azarbarzin A, Marques M, Edwards BA, Eckert DJ, White DP, Wellman A. The combination of Atomoxetine and oxybutynin greatly reduces obstructive sleep apnea severity. A randomized, placebo-controlled, double-blind crossover trial. Am J Respir Crit Care Med. 2019;199(10):1267–76. https://doi.org/10.1164/rccm.201808-1493OC. PMID: 30395486; PMCID: PMC6519859.
30. Lim R, Carberry JC, Wellman A, Grunstein R, Eckert DJ. Reboxetine and hyoscine butylbromide improve upper airway function during nonrapid eye movement and suppress rapid eye movement sleep in healthy individuals. Sleep. 2019;42:zsy261.
31. Schweitzer PK, Taranto-Montemurro L, Ojile JM, Thein SG, Drake CL, Rosenberg R, Corser B, Abaluck B, Sangal RB, Maynard J. The combination of Aroxybutynin and Atomoxetine in the treatment of obstructive sleep apnea (MARIPOSA): a randomized controlled trial. Am J Respir Crit Care Med. 2023;208(12):1316–1327. https://doi.org/10.1164/rccm.202306-1036OC. Erratum in: Am J Respir Crit Care Med. 2024;209(6):767. Erratum in: Am J Respir Crit Care Med 2024 Mar 15;209(6):767. PMID: 37812772; PMCID: PMC10765395.
32. Fenik P, Veasey SC. Pharmacological characterization of serotonergic receptor activity in the hypoglossal nucleus. Am J Respir Crit Care Med. 2003;167:563–9.
33. Carley DW, Radulovacki M. Role of peripheral serotonin in the regulation of central sleep apneas in rats. Chest. 1999;11:1397–401.
34. Mendelson WB, Maczaj M, Holt J. Buspirone administration to sleep apnea patients. J Clin Psychopharmacol. 1991;11:71.

35. Berry RB, Yamaura EM, Gill K, Reist C. Acute effects of paroxetine on genioglossus activity in obstructive sleep apnea. Sleep. 1999;22:1087–92.
36. Prasad B, Radulovacki M, Olopade C, Herdegen JJ, Logan T, Carley DW. Prospective trial of efficacy and safety of ondansetron and fluoxetine in patients with obstructive sleep apnea syndrome. Sleep. 2010;33:982–9.
37. Marshall NS, Yee BJ, Desai AV, Buchanan PR, Wong KK, Crompton R, Melehan KL, Zack N, Rao SG, Gendreau RM, Kranzler J, Grunstein RR. Two randomized placebo-controlled trials to evaluate the efficacy and tolerability of mirtazapine for the treatment of obstructive sleep apnea. Sleep. 2008;31(6):824–31. https://doi.org/10.1093/sleep/31.6.824. PMID: 18548827; PMCID: PMC2442407.
38. Schmidt HS. L-tryptophan in the treatment of impaired respiration in sleep. Bull Eur Physiopathol Respir. 1983;19:625–9.
39. Taranto-Montemurro L, Sands SA, Azarbarzin A, Marques M, de Melo CM, Edwards BA, Eckert DJ, Messineo L, White DP, Wellman A. Effect of 4-aminopyridine on genioglossus muscle activity during sleep in healthy adults. Ann Am Thorac Soc. 2017;14:1177–83.
40. Carley DW, Prasad B, Reid KJ, Malkani R, Attarian H, Abbott SM, Vern B, Xie H, Yuan C, Zee PC. Pharmacotherapy of apnea by Cannabimimetic enhancement, the PACE clinical trial: effects of Dronabinol in obstructive sleep apnea. Sleep. 2018;41:zsx184.
41. Edwards BA, Sands SA, Eckert DJ, White DP, Butler JP, Owens RL, Malhotra A, Wellman A. Acetazolamide improves loop gain but not the other physiological traits causing obstructive sleep apnoea. J Physiol. 2012;590:1199–211.
42. Schmickl CN, Orr JE, Alex RM, Gruenberg E, Parra G, White S, Spenceley A, DeSarkar T, Kong M, DeYoung PN, Sands SA, Owens RL, Malhotra A. Combination drug therapy with acetazolamide, eszopiclone +/− venlafaxine for obstructive sleep apnea (RESCUE-combo): a randomized, double-blind, placebo-controlled trial. Ann Am Thorac Soc. 2024; https://doi.org/10.1513/AnnalsATS.202407-736OC. Epub ahead of print.
43. Walsh J, Rankin T, Mehra S, Naughton MT, Roebuck T, McDermott E, Pattichis A, Smith R, Walsh R, Bleackley M, Maddison K, O'Brien TJ. A randomised dose finding study of combination dronabinol and acetazolamide for the treatment of obstructive sleep apnoea. Sleep. 2024;48:zsae276. https://doi.org/10.1093/sleep/zsae276. Epub ahead of print.
44. Hedner J, Stenlöf K, Zou D, Hoff E, Hansen C, Kuhn K, Lennartz P, Grote L. A randomized controlled clinical trial exploring safety and tolerability of Sulthiame in sleep apnea. Am J Respir Crit Care Med. 2022;205(12):1461–9. https://doi.org/10.1164/rccm.202109-2043OC.
45. Moraes W, Poyares D, Sukys-Claudino L, Guilleminault C, Tufik S. Donepezil improves obstructive sleep apnea in Alzheimer disease: a double-blind, placebo-controlled study. Chest. 2008;133:677–83.
46. Li Y, Owens RL, Sands S, Orr J, Moraes W, Deyoung P, Smales E, Jen R, Malhotra A. The effect of donepezil on arousal threshold and apnea–hypopnea index. A randomized, double-blind, cross-over study. Ann Am Thorac Soc. 2016;13:2012–8.
47. Hedner J, Kraiczi H, Peker Y, Murphy P. Reduction of sleep-disordered breathing after physostigmine. Am J Respir Crit Care Med. 2003;168:1246–51.
48. Carberry JC, Fisher LP, Grunstein RR, Gandevia SC, McKenzie DK, Butler JE, Eckert DJ. Role of common hypnotics on the phenotypic causes of obstructive sleep apnoea: paradoxical effects of zolpidem. Eur Respir J. 2017;50:1701344.
49. Berry RB, Kouchi K, Bower J, Prosise G, Light RW. Triazolam in patients with obstructive sleep apnea. Am J Respir Crit Care Med. 1995;151:450–4.
50. Camacho ME, Morin CM. The effect of Temazepam on respiration in elderly insomniacs with mild sleep apnea. Sleep. 1995;18:644–5.
51. Carberry JC, Grunstein RR, Eckert DJ. The effects of zolpidem in obstructive sleep apnea—an open-label pilot study. J Sleep Res. 2019;28:e12853.
52. Carter SG, Berger MS, Carberry JC, Bilston LE, Butler JE, Tong BK, Martins RT, Fisher LP, McKenzie DK, Grunstein RR, et al. Zopiclone increases the arousal threshold without impairing genioglossus activity in obstructive sleep apnea. Sleep. 2016;39:757–66.

53. Carter SG, Carberry JC, Cho G, Fisher LP, Rollo CM, Stevens DJ, D'Rozario AL, McKenzie DK, Grunstein RR, Eckert DJ. Effect of 1 month of zopiclone on obstructive sleep apnoea severity and symptoms: a randomised controlled trial. Eur Respir J. 2018;52:1800149.
54. Eckert DJ, Owens RL, Kehlmann GB, Wellman A, Rahangdale S, Yim-Yeh S, White DP, Malhotra A. Eszopiclone increases the respiratory arousal threshold and lowers the apnoea/hypopnoea index in obstructive sleep apnoea patients with a low arousal threshold. Clin Sci. 2011;120:505–14.
55. George CF, Feldman N, Zheng Y, Steininger TL, Grzeschik SM, Lai C, Inhaber N. A 2-week,polysomnographic, safety study of sodium oxybate in obstructive sleep apnea syndrome. Sleep Breath. 2011;15:13–20.
56. Smales ET, Edwards BA, Deyoung PN, McSharry DG, Wellman A, Velasquez A, Owens R, Orr JE. Malhotra, a trazodone effects on obstructive sleep apnea and non-REM arousal threshold. Ann Am Thorac Soc. 2015;12:758–64.
57. Wang SH, Keenan BT, Wiemken A, Zang Y, Staley B, Sarwer DB, Torigian DA, Williams N, Pack AI, Schwab RJ. Effect of weight loss on upper airway anatomy and the apnea–hypopnea index. the importance of tongue fat. Am J Respir Crit Care Med. 2020;201:718–27.
58. Hudgel DW, Patel SR, Ahasic AM, Bartlett SJ, Bessesen DH, Coaker MA, Fiander PM, Grunstein RR, Gurubhagavatula I, Kapur VK, et al. The role of weight management in the treatment of adult obstructive sleep apnea. An Official American Thoracic Society Clinical Practice Guideline. Am J Respir Crit Care Med. 2018;198:e70–87.
59. Winslow DH, Bowden CH, DiDonato KP, McCullough PA. A randomized, double-blind, placebo-controlled study of an oral, extended-release formulation of phentermine/topiramate for the treatment of obstructive sleep apnea in obese adults. Sleep. 2012;35:1529–39.
60. Blackman A, Foster GD, Zammit G, Rosenberg R, Aronne L, Wadden T, Claudius B, Jensen CB, Mignot E. Effect of liraglutide 3.0mg in individuals with obesity and moderate or severe obstructive sleep apnea: the SCALE sleep apnea randomized clinical trial. Int J Obes. 2016;40:1310–9.
61. Le KDR, Le K, Foo F. The impact of glucagon-like peptide 1 receptor agonists on obstructive sleep apnoea: a scoping review. Pharmacy (Basel). 2024;12(1):11. https://doi.org/10.3390/pharmacy12010011. PMID: 38251405; PMCID: PMC10801460.
62. Malhotra A, Grunstein RR, Fietze I, Weaver TE, Redline S, Azarbarzin A, Sands SA, Schwab RJ, Dunn JP, Chakladar S, Bunck MC, Bednarik J, SURMOUNT-OSA Investigators. Tirzepatide for the treatment of obstructive sleep apnea and obesity. N Engl J Med. 2024;391(13):1193–1205. https://doi.org/10.1056/NEJMoa2404881. Epub 2024 Jun 21. Erratum in: N Engl J Med. 2024 Oct 17;391(15):1464. PMID: 38912654; PMCID: PMC11598664.
63. Ji T, Lu T, Qiu Y, Li X, Liu Y, Tai J, Guo Y, Zhang J, Wang S, Zhao J, Ni X, Xu Z. The efficacy and safety of montelukast in children with obstructive sleep apnea: a systematic review and meta-analysis. Sleep Med. 2021;78:193–201. https://doi.org/10.1016/j.sleep.2020.11.009. Epub 2020 Nov 10.

Positional Therapy: What Can We Do

8

Li-Ang Lee, Ying-Shuo Hsu, Wen-Te Liu, Tsung-Te Chung, Chih-Yuan Liu, Meiho Nakayama, and Hsueh-Yu Li

Contents

L.-A. Lee (✉)
Department of Otorhinolaryngology, Head and Neck Surgery, Sleep Center, Linkou Medical Center, Chang Gung Memorial Hospital, Chang Gung University, Taoyuan, Taiwan

School of Medicine, College of Medicine, Chang Gung University, Taoyuan, Taiwan

School of Medicine, College of Life Science and Medicine, National Tsing Hua University, Hsinchu, Taiwan

Institute of Brain Science, National Yang Ming Chiao Tung University, Taipei, Taiwan

Sleep Research Center, National Yang Ming Chiao Tung University, Taipei, Taiwan
e-mail: 5738@cgmh.org.tw

Y.-S. Hsu
Institute of Brain Science, National Yang Ming Chiao Tung University, Taipei, Taiwan

Sleep Research Center, National Yang Ming Chiao Tung University, Taipei, Taiwan

Department of Otolaryngology, Shin Kong Wu-Ho-Su Memorial Hospital, Taipei, Taiwan

School of Medicine, Fu Jen Catholic University, New Taipei City, Taiwan

H.-Y. Li (ed.), *Snoring and Obstructive Sleep Apnea*,
https://doi.org/10.1007/978-981-95-2648-2_8

Abstract

This chapter explores positional therapy as an effective management strategy for sleep-disordered breathing, particularly in cases with positional obstructive sleep apnea. It discusses the mechanisms, types, and clinical implications of positional therapy, including the use of vibratory devices and integration with continuous positive airway pressure therapy. The chapter also reviews diagnostic techniques, surgical interventions, and nasopharyngoscopy's role in evaluating airway status. Future directions emphasize enhancing patient-centered outcomes, with research needed on long-term adherence, cost-effectiveness, and impacts on broader health conditions such as cardiovascular disease, metabolic syndrome, and neurocognitive disorder.

W.-T. Liu
School of Respiratory Therapy, College of Medicine, Taipei Medical University, Taipei, Taiwan

Division of Pulmonary Medicine, Department of Internal Medicine, Taipei Medical University-Shuang Ho Hospital, New Taipei City, Taiwan

Sleep Center, Taipei Medical University-Shuang Ho Hospital, New Taipei City, Taiwan

Research Center of Sleep Medicine, College of Medicine, Taipei Medical University, Taipei, Taiwan

T.-T. Chung
Department of Otolaryngology, Show Chwan Memorial Hospital, Changhua, Taiwan

Sleep Center, Chang Bing Show Chwan Memorial Hospital, Changhua, Taiwan

C.-Y. Liu
Power Sleep Inc, New Taipei City, Taiwan

M. Nakayama
Meiho Sleep & Balance Clinic, Nagoya, Japan

H.-Y. Li
Department of Otorhinolaryngology, Head and Neck Surgery, Sleep Center, Linkou Medical Center, Chang Gung Memorial Hospital, Chang Gung University, Taoyuan, Taiwan

School of Medicine, College of Medicine, Chang Gung University, Taoyuan, Taiwan

Keywords

Sleep-disordered breathing · Obstructive sleep apnea · Positional dependence Positional therapy · Head positional pillow · Vibratory device · Chest-worn belt Adjustable bed · Nasopharyngoscopy · Drug-induced sleep endoscopy

8.1 Introduction

Sleep is a critical physiological need across all species, crucial for the restoration and maintenance of various bodily functions [1]. The relationship between sleep positions and respiratory dynamics has long been observed, particularly in individuals who snore and their partners [2]. It is well documented that sleep position profoundly influences disorders like obstructive sleep apnea (OSA). The supine position, or lying on one's back, tends to exacerbate OSA by promoting the collapse of the upper airway, leading to more severe apnea episodes [3]. Conversely, lateral positions (sleeping on one's side) are associated with reduced severity of these episodes [4]. Positional therapy, which employs techniques to avoid the supine posture during sleep, has shown significant efficacy in mitigating apnea–hypopnea index (AHI) in many patients, highlighting its potential as a therapeutic strategy in sleep medicine [5]. However, positional therapy is not recommended for patients with shoulder problems or any other physical disability that interferes with their sleep in the lateral position, cannot avoid the supine posture during sleep [6]. This chapter will explore positional therapy's mechanisms, types, and clinical implications in managing sleep-disordered breathing (SDB), including its integration into multimodal treatment approaches [7]. Follow-up is crucial, as changes in a patient's condition, like weight gain, may alter the effectiveness of positional therapy.

8.2 Understanding Positional Dependency

Positional dependency significantly affects the pathophysiology of SDB, particularly in conditions like habitual snoring and OSA. This term denotes the variations in the severity and frequency of snoring and apnea episodes that are influenced by the sleeper's posture. Typically, patients exhibit different symptoms based on whether they are in a supine (lying on the back) or non-supine (side sleeping, elevated head, or prone sleeping) position.

8.2.1 Mechanisms Influencing Positional Dependency

In positional OSA (POSA), lying on the back often exacerbates airway collapse [8]. Gravity impacts the upper airway negatively, causing it to collapse more readily, which leads to a higher apnea–hypopnea index (AHI) [9]. Conversely, non-supine positions such as lateral sleeping or using a head-of-bed elevation (HOBE) mitigate these obstructions. For instance, elevating the head and the trunk to a 30-degree angle, known as the head-of-bed elevation (HOBE), has been shown to significantly decrease airway collapses and improve respiratory outcomes during sleep [10].

8.2.2 Epidemiological and Physiological Considerations

A considerable portion of OSA patients (50–60%) are affected by POSA [11], and about 26% experience symptoms exclusively when supine [12]. Typically, POSA patients are younger, have a lower body mass index, and lower AHI compared to those with non-positional OSA (NPOSA) [3]. Research indicates that nocturnal shifts in body composition, such as variations in fat distribution across different body regions, may influence the likelihood of developing POSA rather than NPOSA [13]. Furthermore, compared to NPOSA sufferers, those with POSA often show characteristics indicative of milder OSA, including less severe airway collapsibility, more effective upper-airway muscle compensation, lower ventilatory control instability (loop gain), and a reduced arousal threshold [14].

8.2.3 Clinical Implications and Management Strategies

Understanding positional dependency is crucial for tailoring effective treatment strategies for sleep apnea patients. This includes lifestyle modifications, mechanical interventions such as positional therapy devices, and, in some cases, surgical interventions [15]. The management plan should be flexible and responsive to the individual's specific positional influences on sleep apnea severity.

Incorporating positional therapy into the management of OSA can significantly enhance patient outcomes by reducing the severity of apnea incidents and improving overall sleep quality [16]. By leveraging a combination of lifestyle changes, mechanical aids, and medical interventions tailored to positional dependencies, clinicians can offer more effective and personalized treatment plans for their patients.

8.3 Types of Positional Therapy

Positional therapy plays a pivotal role in managing habitual snoring and OSA by addressing positional dependency. This therapeutic approach involves various devices designed to alter sleep positions to mitigate snoring events, apnea episodes, enhancing sleep quality and reducing the AHI. Here, we illustrate the different types of positional therapy (Table 8.1):

8.3.1 Positional Devices

8.3.1.1 Head Positional Pillows

These are specially designed to encourage side sleeping or elevate the head and upper body. Pillows like the cervical pillow increase upper airway space by extending the neck, which can reduce respiratory disturbances. Specifically, our previous

Table 8.1 Various types of positional devices used in therapy for habitual snoring and obstructive sleep apnea

Device Type	Description	Suitability
Head positional pillows	Designed to encourage side sleeping or elevate the head, increasing upper airway space	Mild-to-moderate OSA; individuals who prefer non-intrusive solutions
Chest-worn belts	Belts with attachments that make back sleeping uncomfortable, promoting side sleeping	Mild-to-moderate OSA; effective for individuals who primarily sleep on their back
Adjustable beds	Beds that can be angled to elevate the head and the trunk to reduce airway collapse	All severity levels of OSA; particularly beneficial for elderly or those with physical limitations
Smart vests	Wearable vests with sensors that prevent back sleeping and provide data on sleep patterns	Monitoring and adjustment of sleep posture; tech-savvy individuals
Biofeedback devices	Devices that vibrate to prompt position change when the user rolls onto their back	All severity levels of OSA; individuals sensitive to minor disturbances
Smart antisnore pillows	Pillows that adjust head position automatically based on snoring sound	Mild-to-moderate OSA; ideal for users needing dynamic adjustment throughout the night
Integrated systems	Combines positional therapy with other treatments like CPAP or oral appliances	Complex or severe OSA; patients using multiple therapeutic approaches

Abbreviations: *CPAP* continuous positive airway pressure, *OSA* obstructive sleep apnea

study has shown that after transitioning to head positional pillows, patients experienced a reduction in subjective snoring severity from a score of 5–4 [17]. Additionally, the objective snoring index decreased from 218 events per hour to 115 events per hour. Moreover, there was an observed increase in the minimum peripheral oxygen saturation from 85% to 88%, highlighting the pillows' efficacy in enhancing overall respiratory function during sleep. These results underscore the potential benefits of head positional pillows in managing snoring and improving oxygenation in individuals with SDB.

8.3.1.2 Chest-Worn Belts

These belts have attachments that make sleeping on the back uncomfortable, thereby promoting side sleeping. Clinical studies have documented the effectiveness of these belts, demonstrating a significant reduction in the AHI from 11 events per hour to just 2 events per hour [18]. Additionally, these belts have proven effective in altering sleeping behaviors, with documented reductions in the percentage of total sleep time spent in the supine position from 40% to virtually 0%. These findings highlight the potential of chest-worn belts as a simple yet effective therapeutic tool for managing and significantly improving conditions like POSA.

8.3.2 Advanced Positional Devices

8.3.2.1 Adjustable Beds

Beds that can be angled to elevate the head and trunk (similar to the HOBE position) offer a static, yet effective means of reducing airway collapse associated with supine sleeping [19].

8.3.3 Wearable Devices

8.3.3.1 Smart Vests

Incorporating sensor technology, these vests not only prevent back sleeping but also provide feedback and data on sleep patterns and posture, which can be useful for ongoing monitoring and adjustments in therapy [20]. It can effectively diminish the percentage of supine sleep and subjective sleepiness and improve sleep-related quality of life in patients with mild-to-moderate POSA [21].

8.3.4 Innovative Solutions

8.3.4.1 Biofeedback Devices

Some devices provide real-time feedback to the user about their sleeping position [22]. If the user rolls onto their back, the device gently vibrates to prompt them to change positions without significantly disrupting sleep.

8.3.4.2 Smart Antisnore Pillows

These pillows automatically adjust the head's position in response to detected snoring sound [23]. Equipped with sensors, they change the head's orientation to optimize airway openness without waking the patient, making them suitable for managing mild-to-moderate OSA.

8.3.5 Combination Therapies

8.3.5.1 Integrated Systems

Combining positional therapy with other treatments like continuous positive airway pressure (CPAP) [24] or oral appliances [25] can enhance therapeutic outcomes, especially in patients with complex or severe OSA.

Each of these therapies offers unique advantages and can be tailored to meet the patient's specific needs, preferences, and the severity of their condition. Initially, traditional positional therapies such as head positional pillows, chest-worn belts, and adjustable beds have been shown to significantly improve AHI and reduce the proportion of time spent in the supine position [15, 26, 27]. Newer positional therapy devices like electrically operated pillows, smart vests, and biofeedback devices have demonstrated efficacy and are often better tolerated by patients compared to traditional methods [28]. However, these vibrotactile positional therapy devices have a limited impact on sleep quality, excessive daytime sleepiness, and disease-specific quality of life [27].

Although the effectiveness of positional therapy generally falls slightly short of CPAP [29], combining positional therapy with a CPAP device can be an effective titration method for patients with difficult-to-treat OSA [24]. Similarly, while positional therapy is slightly less effective than oral appliance therapy [30], combining it with oral appliance therapy has been shown to be more effective than either treatment alone and may represent a viable option for treating POSA [25]. The success of positional therapy often hinges on patient adherence and the extent of positional dependency in their OSA, underscoring the importance of personalized treatment plans.

8.4 Polysomnographic Considerations

POSA is distinguished by a higher frequency of respiratory events when the patient is in the supine position. Identifying this polysomnographic phenotype is crucial as it opens the possibility for significant improvement in OSA symptoms through positional therapy, which involves strategic alterations in sleeping posture.

8.4.1 Defining Supine-Related OSA

Historically, supine-related OSA was defined based on a notable reduction in respiratory events when transitioning from supine to lateral positions (Fig. 8.1).

This was quantified by a ratio where the supine AHI was at least double the non-supine AHI [9], with the requirement that both supine and non-supine durations exceed 20 minutes [31]:

$$AHI_{total} \geq 5 \text{ and } \frac{AHI_{supine}}{AHI_{non-supine}} \geq 2$$

This definition has been broadly utilized in various clinical studies, influencing the metrics for evaluating treatment outcomes, particularly those involving surgical or pharmacological interventions.

To calculate the supine AHI, the following simple formula is used, considering the total sleep time (TST) spent in the supine position:

$$\frac{AHI_{face} \times TST_{face}}{TST_{face}}$$

This calculation directly equates the AHI value for the face-down (supine) position to the AHI recorded while the patient is face down, as the time factors cancel out.

Additionally, the non-supine AHI is determined by aggregating AHIs from different non-supine positions weighted by the time spent in each, as follows:

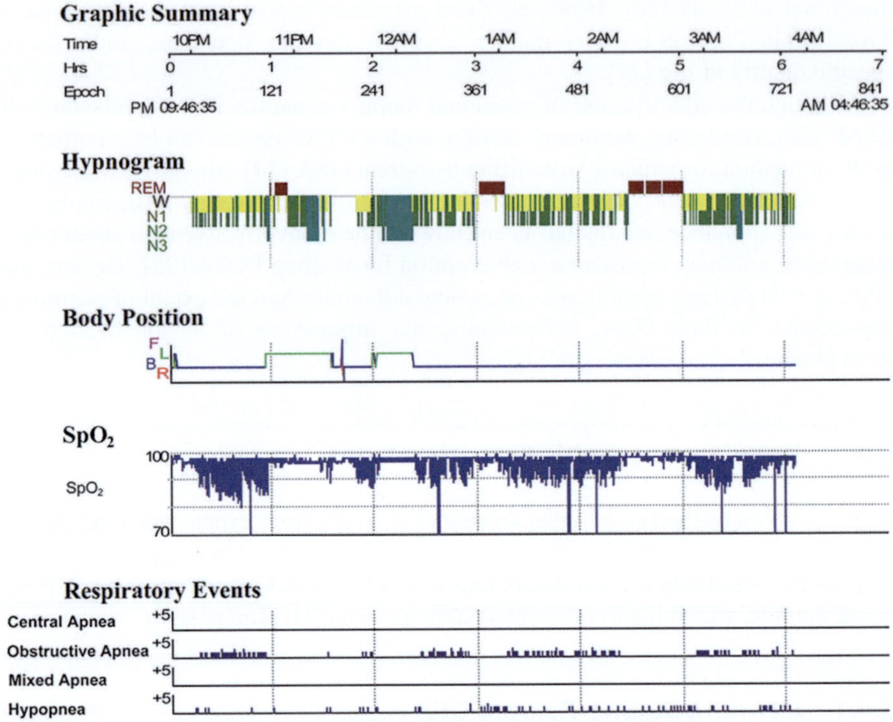

Fig. 8.1 Demonstration of polysomnographic findings in positional obstructive sleep apnea. This figure demonstrates the influence of different sleep positions on various polysomnographic parameters for a patient with positional obstructive sleep apnea. It includes a series of charts that correlate specific body positions—back (B), face (F), left side (L), and right side (R)—with key sleep metrics. These metrics include stages of sleep (rapid eye movement (REM), wake (W), N1, N2, and N3 stages), peripheral oxygen saturation (SpO_2), and respiratory events such as central apnea, obstructive apnea, mixed apnea, and hypopnea throughout the night. Each graph aims to visually represent how changes in sleeping position can affect sleep quality, oxygen levels, and the frequency and type of apnea/hypopnea events experienced by the patient

$$\frac{\left(\mathrm{AHI}_{\mathrm{back}} \times \mathrm{TST}_{\mathrm{back}}\right)+\left(\mathrm{AHI}_{\mathrm{left}} \times \mathrm{TST}_{\mathrm{left}}\right)+\left(\mathrm{AHI}_{\mathrm{right}} \times \mathrm{TST}_{\mathrm{tright}}\right)}{\mathrm{TST}_{\mathrm{back}}+\mathrm{TST}_{\mathrm{left}}+\mathrm{TST}_{\mathrm{tright}}}$$

This formula provides a comprehensive assessment of the AHI across all non-supine sleeping positions, thereby enhancing the understanding of how sleep apnea severity varies with different sleeping postures.

This nuanced approach helps clinicians tailor treatment strategies more effectively by considering the impact of sleep posture on OSA severity. These methodologies help clinicians devise more effective treatment strategies by considering the significant impact of sleep posture on the severity of OSA. This detailed approach ensures that therapeutic interventions are tailored to individual patient needs, potentially improving clinical outcomes.

8.4.2 Recent Advances in Definition

More contemporary definitions of supine-related OSA incorporate broader clinical considerations. For instance, if an OSA patient avoids sleeping on their back, their overall AHI may approximate their non-supine AHI. Complete resolution of OSA is considered achievable for individuals whose non-supine AHI is less than five events per hour. This concept, first introduced by Mador and colleagues [32], has been increasingly acknowledged and applied in subsequent research [33].

8.4.3 Consistency and Clinical Implications

An essential aspect for the clinical effectiveness of positional therapy is the consistency of the supine OSA phenotype across multiple nights. Interventions aimed at modifying sleep position must rely on the persistent occurrence of respiratory events in the supine position and their absence in non-supine positions. Current research indicates that males are more likely to demonstrate consistent supine-related OSA. Moreover, the predictability of this condition seems to correlate with a lower non-supine AHI [34]. Despite its importance, many existing definitions of supine-related OSA do not fully consider the variability and repeatability of these findings, underscoring a need for more nuanced criteria that reflect the complex dynamics of sleep and breathing patterns.

This expanded understanding and redefinition of supine-related OSA are critical for tailoring effective treatment strategies, particularly in leveraging positional therapy as a non-invasive option to manage and potentially resolve POSA.

8.5 Specific Endoscopic Considerations

Nasopharyngoscopy is a valuable office-based tool used in the Otorhinolaryngology, Head and Neck Surgery Department to assess the upper airway. Utilizing awake endoscopy with Müller's maneuver allows examiners to evaluate patients with SDB, prioritizing them for further polysomnographic assessment and potential surgical interventions [35]. Our research has shown that measurements of airway dimensions at the uvula base during Müller's maneuver are consistent across different raters [36]. Figure 8.2 illustrates these findings, contrasting positional and non-positional dependencies. Importantly, a higher baseline transverse retropalatal collapsibility is associated with an elevated AHI, and lower postoperative values correlate with better surgical outcomes, though these measures do not significantly impact positional dependency.

Moreover, nasopharyngoscopy can guide the selection of tailored treatments for POSA, such as determining the optimal height of a head positional pillow [17]. This adjustment aims to achieve a sniff position, aligning the upper respiratory tract and opening the airway during supine sleep. Theoretically, both head positions

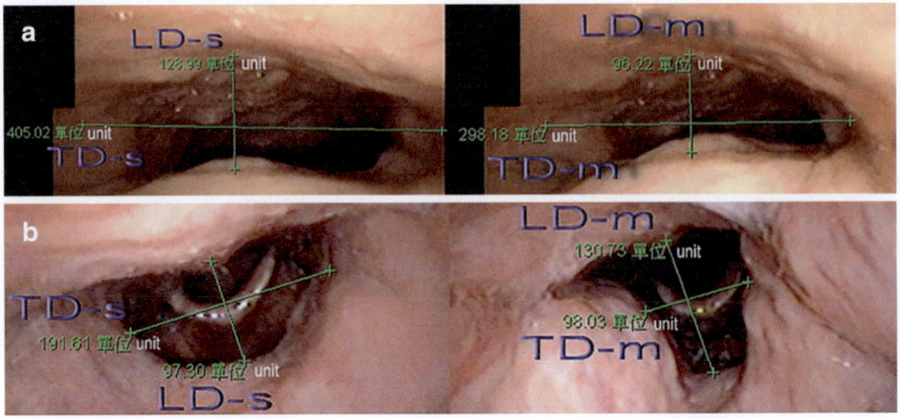

Fig. 8.2 Nasopharyngoscopic measurements of the retropalatal airway space. (**a**) This section displays the transverse and longitudinal dimensions of the retropalatal airway at the uvular base during both the static phase (left panel) and Müller's phase (right panel). (**b**) Patient with on-positional dependency: Here, the same measurements are shown for a patient whose airway dimensions change in both the static phase (left panel) and during Müller's phase (right panel). Abbreviations: LD longitudinal dimension, m Müller's phase, s static phase, TD transverse dimension

promoted by the pillow—median for head extension and lateral for rotation—help maintain airway patency and reduce snoring.

Endoscopic findings in POSA reveal that sleep position significantly influences upper airway dynamics. Drug-induced sleep endoscopy (DISE) studies show that the supine position typically exacerbates airway obstruction more than lateral positions. Specifically, anteroposterior airway obstructions are more pronounced when patients sleep on their backs, whereas lateral positions often lead to transverse airway obstructions. Patients with OSA exhibiting complete concentric retropalatal collapse in the supine position, head rotation, and oral appliance have a severer disease severity than those without such collapses [37]. Understanding these patterns helps tailor treatment strategies, such as positional therapy, oral appliance, surgical interventions, and weight reduction, to manage POSA effectively.

8.6 Postoperative Considerations and Positional Shifting

Postoperative fluctuations in sleep positions can impact the outcomes of OSA surgeries like uvulopalatopharyngoplasty (UPPP), which generally improves AHI in lateral positions for NPOSA patients [38–40]. However, positional dependency does not significantly alter UPPP's success rates [38]. Relocation pharyngoplasty, which advances and tensions the soft palate and reinforces the lateral pharyngeal

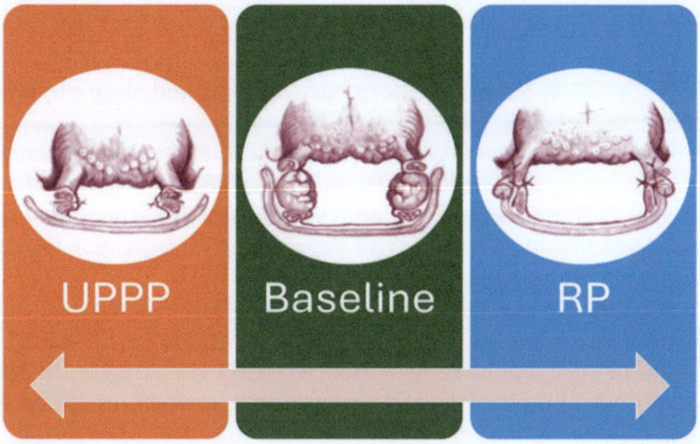

Fig. 8.3 Reconstruction of the lateral pharyngeal wall. This figure presents the pharyngeal structure, including the palatine tonsils, tongue base, and lateral pharyngeal wall as seen in the horizontal plane (middle panel). The left panel demonstrates the impact of traditional uvulopalatopharyngoplasty (UPPP), which involves direct suturing of the tonsillar fossa after tonsil removal, resulting in a narrowing of the retroglossal dimension. The right panel shows how relocation pharyngoplasty (RP) employs plication of the superior constrictor muscle and posterior pillar muscle to support and maintain the lateral pharyngeal wall, thereby preserving the retroglossal space [41, 42]

wall, has shown to help stabilize these structures against gravity in the supine position (Fig. 8.3) [41, 42]. Research indicates that positional dependency at baseline predicts surgical success in severe OSA cases treated with relocation pharyngoplasty [42]. This surgery also tends to increase positional dependency [43], offering additional therapeutic opportunities for managing POSA postoperatively.

8.7 Summary and Future Directions

Positional therapy plays a pivotal role in managing SDB, with its effectiveness highlighted across various treatment modalities. As a potentially cost-effective first-line therapy [44], especially for POSA, positional therapy is recommended for certain patient groups by the European Respiratory Society [45]. The therapy's integration into treatment plans ranges from mild-to-severe OSA cases, adapting to individual needs with devices like vibratory neck-based systems or combining with CPAP [6]. Current findings show promising long-term efficacy and adherence for mild-to-moderate OSA [46]. Advances in diagnostics, surgery, and device innovation are expected to enhance precision and outcomes in OSA management. Future research will focus on enhancing patient-centered outcomes and exploring the therapy's impact on broader health issues like cardiovascular disease, metabolic syndrome, neurocognitive disorders.

References

1. Troynikov O, Watson CG, Nawaz N. Sleep environments and sleep physiology: a review. J Therm Biol. 2018;78:192–203.
2. Tiotiu A, Mairesse O, Hoffmann G, Todea D, Noseda A. Body position and breathing abnormalities during sleep: a systematic study. Pneumologia. 2011;60(4):216–21.
3. Landry SA, Beatty C, Thomson LDJ, Wong AM, Edwards BA, Hamilton GS, Joosten SA. A review of supine position related obstructive sleep apnea: classification, epidemiology, pathogenesis and treatment. Sleep Med Rev. 2023;72:101847.
4. Oksenberg A, Gadoth N. Are we missing a simple treatment for most adult sleep apnea patients? The avoidance of the supine sleep position. J Sleep Res. 2014;23(2):204–10.
5. Joosten SA, Edwards BA, Wellman A, Turton A, Skuza EM, Berger PJ, Hamilton GS. The effect of body position on physiological factors that contribute to obstructive sleep apnea. Sleep. 2015;38(9):1469–78.
6. Verbraecken J, Dieltjens M, Op de Beeck S, Vroegop A, Braem M, Vanderveken O, Randerath W. Non-CPAP therapy for obstructive sleep apnoea. Breathe (Sheff). 2022;18(3):220164.
7. Dieltjens M, Vroegop AV, Verbruggen AE, Wouters K, Willemen M, De Backer WA, Verbraecken JA, Van de Heyning PH, Braem MJ, de Vries N, et al. A promising concept of combination therapy for positional obstructive sleep apnea. Sleep Breath. 2015;19(2):637–44.
8. Edwards BA, Joosten SA, Sands SA. A detailed mechanistic understanding of positional and nonpositional OSA: what side are you on? Chest. 2024;166(1):7–9.
9. Cartwright RD, Diaz F, Lloyd S. The effects of sleep posture and sleep stage on apnea frequency. Sleep. 1991;14(4):351–3.
10. Souza F, Genta PR, de Souza Filho AJ, Wellman A, Lorenzi-Filho G. The influence of head-of-bed elevation in patients with obstructive sleep apnea. Sleep Breath. 2017;21(4):815–20.
11. Omobomi O, Quan SF. Positional therapy in the management of positional obstructive sleep apnea-a review of the current literature. Sleep Breath. 2018;22(2):297–304.
12. Heinzer R, Petitpierre NJ, Marti-Soler H, Haba-Rubio J. Prevalence and characteristics of positional sleep apnea in the HypnoLaus population-based cohort. Sleep Med. 2018;48:157–62.
13. Tung NT, Lin SY, Dung HB, Thuy TPC, Kuan YC, Tsai CY, Lo CC, Lo K, Liu WT, Chuang HC. Associations of overnight changes in body composition with positional obstructive sleep apnea. Sleep Breath. 2023;27(2):631–40.
14. Wang X, Zhou T, Huang W, Zhang J, Zou J, Guan J, Yi H, Yin S. Differences in physiologic endotypes between nonpositional and positional OSA: results from the Shanghai Sleep Health Study Cohort. Chest. 2024;166(1):212–25.
15. Srijithesh PR, Aghoram R, Goel A, Dhanya J. Positional therapy for obstructive sleep apnoea. Cochrane Database Syst Rev. 2019;5(5):CD010990.
16. Lee K-I, Choi JH. Positional therapy for obstructive sleep apnea: therapeutic modalities and clinical effects. Sleep Med Res. 2023;14(3):129–34.
17. Chen WC, Lee LA, Chen NH, Fang TJ, Huang CG, Cheng WN, Li HY. Treatment of snoring with positional therapy in patients with positional obstructive sleep apnea syndrome. Sci Rep. 2015;5:18188.
18. Permut I, Diaz-Abad M, Chatila W, Crocetti J, Gaughan JP, D'Alonzo GE, Krachman SL. Comparison of positional therapy to CPAP in patients with positional obstructive sleep apnea. J Clin Sleep Med. 2010;6(3):238–43.
19. Iannella G, Cammaroto G, Meccariello G, Cannavicci A, Gobbi R, Lechien JR, Calvo-Henriquez C, Bahgat A, Di Prinzio G, Cerritelli L, et al. Head-Of-Bed Elevation (HOBE) for improving Positional Obstructive Sleep Apnea (POSA): an experimental study. J Clin Med. 2022;11(19)
20. van Maanen JP, Meester KA, Dun LN, Koutsourelakis I, Witte BI, Laman DM, Hilgevoord AA, de Vries N. The sleep position trainer: a new treatment for positional obstructive sleep apnoea. Sleep Breath. 2013;17(2):771–9.

21. van Maanen JP, de Vries N. Long-term effectiveness and compliance of positional therapy with the sleep position trainer in the treatment of positional obstructive sleep apnea syndrome. Sleep. 2014;37(7):1209–15.
22. Ravesloot MJL, White D, Heinzer R, Oksenberg A, Pepin JL. Efficacy of the new generation of devices for positional therapy for patients with positional obstructive sleep apnea: a systematic review of the literature and meta-analysis. J Clin Sleep Med. 2017;13(6):813–24.
23. Chung TT, Lee MT, Ku MC, Yang KC, Wei CY. Efficacy of a smart antisnore pillow in patients with obstructive sleep apnea syndrome. Behav Neurol. 2021;2021:8824011.
24. Goyal A, Pakhare A, Subhedar R, Khurana A, Chaudhary P. Combination of positional therapy with positive airway pressure for titration in patients with difficult to treat obstructive sleep apnea. Sleep Breath. 2021;25(4):1867–73.
25. Huang W, Li C, Zou J, Wang X, Zhang J, Guan J, Yi H, Yin S. Effects of the combination of novel eye mask sleep position therapy device and oral appliance on positional OSA: a multi-arm, parallel-group randomized controlled trial. Sleep Med. 2023;102:52–63.
26. Barnes H, Edwards BA, Joosten SA, Naughton MT, Hamilton GS, Dabscheck E. Positional modification techniques for supine obstructive sleep apnea: a systematic review and meta-analysis. Sleep Med Rev. 2017;36:107–15.
27. AlQarni AS, Turnbull CD, Morrell MJ, Kelly JL. Efficacy of vibrotactile positional therapy devices on patients with positional obstructive sleep apnoea: a systematic review and meta-analysis. Thorax. 2023;78(11):1126–34.
28. Yingjuan M, Siang WH, Leong Alvin TK, Poh HP. Positional therapy for positional obstructive sleep apnea. Sleep Med Clin. 2020;15(2):261–75.
29. Ha SC, Hirai HW, Tsoi KK. Comparison of positional therapy versus continuous positive airway pressure in patients with positional obstructive sleep apnea: a meta-analysis of randomized trials. Sleep Med Rev. 2014;18(1):19–24.
30. Mohamed AM, Mohammed OM, Liu S, Al-Balaa M, Al-Warafi LA, Peng SJ, Qiao YQ. Oral appliance therapy vs. positional therapy for managing positional obstructive sleep apnea; a systematic review and meta-analysis of randomized control trials. BMC Oral Health. 2024;24(1):666.
31. Bignold JJ, Mercer JD, Antic NA, McEvoy RD, Catcheside PG. Accurate position monitoring and improved supine-dependent obstructive sleep apnea with a new position recording and supine avoidance device. J Clin Sleep Med. 2011;7(4):376–83.
32. Mador MJ, Kufel TJ, Magalang UJ, Rajesh SK, Watwe V, Grant BJ. Prevalence of positional sleep apnea in patients undergoing polysomnography. Chest. 2005;128(4):2130–7.
33. Joosten SA, O'Driscoll DM, Berger PJ, Hamilton GS. Supine position related obstructive sleep apnea in adults: pathogenesis and treatment. Sleep Med Rev. 2014;18(1):7–17.
34. Joosten SA, O'Donoghue FJ, Rochford PD, Barnes M, Hamza K, Churchward TJ, Berger PJ, Hamilton GS. Night-to-night repeatability of supine-related obstructive sleep apnea. Ann Am Thorac Soc. 2014;11(5):761–9.
35. Aksoy EA, Serin GM, Polat S, Unal OF, Tanyeri H. The morphology of the nasopharyngeal inlet in obstructive sleep apnea. Eur Arch Otorrinolaringol. 2014;271(4):771–5.
36. Chen HC, Lee LA, Hsin LJ, Lin WN, Fang TJ, Huang CG, Li HY. Transverse retropalatal collapsibility is associated with obstructive sleep apnea severity and outcome of relocation pharyngoplasty. Otolaryngol Head Neck Surg. 2015;153(6):1056–63.
37. Lin HC, Wang CH, Kuo TBJ, Yang CCH, Lee JC, Chiu FS, Chang Y, Jacobowitz O, Chu CM, Hsu YS. Upper airway surgery or weight control? Modified drug-induced sleep endoscopy for obstructive sleep apnea. Otolaryngol Head Neck Surg. 2023;169(5):1345–55.
38. Lee CH, Kim SW, Han K, Shin JM, Hong SL, Lee JE, Rhee CS, Kim JW. Effect of uvulopalatopharyngoplasty on positional dependency in obstructive sleep apnea. Arch Otolaryngol Head Neck Surg. 2011;137(7):675–9.
39. Fujita S, Conway W, Zorick F, Roth T. Surgical correction of anatomic azbnormalities in obstructive sleep apnea syndrome: uvulopalatopharyngoplasty. Otolaryngol Head Neck Surg. 1981;89(6):923–34.

40. Katsantonis GP, Miyazaki S, Walsh JK. Effects of uvulopalatopharyngoplasty on sleep architecture and patterns of obstructed breathing. Laryngoscope. 1990;100(10 Pt 1):1068–72.
41. Li HY, Lee LA. Relocation pharyngoplasty for obstructive sleep apnea. Laryngoscope. 2009;119(12):2472–7.
42. Li HY, Cheng WN, Chuang LP, Fang TJ, Hsin LJ, Kang CJ, Lee LA. Positional dependency and surgical success of relocation pharyngoplasty among patients with severe obstructive sleep apnea. Otolaryngol Head Neck Surg. 2013;149(3):506–12.
43. Ruiz M, de Apodaca P, Carrasco Llatas M, Matarredona Quiles S, Dalmau Galofre J. Development of positional obstructive sleep apnea (POSA) after upper airway surgery in OSA patients. Sleep Breath. 2020;24(3):849–56.
44. Vega ME, Diaz-Abad M, Jaffe F, Yu D, Chatila W, D'Alonzo GE, Krachman S. Use of positional therapy when incorporated into a diagnosis-treatment algorithm for obstructive sleep apnea. Sleep Sci. 2019;12(1):15–20.
45. Randerath W, Verbraecken J, de Raaff CAL, Hedner J, Herkenrath S, Hohenhorst W, Jakob T, Marrone O, Marklund M, McNicholas WT, et al. European Respiratory Society guideline on non-CPAP therapies for obstructive sleep apnoea. Eur Respir Rev. 2021;30(162):210200.
46. De Corso E, Mastrapasqua RF, Fiorita A, Settimi S, Mele DA, Picciotti PM, Loperfido A, Marrone S, Rizzotto G, Paludetti G, et al. Efficacy and long-term follow-up of positional therapy by vibrotactile neck-based device in the management of positional OSA. J Clin Sleep Med. 2020;16(10):1711–9.

Mouth Taping

9

Yi-Chieh Lee and Chun-Ting Lu

Contents

Abstract

Mouth breathing during sleep is commonly observed in patients with obstructive sleep apnea (OSA) and is associated with increased upper airway collapsibility. Studies suggest that mouth breathing alters upper airway anatomy and aerodynamics, potentially worsening OSA severity. Research on mouth taping as a therapeutic approach for OSA has shown mixed results. Some studies report reduced apnea–hypopnea index (AHI) and improved inspiratory flow in selected patients, while others suggest that mouth taping may exacerbate symptoms in individuals with velopharyngeal obstruction. Oral appliances like mouth tape and shields have demonstrated potential in improving continuous positive airway pressure compliance and also reducing OSA severity. However, patient selection is crucial, as mouth taping may not be suitable for obese individuals or those with severe OSA based on current understanding. Treating nasal obstruction and pre-treatment drug-induced sleep endoscopy (DISE) are recommended to identify suitable candidates.

Y.-C. Lee (✉) · C.-T. Lu
Department of Otolaryngology-Head and Neck Surgery, New Taipei Municipal Tucheng Hospital (Built and Operated by Chang Gung Medical Foundation), New Taipei City, Taiwan

H.-Y. Li (ed.), *Snoring and Obstructive Sleep Apnea*,
https://doi.org/10.1007/978-981-95-2648-2_9

Keywords

Mouth breathing · Velopharyngeal obstruction

9.1 Introduction

Humans generally prefer nasal breathing for optimal physiological function, both during the day and while sleeping. [1, 2] However, if there is increased nasal resistance or any obstructions along the nasal breath route, the nasal breathing pattern may change to a mouth breathing pattern to compensate for decreased nasal flow and to maintain adequate respiration [3, 4].

Mouth breathing during sleep is a common phenomenon in adults with obstructive sleep apnea (OSA). Several studies have explored the impact of mouth breathing on OSA patients, focusing on anatomical and aerodynamic changes, as well as strategies for managing mouth breathing.

9.2 Aerodynamic Change During Different Breathing Pattern

In healthy adults, the breathing route can influence the upper airway. A study involving healthy participants has shown that during wakefulness, mouth breathing is associated with a reduction in retropalatal and retroglossal distances, as well as a lengthening of the pharynx [5]. During sleep, mouth opening increases upper airway collapsibility, as measured by upper airway pressure [6]. Furthermore, upper airway resistance is significantly higher during oral breathing compared to nasal breathing [7].

In patients with OSA, mouth breathing may predispose individuals to upper airway collapse. Using computational fluid dynamics simulations, a study found that during wakefulness, mouth breathing, compared to nasal breathing, was associated with higher airflow velocity and static pressure, which may increase the risk of pharyngeal airway collapse [2]. Similarly, a three-dimensional (3-D) computed tomography (CT) study revealed that mouth breathing is associated with a more elongated and narrower upper airway [8]. Furthermore, a study using drug-induced sleep endoscopy (DISE) found that mouth breathing is linked to a higher degree and prevalence of both lateral pharyngeal wall collapse and tongue base collapse [9].

In the OSA population, in addition to nasal obstruction, other factors appear to be related to mouth breathing. The proportion of mouth breathing is positively correlated with the apnea–hypopnea index (AHI) and body mass index (BMI) during natural sleep [1]. Findings from DISE also showed that patients who breathe through the mouth exhibit greater OSA severity, poorer oximetric variables, and higher BMI compared to those utilizing other breathing routes [9].

Although the mechanisms underlying mouth breathing in OSA are not fully understood, the soft palate appears to play a significant role. Research indicates that velopharyngeal obstruction correlates with a higher proportion of mouth breathing [10]. Some studies propose a novel concept suggesting that prolapse or extrusion of the soft palate during sleep may contribute to mouth breathing. The soft palate may prolapse into the nasopharynx during expiration, act like a valve, thereby restricting expiratory flow and causing a shift from nasal to mouth breathing [11, 12]. A bidirectional relationship may exist between OSA and mouth breathing. Mouth breathing could result from velopharynx obstruction during sleep, while the anatomical and aerodynamic changes associated with mouth breathing may increase the likelihood of upper airway collapse.

9.3 Mouth Taping in OSA Patients

Mouth breathing is a common symptom in patients with OSA during sleep, often due to nasal obstruction or habitual mouth breathing [13]. Researches suggest a connection between mouth breathing and OSA [1, 2, 5], and additionally, mouth breathing may compromise adherence to nasal continuous positive airway pressure (CPAP) treatment [14].

9.3.1 Mouth-Taping Effect

Previous literature has highlighted the negative impact of mouth breathing on OSA, prompting numerous studies to explore the effect of mouth closure on treating mouth breathers with OSA. The impact of mouth closure can be assessed by improvements in anatomical obstruction sites or via sleep tests.

A study by Koo et al. demonstrated improvement in the obstruction site in 40% (7 out of 18) of mouth-breathing patients after mouth closure, as measured by DISE [15]. In a recent non-randomized controlled trial, Yang et al. showed heterogeneous results regarding the improvement of inspiratory flow after mouth closure in mouth-breathing OSA patients, as measured by DISE [16]. They stratified mouth-breathing patients into three groups based on the level of oral breathing: mild, moderate, and high. Inspiratory flow improved in the moderate-level group but worsened in the high-level group. The study found that patients with velopharyngeal obstruction may not benefit from mouth closure, emphasizing the need for a personalized approach. The mouth-puffing phenomenon while the mouth was taped was also investigated in the study, which found that mouth taping worsened OSA severity in certain groups of OSA patients [17].

Our team conducted a preliminary study using 3 M seal-off mouth tape in mild OSA patients with open-mouth breathing, showing that OSA severity (measured by AHI levels) was reduced by half after mouth taping during sleep [18]. However, a study by Bhat et al. reported different findings, as they did not observe a significant reduction in AHI after chinstrap use in OSA patients [19]. The efficacy of mouth taping in reducing the AHI score remains controversial.

Recent advancements have led to the development of oral appliances such as mouth tape and oral shield devices, which have demonstrated clinical effectiveness in either reducing the severity of OSA or improving CPAP treatment compliance. Foellner et al. demonstrated the positive effects of using an oral shield device alongside nasal CPAP. This effect was achieved by preventing oral leaks during nasal CPAP use, resulting in better compliance and comfort [20]. Huang et al. also reported promising results for novel oral patches in treating mild OSA and habitual mouth breathers, with a moderate reduction in both the snore index and AHI [21].

Based on the current literature, mouth breathers with OSA may benefit from mouth taping as an individualized treatment approach, either to reduce the severity of OSA or to enhance tolerance and compliance with CPAP therapy.

9.3.2 Patient Selection

Current literature suggests that mouth taping may be beneficial for specific populations with OSA. However, treating mouth-breathing OSA patients with mouth taping requires careful consideration and selection, as the condition may worsen in some individuals.

First, most studies on the effects of mouth taping treatment have focused on non-obese (BMI < 30) patients with mild-to-moderate OSA, with limited research on its effects in those obese patients and those with severe OSA. Therefore, it is recommended that mouth taping should be applied to non-obese patients with mild-to-moderate OSA initially. Additionally, patients should be treated for nasal obstruction through medication or surgical approaches, as nasal obstruction is a contributing factor to mouth breathing during sleep and plays a role in mild OSA and sleep-disordered breathing [13]. DISE is also recommended before implementing this treatment. For those primarily breathes through mouth and with velopharyngeal obstruction observed in DISE, the condition may be worse [16].

9.4 Summary

Mouth breathing during sleep may compromise airway patency and, therefore, exacerbate OSA. Mouth taping could benefit specific OSA populations, particularly those with mild-to-moderate OSA, as studies on severe OSA remain limited. Prior treatment for nasal obstruction and a DISE evaluation are recommended before applying this therapy. Patients with velopharyngeal obstruction identified during DISE or those who primarily breathe through the mouth may experience worsened conditions with mouth taping. Care should be taken when implementing the treatment.

References

1. Koutsourelakis I, Vagiakis E, Roussos C, Zakynthinos S. Obstructive sleep apnoea and oral breathing in patients free of nasal obstruction. Eur Respir J. 2006;28(6):1222–8. https://doi.org/10.1183/09031936.00058406.
2. Suzuki M, Tanuma T. The effect of nasal and oral breathing on airway collapsibility in patients with obstructive sleep apnea: computational fluid dynamics analyses. PLoS One. 2020;15(4):e0231262. https://doi.org/10.1371/journal.pone.0231262.
3. Harvold EP, Tomer BS, Vargervik K, Chierici G. Primate experiments on oral respiration. Am J Orthod. 1981;79(4):359–72. https://doi.org/10.1016/0002-9416(81)90379-1.
4. McLean HA, Urton AM, Driver HS, et al. Effect of treating severe nasal obstruction on the severity of obstructive sleep apnoea. Eur Respir J. 2005;25(3):521–7. https://doi.org/10.1183/09031936.05.00045004.
5. Lee SH, Choi JH, Shin C, Lee HM, Kwon SY, Lee SH. How does open-mouth breathing influence upper airway anatomy? Laryngoscope. 2007;117(6):1102–6. https://doi.org/10.1097/MLG.0b013e318042aef7.
6. Meurice JC, Marc I, Carrier G, Sériès F. Effects of mouth opening on upper airway collapsibility in normal sleeping subjects. Am J Respir Crit Care Med. 1996;153(1):255–9. https://doi.org/10.1164/ajrccm.153.1.8542125.
7. Fitzpatrick MF, McLean H, Urton AM, Tan A, O'Donnell D, Driver HS. Effect of nasal or oral breathing route on upper airway resistance during sleep. Eur Respir J. 2003;22(5):827–32. https://doi.org/10.1183/09031936.03.00047903.
8. Kim EJ, Choi JH, Kim KW, et al. The impacts of open-mouth breathing on upper airway space in obstructive sleep apnea: 3-D MDCT analysis. Eur Arch Otorrinolaringol. 2011;268(4):533–9. https://doi.org/10.1007/s00405-010-1397-6.
9. Hsu YB, Lan MY, Huang YC, Kao MC, Lan MC. Association between breathing route, oxygen desaturation, and upper airway morphology. Laryngoscope. 2021;131(2):E659–e664. https://doi.org/10.1002/lary.28774.
10. Yang H, Huyett P, Wang TY, et al. Mouth closure and airflow in patients with obstructive sleep apnea: a nonrandomized clinical trial. JAMA Otolaryngol Head Neck Surg. 2024;150(11):1012–9. https://doi.org/10.1001/jamaoto.2024.3319.
11. Azarbarzin A, Sands SA, Marques M, et al. Palatal prolapse as a signature of expiratory flow limitation and inspiratory palatal collapse in patients with obstructive sleep apnoea. Eur Respir J. 2018;51(2). https://doi.org/10.1183/13993003.01419-2017.
12. Hsu YS, Ke YK, Kuo TBJ, et al. Expiratory velopharyngeal obstruction: sleep endoscopy-guided treatment strategies to prevent Oral breathing during sleep. Otolaryngology. 2024;171(5):1591–600. https://doi.org/10.1002/ohn.902.
13. Georgalas C. The role of the nose in snoring and obstructive sleep apnoea: an update. Eur Arch Otorrinolaringol. 2011;268(9):1365–73. https://doi.org/10.1007/s00405-010-1469-7.
14. Bachour A, Maasilta P. Mouth breathing compromises adherence to nasal continuous positive airway pressure therapy. Chest. 2004;126(4):1248–54. https://doi.org/10.1378/chest.126.4.1248.
15. Koo SK, Park GH, Koh TK, Jung SH, Lee HB, Ji CL. Effect of mouth closure on upper airway obstruction in patients with obstructive sleep apnoea exhibiting mouth breathing: a drug-induced sleep endoscopy study. Eur Arch Otorrinolaringol. 2020;277(6):1823–8. https://doi.org/10.1007/s00405-020-05904-0.
16. Yang H, Huyett P, Wang TY, et al. Mouth closure and airflow in patients with obstructive sleep apnea: a nonrandomized clinical trial. JAMA Otolaryngol Head Neck Surg. 2024; https://doi.org/10.1001/jamaoto.2024.3319.
17. Jau JY, Kuo TBJ, Li LPH, et al. Mouth puffing phenomenon and upper airway features may be used to predict the severity of obstructive sleep apnea. Nat Sci Sleep. 2023;15:165–74. https://doi.org/10.2147/nss.S384387.

18. Lee YC, Lu CT, Cheng WN, Li HY. The impact of mouth-taping in mouth-breathers with mild obstructive sleep apnea: a preliminary study. Healthcare (Basel). 2022;10(9) https://doi.org/10.3390/healthcare10091755.
19. Bhat S, Gushway-Henry N, Polos PG, et al. The efficacy of a chinstrap in treating sleep disordered breathing and snoring. J Clin Sleep Med. 2014;10(8):887–92. https://doi.org/10.5664/jcsm.3962.
20. Foellner S, Guth P, Jorde I, et al. Prevention of leakage due to mouth opening through applying an oral shield device (Sominpax™) during nasal CPAP therapy of patients with obstructive sleep apnea. Sleep Med. 2020;66:168–73. https://doi.org/10.1016/j.sleep.2019.06.023.
21. Huang TW, Young TH. Novel porous oral patches for patients with mild obstructive sleep apnea and mouth breathing: a pilot study. Otolaryngol Head Neck Surg. 2015;152(2):369–73. https://doi.org/10.1177/0194599814559383.

Myofunctional Therapy in OSA

10

Yi-Ju Lai, Jun-Hui Ong, Ching-Hsia Hung,
and Cheng-Yu Lin

Contents

Y.-J. Lai
Institute of Physical Education, Health and Leisure Studies, College of Management, National Cheng Kung University, Tainan, Taiwan

J.-H. Ong
Institute of Allied Health Sciences, College of Medicine, National Cheng Kung University, Tainan, Taiwan

C.-H. Hung
Institute of Allied Health Sciences, College of Medicine, National Cheng Kung University, Tainan, Taiwan

Department of Physical Therapy, College of Medicine, National Cheng Kung University, Tainan, Taiwan

C.-Y. Lin (✉)
Department of Otolaryngology, National Cheng Kung University, Hospital, College of Medicine, National Cheng Kung University, Tainan, Taiwan

Sleep Medicine Center, National Cheng Kung University Hospital, College of Medicine, National Cheng Kung University, Tainan, Taiwan
e-mail: yu621109@ms48.hinet.net

H.-Y. Li (ed.), *Snoring and Obstructive Sleep Apnea*,
https://doi.org/10.1007/978-981-95-2648-2_10

Abstract

Obstructive sleep apnea (OSA) is influenced by four key pathophysiological factors: narrow collapsible upper-airway, high loop gain, low arousal threshold, and poor muscle responsiveness. Myofunctional therapy is a non-invasive treatment for obstructive sleep apnea. It focuses on strengthening the muscles of the retropalatal region, retroglossal area, hypopharynx, temporomandibular joint (TMJ), and facial muscles, with a treatment duration of 6 weeks to 3 months. In this section, we have compiled the training patterns (phases), recording types, and primary muscle groups from previous literature into tables. Current evidence suggests that myofunctional therapy alone is more effective for individuals with an apnea–hypopnea index (AHI) below 30, whereas those with an AHI above 30 may require additional interventions, such as continuous positive airway pressure (CPAP) or surgical treatment, to achieve optimal results.

Keywords

Myofunctional Therapy · Upper Airway · Soft palate · Tongue · Facial Muscles

10.1 Definition of Myofunctional Therapy

Myofunctional therapy (MT) is a therapeutic approach that involves exercises to improve the strength, coordination, and function of the orofacial and upper airway muscles. MT aims to support optimal tongue posture, nasal breathing, and swallowing, contributing to improved airway patency and reduced symptoms of obstructive sleep apnea (OSA) [1].

10.2 The Role of Upper Airway Muscle in OSA / Pathophysiological Changes in the Upper Airway Muscle of OSA Patients

In OSA, repetitive episodes of partial or complete airway obstruction occur during sleep, primarily due to altered neuromuscular control and anatomical narrowing of the upper airway. The muscles that maintain airway patency can fail, leading to airway collapse. The palate was the most frequent site of the collapse, occurring in

58.8–98% of cases, followed by the oropharynx at 29.9–88.2%, the tongue base at 43.2–60.8%, and the epiglottis at 22.4–31.4% [2, 3]. These pathophysiological changes vary at different levels of the upper airway:

10.2.1 Retropalatal Level: Soft Palate, Uvula, Palate Muscles

The soft palate and associated muscles, such as the levator veli palatini and tensor veli palatini, play a crucial role in maintaining upper airway patency by tensing and elevating the soft palate [4]. During the wakeful stage, the tensor veli palatini muscle activity in OSA patients is greater than in the control group. However, it gradually decreases by up to 30% during the non-rapid eye movement (NREM) stage compared to its activity during wakefulness [4].The reduction in muscle tone or delayed activation contributes to the collapse of the retropalatal region.

10.2.2 Retroglossal Level: Tongue, Genioglossus (GG) Muscles

The genioglossus is the primary upper airway dilator, helps in protracts and depresses the tongue to maintain airway patency [4]. In a narrowed airway, greater dilating muscle activation will be required to counter increased inspiratory negative pressure [5]. However, this compensatory activity decreases during sleep, particularly in REM. Studies showed GG muscle strength and conduction velocity reduced significantly in OSA patients [5–7]. Additionally, OSA patients have a greater tongue volume, especially over the tongue base, as shown by volumetric magnetic resonance imaging (MRI) [8, 9]. Similar findings of increased tongue base thickness were observed using sonography [10]. These anatomical changes contribute to airway obstruction at the retroglossal level.

10.2.3 Hypopharynegal Level: Lateral Pharyngeal Wall

The lateral pharyngeal wall muscles, such as the superior constrictor muscles, which constrict the pharynx, and the palatopharyngeus, which elevates the pharynx, contribute to airway collapse during sleep. Studies indicate that increased pharyngeal fat deposition results in the thickening of the lateral walls, while reduced neuromuscular compensation in these muscles contributes to airway collapsibility in OSA [8, 11].

10.2.4 TMJ and Facial Muscles: Suprahyoid Muscles, Pterygoid Muscles, Masseter, Temporalis, Buccinators, and Orbicularis Oris

The suprahyoid muscles (e.g., mylohyoid and geniohyoid) stabilize the hyoid bone and support airway patency, while dysfunction or fatigue can impair the upper

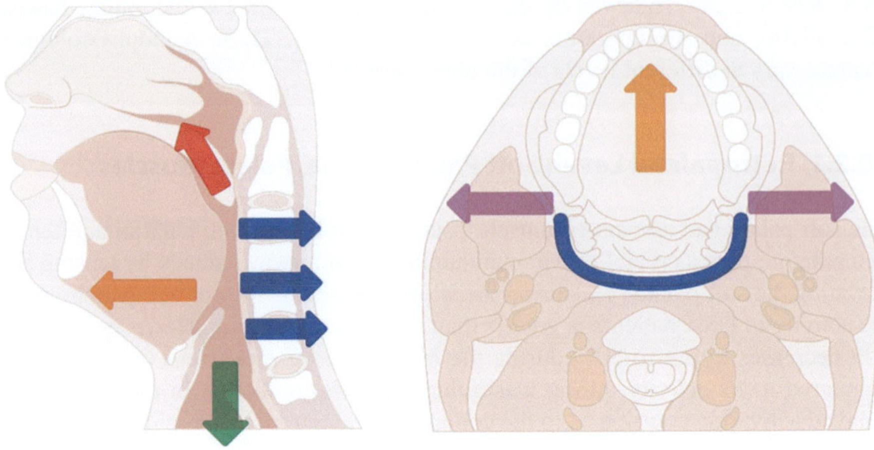

Fig. 10.1 A subject in sagittal (left image) and coronal (right image) views. Arrows represent force vectors from the upper airway dilator muscles in maintaining airway patency during wakefulness. Red arrow: tensor veli palatini, levator veli palatini; orange arrow: genioglossus, geniohyoid, anterior digastric; green arrow: sternohyoid, omohyoid, sternothyroid, mylohyoid; blue arrow and u-shaped area: pharyngeal constrictors; purple arrow: geniohyoid, stylopharyngeus

airway [12, 13]. Additionally, the masseter and temporalis muscles, which regulate jaw positioning, may influence the airway by altering mandibular posture, further contributing to obstruction [13]. Unlike the increased fat composition observed in Kim et al.'s study, demonstrated that the fat composition of the masseter muscle, analyzed using MRI, did not differ significantly between individuals with OSA and control subjects [9]. Facial muscles, including the buccinator and orbicularis oris, maintain oral seal and muscle balance between the front and the rear mouth [12]. Weakness in these muscles can lead to mouth breathing, worsening OSA symptoms by increasing airway resistance [1].

Although the genioglossus muscle is crucial for upper airway stability, its activation alone is insufficient. Other muscles, innervated by the trigeminal, facial, glossopharyngeal, vagus, and hypoglossal nerves, may significantly contribute to reducing pharyngeal collapse. However, limited studies have been performed on these muscles, indicating that further investigation is required (Fig. 10.1).

10.3 Principles and Objectives of Myofunctional Therapy

10.3.1 Principles and Objectives

Several invasive and non-invasive treatments exist to treat obstructive sleep apnea (OSA) [14–16]. Four pathophysiology traits were found as contributors to OSA such as narrow collapsible upper-airway, high loop gain, low arousal threshold, and poor muscle responsiveness [17]. Due to the critical role of the upper airway dilator muscle, researchers designed several exercise protocols that target the oral cavity and oropharyngeal muscles to treat OSA [18, 19]. Guimaraes had proposed myofunctional therapy as a treatment for OSA since the 1990s [20]. Myofunctional therapy consists of isometric and isotonic exercises targeting the soft palate, tongue, and facial muscle [18].

10.3.2 Protocol

We divided the training protocol into three categories based on training patterns (phases) and recording types. Participants used a diary in the first and the second categories to record their exercise frequency. The difference between these two categories lies in the specific training phases. In the first category, several articles discussing the training protocols for the soft palate, tongue, and facial muscles, as well as their relevance to stomatognathic functions, were summarized below [18, 21–25] (Table 10.1). Moreover, the training duration in the first category ranges from 6 weeks to 3 months, but without clearly defined phases. In the second category, the protocol includes different movements of the lips, tongue, jaw, soft palate, and cheeks, which are divided into three training phases, with each phase practiced for one month [26] (Table 10.2). In the third category, the author developed an app called the AirwayGym App, which includes exercises for the tongue, cheeks, and the jaw. Furthermore, participants were required to interact with the app [27] (Table 10.3). The app recorded the exercise frequency, and the training protocol consisted of a 3-month phase.

10.3.3 Outcome

We summarized the outcomes of the above articles, as shown in Table 10.4. Most of the articles demonstrated an improvement in the AHI following myofunctional therapy. However, Poncin et al. did not show an improvement in the AHI after 6 weeks of training [24]. The difference between Poncin et al. and other articles is that they focused solely on tongue elevation tasks. In contrast, the myofunctional therapy in the other studies combined exercises targeting various oropharyngeal structures.

10.3.4 Current Gaps in Research

The lack of a standardized exercise protocol is a major limitation of myofunctional therapy. The exercise protocols in the above articles varied in terms of exercise type, duration, and frequency. Since myofunctional therapy often involves multiple types of exercises, it is difficult to determine which specific motion contributes most to the observed outcomes, even within individual studies.

Table 10.1 The first category

Exercise	Motion requirement	Times or duration	Muscles
Soft palate	Pronounce an oral vowel intermittently (isotonic exercise) and continuously (isometric exercise) [18, 22, 23, 25]	3 minutes 5 times	Palatopharyngeus Palatoglossus Uvula Tensor veli palatini Levator veli palatini Pharyngeus lateral wall
Tongue	Brushing the superior and lateral surfaces of the tongue while the tongue is positioned in the floor of the mouth [18, 22, 23, 25]	5 times each movement, 3 times a day	
	Place tongue on the roof of the mouth with tip against upper front teeth while sucking it against roof of the mouth [21]		Intrinsic tongue muscle (superior longitudinal fiber) [28] and sternohyoid [29]
	Placing the tip of the tongue against the front of the palate and sliding the tongue backward [18, 22, 23, 25]	2–3 minutes 20 times, three times a day	Intrinsic tongue muscle (Superior longitudinal fiber) [28] and sternohyoid [29]
	Forced tongue sucking upward against the palate, pressing the entire tongue against the palate [18, 22, 23]	2–3 minutes 20 times	Intrinsic tongue muscle (Verticalis fiber) [28]
	Stick the whole tongue on the palate and suck backwards [22]	3 minutes	Intrinsic tongue muscle (superior longitudinal fiber and vericalis fiber) [28]
	Forcing the back of the tongue against the floor of the mouth while keeping the tip of the tongue in contact with the inferior incisive teeth [18, 23, 25]	2–3 minutes 20 times, three times a day	Genioglossus [28]
	Push tongue tip forward just in front of lips without touching teeth or lips [21, 23]	30 seconds or 200 times	Genioglossus [28]
	Spread center of the tongue, so the sides of the tongue touch bottom of upper teeth [21]	30 seconds	Intrinsic tongue muscle (verticalis fiber) [28]
	Tongue lateralization: push tongue to right/left corner of the mouth keep it pointed [21]		Genioglossus [30] and Sternohyoid [29]

(continued)

Table 10.1 (continued)

Exercise	Motion requirement	Times or duration	Muscles
	Holding the tongue tip between teeth anteriorly while trying to swallow [21]		Genioglossus [30] and constrictors of pharynx [31]
	Resistive therapy in which tongue press against hand resistance applied to the check on both sides [21]		Genioglossus and styloglossus [31]
	Palatal elevation with and without yawn (to feel the soft palatal lift) [21]		Tensor veli palatini and levator veli palatini [31]
	Tongue rotation in the oral vestibule [25]	10 times for each side	
	Raise tongue against the hard palate in order to squeeze the Iowa Oral Performance Instrument (IOPI, IOPI Medical LLC) bulb positioned immediately posterior to the central incisors [24]	Week 1: 60% load, 10 times*3 sets Week 2: 65% load, 10 times*3 sets Week 3: 70% load, 10 times*3 sets Week 4: 70% load, 12 times*4 sets Week 5: 75% load, 12 times*4 sets Week 6: 80% load, 12 times*4 sets	Intrinsic tongue muscle (Superior longitudinal fiber) [28] and Sternohyoid [29]
	Maintain an isometric lingual pressure on the IOPI bulb equivalent to 50% of the baseline strength value until a pressure drop >10% for more than 2 s [24]	Week 1: 50% load, 2 sets Week 2: 50% load, 2 sets Week 3: 50% load, 2 sets Week 4: 50% load, 3 sets Week 5: 50% load, 3 sets Week 6: 50% load, 3 sets	Intrinsic tongue muscle (Superior longitudinal fiber) [28] and sternohyoid [29]
Facial muscle	Close with pressure [18, 23]	30 seconds*1–4 times	Orbicularis oris

(continued)

Table 10.1 (continued)

Exercise	Motion requirement	Times or duration	Muscles
	Lips tightly closed (with pressure) to open (isotonic exercise) [22]	30 times	Orbicularis oris
	Sucking cheeks with lips closed [18, 22, 23]	3 minutes 5 s*10 times	Buccinator
	Against the finger that is introduced in the oral cavity and press the buccinator muscle outward [18, 22, 25]	30 s*3times 10 times each side, three times a day	Buccinator
	Alternated elevation of the mouth angle muscle [18, 22]	10 intermittent elevations three times Isometric 30 s*3times	Major zygomaticus Minor zygomaticus Levator labii superioris Levator anguli oris
	Lateral jaw movements with alternating elevation of the mouth angle muscle [18, 22]	30 s*3times	Lateral pterygoid Medial pterygoid
	Gargle without water [23]	200 times	Orbicularis oris
	Suck air from a syringe of 20 ml [25]	5 times, 3 times/day	Orbicularis oris
Stomatognathic function	Forced nasal inspiration and oral expiration in conjunction with phonation of open vowels, while sitting [18, 22, 25]		Levator veli palatini
	Balloon inflation with prolonged nasal inspiration and then forced blowing [18, 22, 25]	5 times	Orbicularis oris
	Suck the yogurt with a narrow straw [25]		Orbicularis oris
Swallowing and chewing	Alternate bilateral chewing and deglutition, using the tongue in the palate, closed teeth, without perioral contraction, whenever feeding [18, 22, 25]		Masseter Lateral pterygoid Medial pterygoid Digastric Genioglossus Mylohyoid Zygomaticus

Most studies recruited fewer than 30 participants in the myofunctional therapy group, with a higher proportion of male participants and an age range below 65 years, limiting the generalizability of the results. The training duration ranged from 6 weeks to 3 months; therefore, the short-term effects (less than 6 weeks) and long-term effects (more than 3 months) of myofunctional therapy in adults could not be determined.

Table 10.2 The second category [26]

Phase 1: 4 lip exercises, 5 tongue exercises, 2 jaw exercises, and 2 soft palate exercises	
Exercise	Motion requirement
Lips	Slowly open and close your mouth to its full extent, making sure the lips meet when closing
	Pucker your lips (as if about to kiss). Hold for a count of 10.
	Spread your lips into a big, exaggerated smile
Tongue	Brushing the superior and lateral surfaces of the tongue while the tongue is positioned in the floor of the mouth
	Placing the tip of the tongue against the front of the palate and sliding the tongue backward
	Open your mouth and stick out your tongue
	Repeatedly stick your tongue in and out as fast as possible
	Flick tongue from corner to corner as quickly as possible
	Move tongue all around your lips in a circle as quickly as possible
Jaw	Open and close mouth as quickly as possible, making sure lips close each time
	Say "Ma-Ma-Ma-Ma" as quickly as possible
	Alternate bilateral chewing and deglutition, using the tongue in the palate, closed teeth, without per oral contraction, whenever feeding
Phase 2: 2 lip exercises, 2 tongue exercises, 2 jaw exercises, 5 soft palate exercises, and 2 cheek exercises	
Exercise	Motion requirement
Lip	Pucker-hold-smile-hold
	Close your lips and press them tightly together
Tongue	Stick out your tongue and try to reach your chin with the tongue tip
	Hold at the farthest extension
Jaw	Repeat with "Ka-Ka-Ka-Ka" as quickly and accurately as possible
	Repeat with "Kala-Kala-Kala-Kala"
Palate	Palatal elevation with and without yawn (to feel the soft palatal lift)
	Pronounce an oral vowel intermittently (isotonic exercise) and continuously (isometric exercise)
Buccinator	Suction movements with repetitions (isotonic) and holding position (isometric)
	finger introduced in the oral cavity, pressing the buccinator muscle outward
Pterygoid	Lateral jaw movements with alternating elevation of the mouth angle muscle
	Forced nasal inspiration and oral expiration in conjunction with phonation of open vowels, while sitting
Phase 3: 2 lip exercises, 2 tongue exercises, 1 jaw exercises, and 2 soft palate exercises	
Buccinator	Balloon inflation with prolonged nasal inspiration and then forced blowing, repeated five times without taking the balloon out of the mouth

(continued)

Table 10.2 (continued)

Pterygoid	Alternated elevation of the jaw muscle forwards and sideways with repetitions
	Sing through the vowel sounds (A-E-I-O-U) as loudly as possible
Tongue	Stick out your tongue. Hold a spoon upright against the tip of extended tongue and try to push it away while hand holds the spoon in place
	Forced tongue sucking upward against the palate, pressing the entire tongue against the palate
Lip	Try to pucker with mouth wide open, without closing jaws together
	Close your lips firmly, and then make a "slurping" noise, as if sipping a drink

Table 10.3 The third category [27]

Exercise	Motion requirement	Times or duration	Muscles
Tongue	Follow the order of the lights and music. Stick your tongue out and press the screen with it following the pattern of the lights.	15 times	Genioglossus [30] and sternohyoid [29]
	Stick your tongue out and press the screen with it for 5 s and release. The tongue should not touch your teeth or lips	15 times	Genioglossus [30] and sternohyoid [29]
	Place your tongue upwards and keep the pressure on the screen for 5 s, release your tongue and then close your mouth and rest	15 times	Intrinsic tongue muscle (superior longitudinal fiber) [30]
	Place your tongue down and keep pressing the screen for 5 s, release your tongue and then close your mouth and rest	15 times	Genioglossus [30] and sternohyoid [29]
Cheek	Press the inside of the left or right cheek with the tip of your tongue. Allow the screen of your phone to touch your cheek gently. Keep the pressure for 5 s.	15 times	Genioglossus [30] and sternohyoid [29]
Jaw	Sitting upright place the phone under chin so there is no contact with the chin. Bend head forward like you are going to drink from a cup and contact your chin on the phone. Keep the contact and move your head from side to side during 10 s with your mouth closed. Lift your head and rest.	15 times	Lateral pterygoid [31]
	Bring the phone closer to the angle of the jaw on the left or right side, as if you were talking on the phone but with the screen to the outside. Turn your jaw toward the phone and keep the pressure with your index finger on the screen for 5 s. Say "I" while doing this.	15 times	Lateral pterygoid [31]

Table 10.4 The improvement of AHI after myofunctional therapy

Study, author, year	Category	N	Age	Gender	BMI	Duration	Frequency	Pre-AHI	Post-AHI	% Change AHI
O'Connor-Reina et al. 2020 [27]	3	18	59.2 (53.7–64.6)[a]	14 male	28.9 (26.8–31.0)[a]	3 months	20 min/day	44.8 (33.8–55.7)[a]	20.9 (14.0–27.7)[a]	−53.3%
Diaféria et al. 2017 [25]	1	27	45.2 (13.0)[b]	27 male	25.0 (7.4)[b]	3 months	20 min/session 3 sessions/day	28.0 (22.7)[b]	13.9 (18.5)[b]	−50.4%
Baz et al. 2012 [21]	1	30	44.1 (7.5)[b]	22 male	33.6 (2.0)[b]	3 months	10 min/session 3–5 sessions/day	22.3 (4.5)[b]	11.5 (5.4)[b]	−48.4%
Guimarães et al. 2009 [18]	1	16	51.5 (6.8)[b]	10 male	29.6 (3.8)[b]	3 months	30 min/day	22.4 (4.8)[b]	13.7 (8.5)[b]	−38.8%
Ye et al. 2018 [23]	1	25	63.4 (9.9)[b]	19 male	26.7 (2.8)[b]	6 weeks	20 min/session 2 sessions/day	20.8 (4.0)[b]	14.3 (4.1)[b]	−31.3%
Erturk et al. 2020 [22]	1	14	53.7 (7.1)[b]	–	31.4 (3.8)[b]	3 months	1 session/day 5 days/week	42.6 (27.1)[b]	40.0 (21.5)[b]	−6.1%
Verma et al. 2016 [26]	2	20	41.1 (10.6)[b]	15 male	25.6 (3.1)[b]	3 months	10 times/exercise 5 session/day	20.1 (9.1)[b]	19.7 (9.4)[b]	−2.0%
Poncin et al. 2022 [24]	1	12	48.0 [44.8–55.5][a]	8 male	26.5 (24.5–29.6)[a]	6 weeks	15 min/session 4 sessions/week	18.9 [10.3–37.5][a]	22.2 [8.9–40.6][a]	17.46%

AHI apnea–hypopnea index; *BMI* body mass index; *N* number of patients; *M* male
[a]Median (IQR)
[b]Mean (SD)

Adherence rate played a critical role in myofunctional therapy, ranging from 46% to 100% in the articles reviewed. Some studies defined an adequate adherence rate as completing at least 75%–85% of all exercise sessions. Participants who did not meet this adherence threshold were excluded by the authors [18, 24, 27]. Most adherence rates were calculated based on participants' diaries, with only one study using an app for monitoring [27]. In studies employing diaries, examiners were often required to conduct weekly face-to-face evaluations or phone consultations with participants to maintain or enhance adherence rates. However, using diaries to record exercise frequency introduces the potential for retrospective completion, thereby increasing the risk of recall bias. In contrast, app-based monitoring offers greater efficiency, enabling exercise data to be automatically recorded and stored on a networked platform upon task completion. Furthermore, app-based systems reduce the frequency of participant-evaluator interactions, necessitating only monthly evaluations [27].

Most of the included studies did not assess the long-term maintenance effects of myofunctional therapy, with only one study incorporating a 3-week washout period [25]. In this study, the AHI after the three-week washout period did not return to baseline levels, indicating the maintenance effect of myofunctional therapy [25]. However, the long-term effect of myofunctional following the cessation of treatment remains unknown [32].

10.4 Integration with Other Therapy

10.4.1 Continuous Positive Airway Pressure (CPAP)

Study by Suzuki, Okamoto [33] demonstrated that 6 months of myofunctional therapy combined with CPAP significantly reduced AHI and average apnea duration. The author reported that this effect might be attributed to increased tongue strength from myofunctional therapy. Another study found that compared to baseline and the placebo group, myofunctional therapy, CPAP, and combined myofunctional therapy with CPAP significantly reduced AHI, ESS, and snoring after three months of treatment. Notably, the standalone myofunctional therapy group maintained these effects after a 3-week washout. Additionally, the combined therapy group demonstrated greater CPAP compliance after one and three weeks of treatment than the standalone CPAP group [25]. However, research specifically evaluating the combined effect of myofunctional therapy and CPAP is limited.

10.4.2 Surgery

In a previous study, the transoral robotic surgery combined with the myofunctional therapy group demonstrated a significant reduction in AHI in the supine position after three months of myofunctional therapy [34]. The combined treatment group also showed greater improvements in tongue protrusion strength than the standalone surgery group. This result suggests that myofunctional therapy could complement surgery by improving post-surgical outcomes through muscle training. However, there is limited direct evidence assessing the integration of myofunctional therapy and surgery.

10.4.3 Mandibular Advancement Device (MAD)

Mandibular advancement devices reposition the jaw to improve airway patency. It had been showed to reduce more than 50% of AHI and arousal index in OSA patients [35]. Combining MAD therapy with myofunctional therapy may enhance outcomes by addressing muscle tone and tongue positioning issues. Research by Siripajana, Chalidapongse [36] found that adding myofunctional therapy for patients with moderate OSA who had mild residual AHI after MAD showed no significant difference in respiratory event index after two months of myofunctional therapy. The author suggests that a longer duration of training may be necessary.

10.4.4 Comprehensive Respiratory Exercise

The previous study involved a 12-week exercise program that included myofunctional therapy, stretching, aerobic, and resistance training for moderate to severe OSA patients. It significantly reduced the AHI and arousal index and improved oxygen saturation compared to the control group [37]. Another study found no change in AHI after eight weeks of comprehensive community programs, including myofunctional therapy, walking, dietary recommendations, and sleep hygiene education. However, post hoc analysis revealed a significant reduction in AHI among participants under 60 years [38]. The author recommends longer and better-supervised training programs in future research.

10.5 Conclusion

Myofunctional therapy has emerged as a promising non-invasive approach for managing obstructive sleep apnea. Evidence to date indicates that myofunctional therapy alone is more effective for individuals with an apnea–hypopnea index below 30, while patients with an apnea–hypopnea index above 30 may require combined interventions, such as CPAP or surgical treatments, for optimal results. Moving forward, research should prioritize evaluating the long-term effects of MT and optimizing its application in clinical practice.

References

1. Camacho M, Certal V, Abdullatif J, Zaghi S, Ruoff CM, Capasso R, et al. Myofunctional therapy to treat obstructive sleep apnea: a systematic review and meta-analysis. Sleep. 2015;38(5):669–75.
2. Hybaskova J, Jor O, Novak V, Zelenik K, Matousek P, Kominek P. Drug-induced sleep endoscopy changes the treatment concept in patients with obstructive sleep Apnoea. Biomed Res Int. 2016;2016:6583216.
3. Van den Bossche K, Van de Perck E, Wellman A, Kazemeini E, Willemen M, Verbraecken J, et al. Comparison of drug-induced sleep endoscopy and natural sleep endoscopy in the assessment of upper airway pathophysiology during sleep: protocol and study design. Front Neurol. 2021;12:768973.
4. Bosi M, De Vito A, Gobbi R, Poletti V, Vicini C. The importance of obstructive sleep apnoea and hypopnea pathophysiology for customized therapy. Eur Arch Otorrinolaringol. 2017;274(3):1251–61.
5. Cori JM, O'Donoghue FJ, Jordan AS. Sleeping tongue: current perspectives of genioglossus control in healthy individuals and patients with obstructive sleep apnea. Nat Sci Sleep. 2018;10:169–79.
6. McSharry D, O'Connor C, McNicholas T, Langran S, O'Sullivan M, Lowery M, et al. Genioglossus fatigue in obstructive sleep apnea. Respir Physiol Neurobiol. 2012;183(2):59–66.
7. O'Connor-Reina C, Plaza G, Garcia-Iriarte MT, Ignacio-Garcia JM, Baptista P, Casado-Morente JC, et al. Tongue peak pressure: a tool to aid in the identification of obstruction sites in patients with obstructive sleep apnea/hypopnea syndrome. Sleep Breath. 2020;24(1):281–6.
8. Schwab RJ, Pasirstein M, Pierson R, Mackley A, Hachadoorian R, Arens R, et al. Identification of upper airway anatomic risk factors for obstructive sleep apnea with volumetric magnetic resonance imaging. Am J Respir Crit Care Med. 2003;168(5):522–30.
9. Kim AM, Keenan BT, Jackson N, Chan EL, Staley B, Poptani H, et al. Tongue fat and its relationship to obstructive sleep apnea. Sleep. 2014;37(10):1639–48.
10. Chen JW, Chang CH, Wang SJ, Chang YT, Huang CC. Submental ultrasound measurement of dynamic tongue base thickness in patients with obstructive sleep apnea. Ultrasound Med Biol. 2014;40(11):2590–8.
11. Isono S. Obstructive sleep apnea of obese adults: pathophysiology and perioperative airway management. Anesthesiology. 2009;110(4):908–21.
12. Lin HY, Su PL, Lin CY, Hung CH. Models of anatomically based oropharyngeal rehabilitation with a multilevel approach for patients with obstructive sleep apnea: a meta-synthesis and meta-analysis. Sleep Breath. 2020;24(4):1279–91.
13. Dempsey JA, Veasey SC, Morgan BJ, O'Donnell CP. Pathophysiology of sleep apnea. Physiol Rev. 2010;90(1):47–112.
14. Sullivan CE, Issa FG, Berthon-Jones M, Eves L. Reversal of obstructive sleep apnoea by continuous positive airway pressure applied through the nares. Lancet. 1981;1(8225):862–5.
15. Camacho M, Certal V, Capasso R. Comprehensive review of surgeries for obstructive sleep apnea syndrome. Braz J Otorhinolaryngol. 2013;79(6):780–8.
16. Randerath WJ, Verbraecken J, Andreas S, Bettega G, Boudewyns A, Hamans E, et al. Non-CPAP therapies in obstructive sleep apnoea. Eur Respir J. 2011;37(5):1000–28.
17. Osman AM, Carter SG, Carberry JC, Eckert DJ. Obstructive sleep apnea: current perspectives. Nat Sci Sleep. 2018;10:21–34.
18. Guimarães KC, Drager LF, Genta PR, Marcondes BF, Lorenzi-Filho G. Effects of oropharyngeal exercises on patients with moderate obstructive sleep apnea syndrome. Am J Respir Crit Care Med. 2009;179(10):962–6.
19. O'Connor Reina C, Plaza G, Ignacio-Garcia JM, Baptista Jardin P, Garcia-Iriarte MT, Casado-Morente JC, et al. New mHealth application software based on myofunctional therapy applied to sleep-disordered breathing in non-compliant subjects. Sleep Sci Pract. 2020;4:1–10.

20. Guimaraes K. Soft tissue changes of the oropharynx in patients with obstructive sleep apnea. J Bras Fonoaudiol. 1999;1(1):69–75.
21. Baz H, Elshafey M, Elmorsy S, Abu-Samra M. The role of oral myofunctional therapy in managing patients with mild to moderate obstructive sleep apnea. PAN Arab J Rhinol. 2012;2(1):5.
22. Erturk N, Calik-Kutukcu E, Arikan H, Savci S, Inal-Ince D, Caliskan H, et al. The effectiveness of oropharyngeal exercises compared to inspiratory muscle training in obstructive sleep apnea: a randomized controlled trial. Heart Lung. 2020;49(6):940–8.
23. Ye D, Chen C, Song D, Shen M, Liu H, Zhang S, et al. Oropharyngeal muscle exercise therapy improves signs and symptoms of post-stroke moderate obstructive sleep apnea syndrome. Front Neurol. 2018;9:912.
24. Poncin W, Correvon N, Tam J, Borel JC, Berger M, Liistro G, et al. The effect of tongue elevation muscle training in patients with obstructive sleep apnea: a randomised controlled trial. J Oral Rehabil. 2022;49(11):1049–59.
25. Diaféria G, Santos-Silva R, Truksinas E, Haddad FLM, Santos R, Bommarito S, et al. Myofunctional therapy improves adherence to continuous positive airway pressure treatment. Sleep Breath. 2017;21(2):387–95.
26. Verma RK, Johnson JJ, Goyal M, Banumathy N, Goswami U, Panda NK. Oropharyngeal exercises in the treatment of obstructive sleep apnoea: our experience. Sleep Breath. 2016;20(4):1193–201.
27. O'Connor-Reina C, Ignacio Garcia JM, Rodriguez Ruiz E, Morillo Dominguez MDC, Ignacio Barrios V, Baptista Jardin P, et al. Myofunctional therapy app for severe apnea-hypopnea sleep obstructive syndrome: pilot randomized controlled trial. JMIR Mhealth Uhealth. 2020;8(11):e23123.
28. Kajee Y, Pelteret JP, Reddy BD. The biomechanics of the human tongue. Int J Numer Method Biomed Eng. 2013;29(4):492–514.
29. Bérzin F. Electromyographic analysis of the sternohyoid muscle and anterior belly of the digastric muscle in head and tongue movements. J Oral Rehabil. 1995;22(11):825–9.
30. Corbin-Lewis K, Liss JM, Sciortino KL. Clinical anatomy & physiology of the swallow mechanism. 2005.
31. Avers D, Brown M. Daniels and Worthingham's muscle testing, First South Asia Edition E Book: techniques of manual examination and performance. Testing: Elsevier; 2018.
32. Clark HM, Solomon NP. Age and sex differences in orofacial strength. Dysphagia. 2012;27(1):2–9.
33. Suzuki M, Okamoto T, Akagi Y, Matsui K, Sekiguchi H, Satoya N, et al. Efficacy of oral myofunctional therapy in middle-aged to elderly patients with obstructive sleep apnoea treated with continuous positive airway pressure. J Oral Rehabil. 2021;48(2):176–82.
34. Lai YJ, Su PL, Li CY, Lin CY, Hung CH, Lin CY. Oropharyngeal rehabilitation for patients with moderate to severe obstructive sleep apnea after transoral robotic surgery. Otolaryngol Head Neck Surg. 2022:1945998221088752.
35. Ramar K, Dort LC, Katz SG, Lettieri CJ, Harrod CG, Thomas SM, et al. Clinical practice guideline for the treatment of obstructive sleep apnea and snoring with oral appliance therapy: an update for 2015. J Clin Sleep Med. 2015;11(7):773–827.
36. Siripajana P, Chalidapongse P, Sanguanwong N, Chaweewannakorn C. Efficacy of oropharyngeal exercises as an adjuvant therapy for obstructive sleep apnea: a randomized controlled trial. J Prosthodont Res. 2024;68(4):540–8.
37. Lin HY, Chang CJ, Chiang CC, Su PL, Lin CY, Hung CH. Effects of a comprehensive physical therapy on moderate and severe obstructive sleep apnea- a preliminary randomized controlled trial. J Formos Med Assoc. 2020;119(12):1781–90.
38. Torres-Castro R, Vilaro J, Marti JD, Garmendia O, Gimeno-Santos E, Romano-Andrioni B, et al. Effects of a combined community exercise program in obstructive sleep apnea syndrome: a randomized clinical trial. J Clin Med. 2019;8(3).

Obesity, OSA, and Management

11

Keng-Hao Liu, I-Wen Chen, and Wey-Ran Lin

Contents

K.-H. Liu (✉)
Division of General Surgery, Department of Surgery, Chang Gung Memorial Hospital, Taoyuan City, Taiwan
e-mail: kenghao@cgmh.org.tw

I.-W. Chen
Division of Endocrinology and Metabolism, Department of Internal Medicine, Chang Gung Memorial Hospital, Taoyuan City, Taiwan

W.-R. Lin
Division of Gastroenterology and Hepatology, Department of Internal Medicine, Chang Gung Memorial Hospital, Taoyuan City, Taiwan

H.-Y. Li (ed.), *Snoring and Obstructive Sleep Apnea*,
https://doi.org/10.1007/978-981-95-2648-2_11

Abstract

Obesity is a global health issue that increases the risk of various diseases and is also strongly associated with obstructive sleep apnea (OSA), with a significant increase in OSA risk linked to weight gain. Weight loss remains the key treatment for OSA, with approaches including diet control, exercise, medications, and surgical options. FDA-approved medications such as Orlistat, Phentermine-topiramate, and GLP-1 receptor agonists are effective for weight management. Bariatric endoscopic procedures like gastric balloon insertion and endoscopic sleeve gastroplasty offer minimally invasive alternatives to surgery. Bariatric surgery, particularly sleeve gastrectomy and Roux-en-Y gastric bypass, provides long-term weight loss and improvement of obesity-related comorbidities. However, some patients may still experience moderate-to-severe OSA, making combined airway and bariatric surgery (CABS) a promising solution for these individuals.

Keywords

Obesity · Obstructive sleep apnea · Medications for obesity · Bariatric endoscopic procedure · Bariatric surgery

11.1 Obesity and OSA

Obesity is a global disease that increases the risk of type 2 diabetes, heart disease, reproductive disorders, bone and respiratory diseases, as well as certain cancers. According to the World Health Organization report, the population of overweight adults increased from 25% in 1990 to 43% in 2022. The prevalence of obstructive sleep apnea (OSA) in the general population is 3–7% for men and 2–5% for women [1]. A 10% weight gain leads to a sixfold increase in the development of OSA, and up to 90% of severely obese patients develop OSA [2].

Obesity contributes to upper airway collapsibility by increasing fat deposits in the airway lumen and muscles (Fig. 11.1), which in turn heightens susceptibility to obstructive sleep apnea. Several mechanisms link obesity to OSA, including increased sympathetic tone, vascular endothelial dysfunction, intermittent hypoxia, inflammation, oxidative stress, and metabolic dysregulation. These factors also contribute to upper airway dysfunction and respiratory control instability [3].

Fig. 11.1 Computed tomographic reconstruction of a male patient with a BMI of 44.2 kg/m^2 showing a very narrow retropalatal and retroglossal space due to increasing fat deposits

11.2 Management of Obesity

Weight loss is crucial in the treatment of OSA. Management of obesity includes dietary control, exercise, behavioral therapy, a limited selection of medications, and surgical or endoscopic interventions.

11.3 FDA-Approved Medications for Obesity [4–7]

11.3.1 Orlistat [4–6]

Orlistat is a gastrointestinal lipase inhibitor that prevents the absorption of approximately 30% of dietary fat, leading to a caloric deficit. Clinical studies have shown that orlistat results in a weight loss 3.1% greater than placebo over one year. In addition to weight reduction, orlistat has been associated with reductions in waist circumference, blood pressure, and low-density lipoprotein (LDL) cholesterol. However, its use is limited by gastrointestinal side effects, such as oily stools and flatulence.

11.3.2 Phentermine and Phentermine–Topiramate [4–6]

Phentermine, a sympathomimetic amine, is approved for short-term use in obesity management, while the combination of phentermine and topiramate is approved for long-term use. Phentermine–topiramate is associated with a weight loss of 8% greater than placebo. In addition to its efficacy in reducing weight, this combination medication has been shown to improve metabolic outcomes, including reductions in blood pressure and LDL cholesterol. Neurological side effects such as dry mouth, paresthesia, and cognitive disturbances might occur.

11.3.3 Naltrexone–Bupropion [4–6]

Naltrexone–bupropion, approved since 2014, has demonstrated moderate efficacy in weight loss. Studies indicate that it achieves 4.1% greater weight reduction than placebo, with improvements rising to 9.3% when combined with intensive behavioral therapy. However, this medication has limited cardiovascular benefits and does not significantly affect blood pressure or major cardiovascular events. While effective, adherence to naltrexone–bupropion is low due to common side effects like nausea, constipation, and headache. It should be used cautiously in individuals with moderate kidney or liver impairment and avoided in young adults with depression or anxiety.

11.3.4 Glucagon-Like Peptide-1 (GLP-1) Receptor Agonists [4–6]

GLP-1 receptor agonists, including liraglutide and semaglutide, are among the most effective agents for long-term weight management. Liraglutide, administered daily, results in a 4.7% greater weight loss than placebo, while semaglutide, a weekly injectable, has demonstrated an impressive 11.4% weight reduction compared to placebo. Both medications provide additional glycemic control and cardiovascular risk reduction, making them particularly suitable for patients with type 2 diabetes. Gastrointestinal side effects, such as nausea and vomiting, can be mitigated through gradual dose escalation.

11.3.5 Tirzepatide [7]

Tirzepatide is a dual GLP-1 and GIP (glucose-dependent insulinotropic polypeptide) receptor agonist that has shown remarkable efficacy in weight loss. In clinical trials, tirzepatide achieved a 12.4% greater weight reduction compared to placebo. It demonstrated significant improvements in weight loss and metabolic health, making it a promising option for obesity treatment.

11.4 Intervention Endoscopic Procedures [8–11]

Bariatric endoscopic techniques offer a potentially reversible, minimally invasive, and more cost-effective method of treating obesity compared to surgery. These techniques can be divided into the following categories:

1. Restriction of stomach volume.
2. Reduction of small intestine absorption.
3. Gastric aspiration.

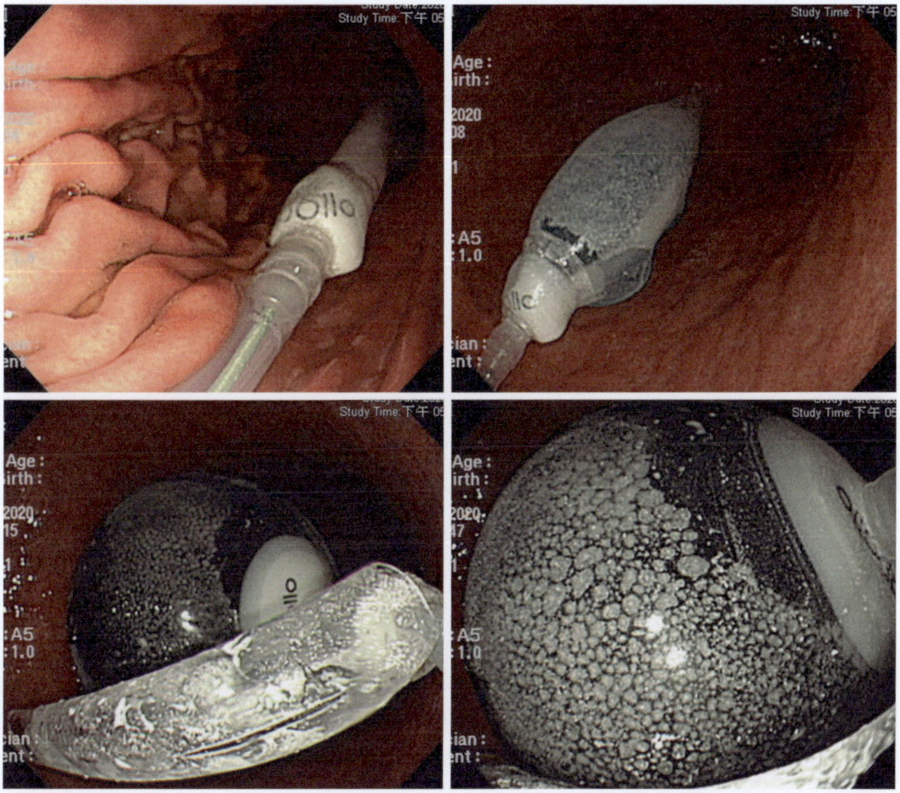

Fig. 11.2 Gastric balloon insertion

11.5 Restriction of Stomach Volume

11.5.1 Gastric Balloon Insertion (GBI)

GBI is recommended for patients with a BMI over 30 kg/m^2. Intragastric balloons are endoscopically placed or swallowed devices that decrease gastric volume and alter motility (Fig. 11.2). These devices typically remain in place for four to twelve months. GBI procedures have shown moderate weight loss benefits, with total body weight loss (TBWL) ranging from 6.8% to 10.2%. Improvements in metabolic markers, such as fasting plasma glucose (FPG), LDL cholesterol, triglycerides, and systolic blood pressure, have also been reported.

11.5.2 Endoscopic Sleeve Gastroplasty (ESG)

ESG uses an endoscopic suturing device to create a tubular gastric sleeve of the stomach (Fig. 11.3). This procedure effectively reduces stomach size and alters

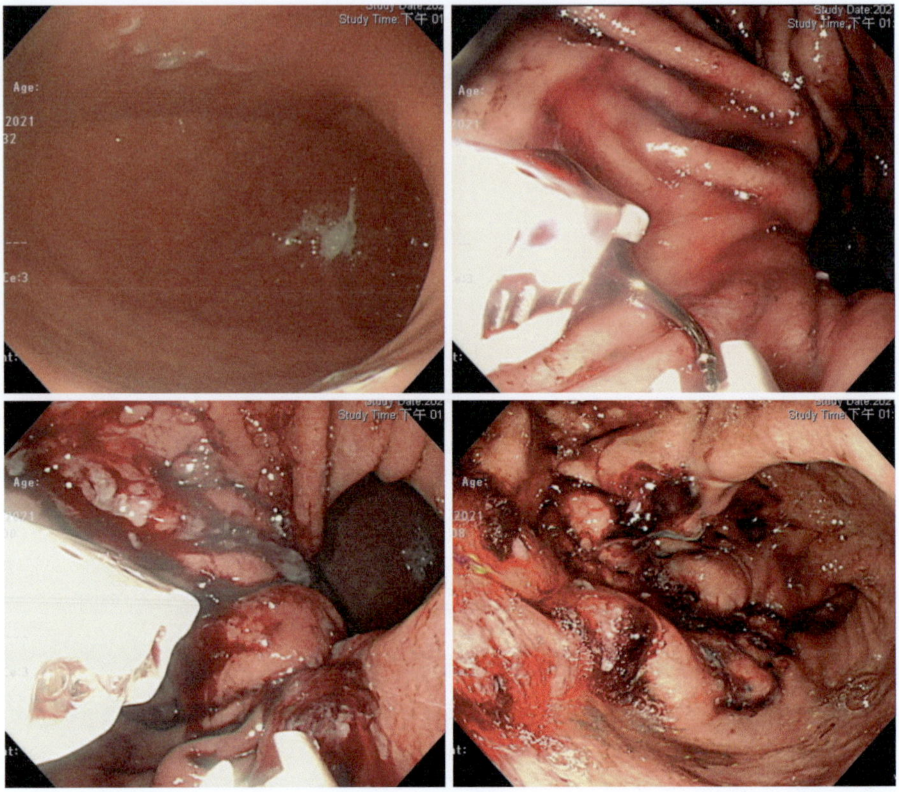

Fig. 11.3 Endoscopic sleeve gastroplasty

motility. ESG has been shown to increase gastric emptying by 90 minutes and reduce caloric intake by 59% during a liquid meal test. Additionally, ESG appears to lower ghrelin levels, although GLP-1 and PYY levels remain largely unchanged. The largest case series on ESG reported TBWL of 15.2% at six months and 18.6% at 24 months. Beyond weight loss, ESG has demonstrated notable cardiometabolic benefits, including reductions in HbA1c, systolic blood pressure, fasting triglycerides, and serum ALT.

11.5.3 Primary Obesity Surgery Endoluminal (POSE)

The POSE procedure uses an incisionless operating platform to create full-thickness plications in the gastric fundus and body, reducing gastric accommodation and delaying gastric emptying. Meta-analyses of studies reported a mean TBWL of 13.5% at 3–6 months and 12.7% at 12–15 months.

11.6 Reduction of Small Intestinal Absorption

11.6.1 Duodenal–Jejunal Bypass Linear (DJBL)

DJBL, marketed as EndoBarrier®, is a thin, flexible sleeve made of biocompatible material. It is endoscopically implanted in the small intestine to create a physical barrier between chyme and the intestinal wall. A recent meta-analysis showed an average HbA1c reduction of 1.3, accompanied by significant weight loss, averaging 11.3 kg (TBWL of 18.9%), with substantial weight reduction sustained one year after device removal.

11.6.2 Duodenal Mucosal Resurfacing (DMR)

DMR uses a single-use balloon catheter, Revita (Fractyl Health), to perform hydrothermal ablation of the superficial duodenal mucosa. The regeneration of the mucosa disrupts pathways involved in the development of type 2 diabetes mellitus (T2DM) and metabolic syndrome. A meta-analysis of four studies involving 127 patients with non-insulin-dependent T2DM showed significant reductions in HbA1c levels, with an average decrease of 1.72% at 3 months and 0.94% at 6 months post-DMR.

11.7 Gastric Aspiration

The AspireAssist (AT) is the first FDA-approved device for treating class II and class III obesity. The system involves endoscopic placement of a gastrostomy tube (A-tube) and a siphon component to aspirate gastric contents approximately 20 minutes after meal consumption, reducing caloric intake by about 30%. This therapy, combined with lifestyle modifications, is approved for long-term use in individuals with a BMI of 35–55 kg/m^2. Clinical studies have demonstrated significant weight loss outcomes with AT. A multicenter study involving 82 participants with a baseline BMI of 41.6 ± 4.5 kg/m^2 reported an average BMI of 34.1 ± 5.4 kg/m^2 and 18.3% ± 8.0% TBWL after the first year. Sustained weight loss of 15.3% TBWL at two years, 16.6% at three years, and 18.7% at four years was observed, all statistically significant ($P < 0.01$). Overall, AT has been validated as a safe and effective treatment for obesity, with notable improvements in quality of life.

11.7.1 Bariatric Surgery

In 2022, more than 300,000 bariatric surgeries were reported globally. The most frequently performed surgical procedures were sleeve gastrectomy (61%), Roux-en-Y gastric bypass (26%), one-anastomosis gastric bypass (4.6%), gastric band (0.9%), and other procedures [12]. Bariatric surgery results in sustained weight loss and can either fully resolve or improve the comorbidities associated with obesity.

11.7.2 Sleeve Gastrectomy

Sleeve gastrectomy (SG, Fig. 11.4) is the most commonly performed metabolic and bariatric surgery (MBS) worldwide, accounting for 67% of procedures [12]. It is popular due to its lower cost, technical simplicity, shorter operating times, and absence of anastomoses and internal hernia risks. While SG has shown 58% excess weight loss at five years and improvements in type 2 diabetes and hypertension, a meta-analysis of follow-up studies of seven years or more found a 28% weight regain and 20% revision rate, primarily due to weight recurrence [13]. SG offers weight loss and metabolic outcomes comparable to Roux-en-Y gastric bypass (RYGB). Post-SG issues, such as gastroesophageal reflux disease (GERD), leaks, and insufficient or recurrent weight loss, are common causes for revision surgeries [14]. Meta-analyses of SG show that 19–32% of patients experience GERD symptoms, and 23–32% develop de novo GERD after surgery. Additionally, 3.1–4.0% of patients may need revisional surgery due to GERD [15]. Another concern is the development of Barrett's esophagus (BE), with 8–11% of SG patients developing the condition over long-term follow-up. The risk of BE is associated with chronic acid exposure caused by GERD [16]. Despite these complications, SG remains a widely used and effective option for bariatric surgery, requiring careful patient management and potential long-term monitoring for GERD and other related conditions [13].

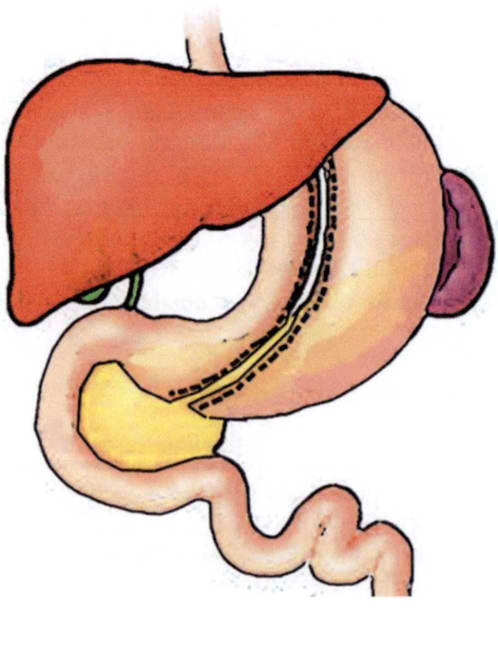

Fig. 11.4 Sleeve Gastrectomy

Fig. 11.5 Roux-en Y Gastric Bypass

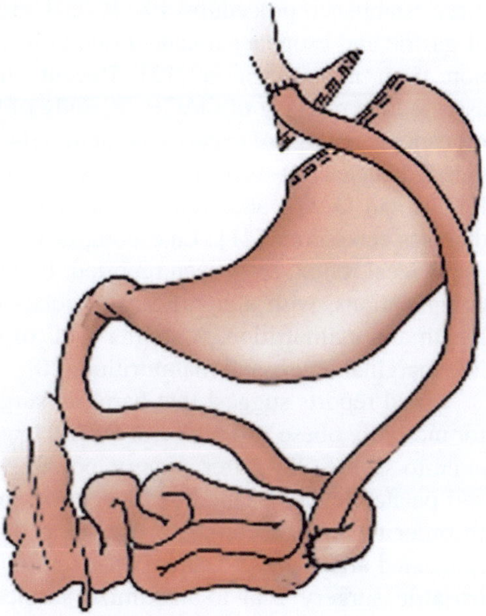

11.7.3 Roux-en-Y Gastric Bypass

Roux-en-Y gastric bypass (RYGB, Fig. 11.5) has been the gold standard surgical procedure for morbid obesity since the 1980s. It still accounts for more than a quarter of all bariatric surgeries [12]. RYGB results in significantly greater weight loss and a lower body mass index (BMI) after up to five years compared to laparoscopic adjustable gastric banding (LAGB). However, RYGB is associated with longer hospital stays and a higher rate of late major complications compared to LAGB. On the other hand, LAGB had a higher rate of reoperation for band removal (40.9%) [17]. The SLEEVEPASS trial [18] revealed that both SG and RYGB lead to significant, sustained long-term weight loss, with similar weight loss patterns over 10 years. Unlike SG, RYGB was associated with lower prevalence of esophagitis, reflux symptoms, and proton-pump inhibitor use. There were no significant differences in long-term complication rates or remission of type 2 diabetes, dyslipidemia, or obstructive sleep apnea between the two procedures. However, remission of hypertension occurred more frequently after RYGB.

11.7.4 One-Anastomosis Gastric Bypass (OAGB)

One-anastomosis gastric bypass (OAGB) has gained popularity, and a Consensus Conference in 2019 concluded that OAGB is effective for weight loss and improves or resolves obesity-related comorbidities such as type 2 diabetes and hypertension [19]. However, there is limited level 1 evidence supporting OAGB compared to more established procedures like RYGB and SG. There are concerns about the risk of gastric and esophageal cancer due to bile reflux and the possibility of malnutrition from malabsorption [12]. Recent studies, including meta-analyses, have assessed the efficacy of OAGB. A study of 95 quantitative studies indicated a significant percentage of excess weight loss (64.0–75.6%) and high resolution rates for type 2 diabetes (60–100%) and hypertension (40–74%) [20]. Comparisons between RYGB and OAGB show similar or better outcomes in terms of weight loss and diabetes resolution [21]. Late complications, including malnutrition and gastroesophageal reflux, have been reported. In one study, 5.46% of patients experienced complications, with a significant correlation between biliopancreatic limb (BPL) length and malnutrition. A shorter BPL of 150 cm may reduce adverse outcomes like marginal ulcers and malnutrition [20].

Varied reports suggest that bariatric surgery reduces adverse respiratory events for morbidly obese patients with OSA. However, up to 71% of patients retain moderate-to-severe OSA after surgery, possibly due to hypertrophic tonsils, redundant soft palate, deviated septal cartilage, or weakness of pharyngeal muscle tone [22]. In order to achieve complete remission of OSA, further airway surgery may be scheduled after bariatric surgery. Lin et al. [23] conducted a combined airway and bariatric surgery trial to maximize surgical outcomes in morbidly obese OSA patients. They enrolled 20 morbidly obese OSA patients (median apnea–hypopnea index (AHI), 63.2 events/h). The combined airway and bariatric surgery (CABS) group ($N = 10$) underwent both bariatric and airway surgeries, while the other group received bariatric surgery only. Postoperatively, both groups showed significant improvements in AHI and BMI. However, the reduction in AHI was significantly greater in the CABS group, with similar BMI loss. In the CABS group, AHI decreased from 75.1 to 4.5 events/h. The classic success rate for OSA in the CABS group was 90%, with a cure rate of 60%. A perioperative assessment of CABS patients revealed significant improvements in the Epworth sleepiness scale, minimal oxygen saturation, snoring index, and heart rate. Overall, CABS leads to a more significant reduction in AHI with high success and cure rates for OSA, significantly enhancing quality of life and general health.

11.8 Conclusion

Weight reduction plays an essential role in the treatment of obese patients with OSA. Anti-obesity medications are critical in the comprehensive treatment of obesity. The choice of medication should be individualized based on patient needs, comorbidities, and tolerability. The GBI and ESG are available in Taiwan and provide effective alternatives for individuals seeking nonsurgical weight loss solutions. When these intervention endoscopic procedures are combined with lifestyle changes and medication, they can help individuals achieve sustainable long-term weight management. Bariatric surgery provides sustained weight loss and improvement in obesity-related comorbidities. Although weight loss significantly improves OSA, a certain proportion of patients retain moderate-to-severe OSA. CABS provides a comprehensive treatment option for selected morbidly obese patients with OSA.

References

1. Lurie A. Obstructive sleep apnea in adults: epidemiology, clinical presentation, and treatment options. Adv Cardiol. 2011;46:1–42. https://doi.org/10.1159/000327660.
2. Senaratna CV, Perret JL, Lodge CJ, Lowe AJ, Campbell BE, Matheson MC, Hamilton GS, Dharmage SC. Prevalence of obstructive sleep apnea in the general population: a systematic review. Sleep Med Rev. 2017;34:70–81. https://doi.org/10.1016/j.smrv.2016.07.002.
3. Javaheri S, Barbe F, Campos-Rodriguez F, et al. Sleep apnea: types, mechanisms, and clinical cardiovascular consequences. Am J Cardiol. 2017;69:841–58. https://doi.org/10.1016/j.jacc.2016.11.069.
4. Gudzune KA, Kushner RF. Medications for obesity: a review. JAMA. 2024;332(7):571–84. https://doi.org/10.1001/jama.2024.10816.
5. Shi Q, Wang Y, Hao Q, et al. Pharmacotherapy for adults with overweight and obesity: a systematic review and mv of randomized controlled trials. Lancet. 2024;403(10434):e21–31. https://doi.org/10.1016/S0140-6736(24)00351-9.
6. Lincoff AM, Brown-Frandsen K, Colhoun HM, et al. SELECT Trial Investigators. Semaglutide and cardiovascular outcomes in obesity without diabetes. N Engl J Med. 2023;389(24):2221–32. https://doi.org/10.1056/NEJMoa2307563.
7. deMesquita YLL, Pera Calvi I, Reis Marques I, et al. Efficacy and safety of the dual GIP and GLP-1 receptor agonist tirzepatide for weight loss: ameta-analysis of randomized controlled trials. Int J Obes (Lond). 2023;47(10):883–92. https://doi.org/10.1038/s41366-023-01337-x.
8. Jirapinyo P, Hadefi A, Thompson CC, Patai ÁV, Pannala R, Goelder SK, et al. American Society for Gastrointestinal Endoscopy-European Society of Gastrointestinal Endoscopy guideline on primary endoscopic bariatric and metabolic therapies for adults with obesity. Endoscopy. 2024;56(6):437–56. https://doi.org/10.1055/a-2292-2494. Epub 2024 Apr 19. PMID: 38641332.
9. Norton BC, Telese A, Papaefthymiou A, Aslam N, Makaronidis J, Murray C, et al. Metabolic and bariatric endoscopy: a mini-review. Life (Basel). 2023;13(9):1905. https://doi.org/10.3390/life13091905.
10. Abdulla M, Mohammed N, AlQamish J. Overview on the endoscopic treatment for obesity: a review. World J Gastroenterol. 2023;29(40):5526–42. https://doi.org/10.3748/wjg.v29.i40.5526. PMID: 37970474; PMCID: PMC10642436.
11. Matteo MV, Bove V, Pontecorvi V, Gualtieri L, Carlino G, Spada C, et al. The evolution and current state of bariatric endoscopy in Western countries. Clin Endosc. 2024;57(6):711–24. https://doi.org/10.5946/ce.2023.253.
12. IFSO-CONSENSUS-CONFERENCE-2023-BOOK.
13. Vitiello A, Abu-Abeid A, Dayan D, Berardi G, Musella M. Long-term results of laparoscopic sleeve gastrectomy: a review of studies reporting 10+ years outcomes. Obes Surg. 2023;33(11):3565–70. https://doi.org/10.1007/s11695-023-06824-8.
14. Yeung KT, Penney N, Ashrafian L, Darzi A, Ashrafian H. Does sleeve gastrectomy expose the distal esophagus to severe reflux? A systematic review and meta-analysis. Ann Surg. 2020;271(2):257–65. https://doi.org/10.1097/SLA.0000000000003275.
15. Sharples AJ, Mahawar K. Systematic review and meta-analysis of randomised controlled trials comparing long-term outcomes of Roux-en-Y gastric bypass and sleeve gastrectomy. Obes Surg. 2020;30(2):664–72. https://doi.org/10.1007/s11695-019-04235-2.
16. Qumseya BJ, Qumsiyeh Y, Ponniah SA, Estores D, Yang D, Johnson-Mann CN, Friedman J, Ayzengart A, Draganov PV. Barrett's esophagus after sleeve gastrectomy: a systematic review

and meta-analysis. Gastrointest Endosc. 2021;93(2):343–52.e2. https://doi.org/10.1016/j.gie.2020.08.008.
17. Colquitt JL, Pickett K, Loveman E, Frampton GK. Surgery for weight loss in adults. Cochrane Database Syst Rev. 2014;8:CD003641. https://doi.org/10.1002/14651858.CD003641.pub4.
18. Salminen P, Grönroos S, Helmiö M, Hurme S, Juuti A, Juusela R, Peromaa-Haavisto P, Leivonen M, Nuutila P, Ovaska J. Effect of laparoscopic sleeve gastrectomy vs Roux-en-Y gastric bypass on weight loss, comorbidities, and reflux at 10 years in adult patients with obesity: the SLEEVEPASS randomized clinical trial. JAMA Surg. 2022;157(8):656–66. https://doi.org/10.1001/jamasurg.2022.2229. PMID: 35731535; PMCID: PMC9218929.
19. Ramos AC, Chevallier J-M, Mahawar K, Brown W, Kow L, White KP, Shikora S, Consensus Conference Contributors IFSO. IFSO (International Federation for Surgery of Obesity and Metabolic Disorders) consensus conference statement on one-anastomosis gastric bypass (OAGB-MGB): results of a modified Delphi study. Obes Surg. 2020;30(5):1625–34. https://doi.org/10.1007/s11695-020-04519-y.
20. De Luca M, Piatto G, Merola G, Himpens J, Chevallier J-M, Carbajo M-A, Mahawar K, Sartori A, Clemente N, Herrera M, Higa K, Brown WA, Shikora S. IFSO update position statement on one anastomosis gastric bypass (OAGB). Obes Surg. 2021;31(7):3251–78. https://doi.org/10.1007/s11695-021-05413-x.
21. Magouliotis DE, Tasiopoulou VS, Tzovaras G. One anastomosis gastric bypass versus Roux-en-Y gastric bypass for morbid obesity: an updated meta-analysis. Obes Surg. 2019;29(9):2721–30. https://doi.org/10.1007/s11695-019-04005-0.
22. Nastałek P, Polok K, Celejewska-Wójcik N, Kania A, Sładek K, Małczak P, Major P. Impact of bariatric surgery on obstructive sleep apnea severity and continuous positive airway pressure therapy compliance: prospective observational study. Sci Rep. 2021;11(1):5003. https://doi.org/10.1038/s41598-021-84570-6.
23. K. Lin C-C, Liu K-H, Lee L-A, Chuang L-P, Lin Y-S, Hsin L-J, Lin W-N, Chiang Y-T, Cheng W-N, Li H-Y. Combined airway and bariatric surgery (CABS) for obstructive sleep apnea patients with morbid obesity: a comprehensive alternative preliminary study. J Clin Med. 2022;11(23):7078. https://doi.org/10.3390/jcm11237078. PMID: 36498653; PMCID: PMC9738588.

Continuous Positive Airway Pressure (CPAP) Ventilator

12

Ning-Hung Chen, Qiao-Lu Huang, Shih-Wei Lin, Hsiung Ying Huang, and Pi-Hung Tung

Contents

N.-H. Chen (✉)
Department of Pulmonary and Critical Care Medicine, Chang Gung Memorial Hospital, Linkou, Taoyuan City, Taiwan

School of Traditinal Chinese Medicine, Chang Gung University, Guishan District, Taiwan

Respiratory Therapy, Chang Gung Memorial Hospital, Taiwan, Taoyuan

Q.-L. Huang

Department of Pulmonary and Critical Care Medicine, Chang Gung Memorial Hospital, Linkou, Taoyuan City, Taiwan

Department of Chest, Chang Gung Memorial Hospital, Xiamen, China

S.-W. Lin
Department of Pulmonary and Critical Care Medicine, Chang Gung Memorial Hospital, Linkou, Taoyuan City, Taiwan

Respiratory Therapy, Chang Gung Memorial Hospital, Taiwan, Taoyuan

H. Y. Huang
Department of Chest, Chang Gung Memorial Hospital, Xiamen, China

P.-H. Tung
Department of Pulmonary and Critical Care Medicine, Chang Gung Memorial Hospital, Linkou, Taoyuan City, Taiwan

H.-Y. Li (ed.), *Snoring and Obstructive Sleep Apnea*,
https://doi.org/10.1007/978-981-95-2648-2_12

Abstract

Continuous positive airway pressure ventilator (CPAP) was designed more than 50 years ago for the treatment of obstructive sleep apnea syndrome (OSA). CPAP owns the best evidence in both short-term and long-term treatment outcomes. Hypertension, atrial fibrillation, cognitive dysfunction, stroke, glucose control, lipid profile, and renal and liver function all have significant improvement after the CPAP treatment. Acceptance and compliance were still the problem for the use of CPAP. New developments including bi-level positive pressure ventilator (BiPAP), auto-adjust positive pressure ventilator (APAP), as well as comfort design on mask and performance were introduced, which make the acceptance and compliance better than before. Although they are minor, some complication were reported, which need physicians to take care. More efforts are needed for the CPAP development on the treatment of OSA.

Keywords

Obstructive sleep apnea syndrome · CPAP · Auto-adjusting PAP · Bi-level PAP · Outcome · Complication

12.1 Introduction

Continuous positive airway pressure (CPAP) is the gold standard of treatment for patients with obstructive sleep apnea (OSA) syndrome. Although the other treatment modalities, such as uvulopalatoplasty (UPPP), oral appliances, and mandible advancement surgery, were also used for the disease, CPAP is always the first choice of treatment with the largest body of evidence supporting its favorable effect.

12.2 History

CPAP was first described by Dr. Sullivan [1] in 1981. In his paper, he described five patients with severe OSA successfully treated by a positive pressure ventilator via a "comfortable" nose mask [1]. CPAP was finally recognized in 1985 as an effective treatment for OSA and commercialized by the ResMed company that named the first CPAP machine after Dr. Sullivan.

The first fixed-pressure CPAP that was commercially available had some initial drawbacks. However, the compliance of CPAP was poor initially due to many complaints such as loud noise, dry mouth, and mask leak. "Comfort" technique was developed later to conquer these problems. Auto-titration of the pressure from CPAP was introduced in 1990 and is widely accepted. The other techniques such as expiratory pressure relief (A-flex, C-flex), sense awake, and the new mask designs make the acceptance of CPAP better.

12.3 The Positive Airway Pressure (PAP) Machine

There are several types of positive airway pressure (PAP) machines, not only the fixed pressure ventilator. The following is a brief introduction:

Continuous Positive Airway Pressure (CPAP) machine: It provides a constant stream of pressurized air. The pressure remains fixed throughout the inspiratory and expiratory phases. CPAP provides a positive pressure to keep the upper airway patent even under the negative intraluminal pressure of upper airway in inspiration. It will also produce small overpressure within the lungs that helps in the gas exchange [2].

Auto-Adjusting Positive Airway Pressure (APAP) machine: Also known as Auto-CPAP, it can gradually increase or decrease the pressure based on respiratory events. Different brands of APAP have their own algorithm facing the respiratory events. It similarly maintains the same pressure throughout the respiratory cycle [3].

Bi-level Positive Airway Pressure (BiPAP) machine: It delivers higher pressure in inhalation and a lower one in exhalation. BiPAP machines may be configured in different modes such as spontaneous mode (S), spontaneous timed mode (S/T), or timed mode (T), and some can also adjust the pressure automatically [4].

Adaptive servo-ventilation (ASV): It alleviates central and obstructive sleep apnea by delivering servo-controlled inspiratory pressure support on top of expiratory positive airway pressure [5].

To improve the comfort of patients wearing PAP machines, the following technical supports are currently available:

Optimizing Mask Design: Material selection: Some studies have shown that choosing soft materials with good fit can improve the wearing comfort for patients [6]. Structural design: Noninvasive PAP devices are designed with a certain degree of leakage, and different devices will adjust in different ways [7]. Reasonable structural design can ensure a better fit between the mask and the patient's face, reducing air leakage and pressure. Some technical concept can be applied to the structural design of PAP machine masks to achieve a more accurate fit and better comfort [6].

Personalized Adjustment: Pressure regulation: Different types of PAP devices use different algorithms to provide pressure to meet the patient needs. For example, devices such as Continuous PAP, Autoadjusting CPAP, and Bi-level PAP use their unique algorithms to dynamically adjust the pressure according to the patient's breathing state and sleep situation. This can avoid discomfort caused by excessive or insufficient pressure and improve wearing comfort. At the same time, some devices can also detect respiratory events and respond to these events to further optimize pressure regulation and provide patients with a more comfortable treatment experience [7].

Improving the ease of use of equipment: Simplifying the operation process: Designing a simple and easy-to-understand operation interface and process so that patients can easily use PAP machines [6]. Providing clear instructions for use: Detailed instructions for use can help patients use PAP machines correctly and avoid discomfort caused by improper operation [6].

Improving the Performance of Equipment: Reducing noise: Low-noise PAP machines can provide patients with a quieter treatment environment, improving the sleep quality and wearing comfort of patients [8]. Enhancing Stability: Stable performance can ensure that PAP machines will not have failures or fluctuations during use, improving the sense of security and comfort of patients [8].

Humidifier and/or Heated Tube: Some devices have heating and humidification functions, which can reduce the irritation of dry air to the respiratory tract and improve patient comfort. The humidifier power depends on the brand of the machine, meaning patients experience different humidifier issues, including dry throat, water condensation in the mask or tube, and nasal discharge. Patient education regarding the adjustment of the humidifier is necessary before CPAP use [9].

Data Collection and Feedback System: Some PAP devices are equipped with data collection systems that can track the patient's usage, including compliance, pressure, leakage, and treatment effect. These data can help doctors better understand the patient's treatment situation, adjust the device settings, and improve the treatment effect and patient comfort. At the same time, patients can also understand their treatment progress through these data and enhance treatment confidence and compliance [7].

12.4 Initiation of CPAP

Optimal pressure should be determined through overnight CPAP titration. While APAP is common, overnight CPAP titration is recommended, particularly for patients with poor CPAP tolerance. The American Academy of Seep Medicine (AASM) recommends that positive airway pressure therapy be initiated using either APAP at home or in-laboratory PAP titration in adults with OSA and no significant comorbidities [10]. The decision regarding whether to initiate PAP therapy in a home environment or laboratory setting requires careful consideration of multiple factors. These factors include, but are not limited to, the accessibility of facilities and equipment, the relative cost-effectiveness of each option, individual patient preferences and comfort levels, the professional assessment and recommendations of the sleep clinician, the complexity of the patient's sleep disorder, the availability of technical support, and the patient's ability to manage the equipment independently. The ultimate choice should be tailored to optimize treatment adherence and therapeutic outcomes while taking into account practical constraints and healthcare resource utilization.

While ideal CPAP therapy involves nightly use, adherence remains suboptimal. Most studies define adherence as CPAP usage $\geq$4 hours per night for 70% of nights, although no formal definition exists [11]. Based on this definition, reported adherence consistently shows rates of compliance between 65% and 90%, while objective data reveal actual usage rates of 40–83%, highlighting the importance of measuring the real-world CPAP use [12]. Research shows that consultation with a sleep specialist can help patients have better results with their CPAP therapy [13]. These consultations can happen either before a sleep study in a lab or while starting and continuing CPAP treatment. When patients work with sleep specialists, they are more likely to use their CPAP machines regularly, feel more satisfied with their care, and receive timely treatment.

Choice of Mask: CPAP masks come in several types: nasal masks, full-face masks (covering nose and mouth), nasal pillows (fitting into nostrils), and oral interfaces. Research on oronasal mask effects on compliance is mixed. These masks may work better for patients with nasal congestion, mouth breathing, or those needing higher pressures. While nasal pillows were traditionally limited to pressures below 12 cmH_2O, newer evidence suggests that some patients can use them at higher settings [14]. Though proper fit is essential, the most effective mask is simply the one the patient finds most comfortable.

Interventions focused on promoting behavioral adaptations and positive changes in patient mindset both before the beginning and throughout the ongoing course of PAP therapy are essential components of successful treatment [10]. Research has demonstrated that motivational interventions outperformed standard care approaches and traditional educational programs to improving patient adherence to CPAP therapy. These interventions, which typically involve personalized counseling sessions and goal-setting strategies, showed measurable benefits in encouraging patients to maintain regular CPAP usage patterns.

12.5 Evidence of CPAP Treatment

12.5.1 Hypertension

The CPAP treatment can significantly reduce blood pressure, especially at night. CPAP treatment reduced 24-hour ambulatory systolic blood pressure (SBP) by 2.32 mmHg and diastolic blood pressure (DBP) by 1.98 mmHg, and the improvement in nocturnal blood pressure was more significant, with nocturnal SBP reduced by 2.74 mmHg and DBP by 1.75 mmHg [15]. Patients with refractory hypertension had a more significant reduction in 24-hour SBP and DBP, with reductions of 3.88 mmHg and 3.65 mmHg, respectively, indicating the importance of CPAP in blood pressure control for this group of patients [16]. Recent review concluded that muscle sympathetic nerve activity (MSNA) is higher in individuals with hypertensive OSA patients. CPAP treatment effectively reduces MSNA and BP [17].

12.5.2 Atrial Fibrillation (AF)

Qureshi's study showed that CPAP treatment reduced the risk of AF recurrence by 42% in patients, and the effect was more significant in younger, obese, and male patients [18]. Nalliah's study showed that CPAP treatment can reverse atrial remodeling and increased atrial conduction velocity, voltage, and reduced the proportion of complex potentials in patients with OSA and AF, suggesting that CPAP has a positive impact on the atrial structure and electrophysiology of AF patients and may help improve the pathophysiological basis of AF [19].

12.5.3 Major Adverse Cardiovascular Events (MACE)

Abuzaid's meta-analysis found that overall CPAP use was not significantly associated with a reduced risk of major adverse cardiovascular events, but in the subgroup with CPAP use for more than 4 hours, the risk was significantly reduced [20].

Cowie et al.'s study on patients with heart failure and central sleep apnea (CSA) shows that 12 months of adaptive servo-ventilation (ASV) treatment has no significant effect on the cardiac structure and function and systemic inflammatory biomarkers. This indicates that although ASV treatment can improve some parameters, the effectiveness does not reach statistical significance [21]. The Sleep Apnea Cardiovascular Endpoints (SAVE) trial and CANPAP study had similar results [22–24]. More research is needed to clarify the relationship between treatment duration and the improvement of cardiovascular diseases.

12.5.4 Cognitive Function Improvement

Stroke patients with OSA had significantly better improvement in attention and executive functions after 4 weeks with CPAP treatment [25]. Similar result is seen in studies focused on the long-term effects of CPAP combined with psychoeducation and lifestyle adjustments [26]. Elderly patients with mild sleep apnea and mild cognitive impairment, especially in executive function and language memory, improved after 1 year of CPAP treatment [27]. The higher the treatment compliance, the more significant in cognitive improvement.

12.5.5 Excessive Daytime Sleepiness (EDS)

Apnea patients have lower snore outcome survey (SOS) scores and higher Epworth sleepiness scale (ESS) scores than simple snorers [28]. A retrospective analysis of data from European Sleep Apnea Database found that about one-fourth of the OSA patients still reported highest EDS in the first 3 months of CPAP treatment and then gradually decreased, which suggested that multiple factors were related to EDS [29].

12.5.6 Stroke

CPAP can reduce the risk of stroke in moderate-to-severe OSA patients [30]. Early CPAP treatment has a positive impact on the long-term prognosis of stroke patients [31].

12.5.7 Diabetes

Multiple studies have shown that CPAP treatment has a positive effect on glucose control and insulin resistance in OSA patients with diabetes. Loffler study concluded that diabetic females have more favorable blood glucose control with CPAP treatment [16].

12.5.8 Renal Function

The prevalence of CKD is relatively high (18%) in severe OSA patients without hypertension or diabetes [32]. A 10-year follow-up study on newly diagnosed OSA patients showed that even without diabetes or hypertension, they have a higher risk of developing CKD in the next 3 years, especially women. AHI and the desaturation index (DI) are independent predictors of the urine albumin-to-creatinine ratio (UACR) and the estimated glomerular filtration rate (eGFR), which indicate that severity of OSA and renal function impairment are related [33]. CPAP treatment has a positive impact on renal function in OSA patients. Good compliance with CPAP therapy is associated with a decrease in microalbuminuria and a significant decline in albumin/creatinine ratio after 3 months of CPAP therapy [34].

12.5.9 Lipid Profile

Sleep-related breathing disorder (SRBD) patients have a relatively high risk of dyslipidemia, and DI may be a potential risk factor. Chou's study found that DI is an important independent factor contributing to dyslipidemia [35]. Martínez-Cerón's study found that after 6 months of CPAP treatment, the concentration of low-density lipoprotein cholesterol was significantly lowered [36]. However, CPAP alone is unlikely to alter nonalcoholic fatty liver disease (NAFLD) activities in patients with concomitant OSA [37].

12.6 Complication

Continuous positive airway pressure (CPAP) is an evidence-based treatment for obstructive sleep apnea (OSA) [38] and is relatively safe. Although CPAP is effective, it is an uncomfortable therapy for some patients. Side effects are one of the reasons for poor CPAP adherence and tend to be sub-therapeutic in the long term [39]. Leakage-related side effects are the most common complication of CPAP [39, 40]. Choosing an appropriate mask type and cleaning the mask daily therefore benefit patients by reducing side effects such as leakage and discomfort, which in turn can improve patient adherence and outcomes [38, 40]. In addition, personalized interventions and ongoing support are important to improve CPAP adherence and treatment efficacy [41].

Table 12.1 provides an overview of various complications of CPAP therapy and their causes.

Table 12.1 Summary of the complication for CPAP usage [38–41]

Category	Complications	Causes
Nasal and airway	Dryness or irritation Nasal congestion or runny nose	Dry nose and throat due to pressurized air, Especially common when starting CPAP therapy
Skin and facial	Skin irritation or sores Skin rash Discomfort or pain	Caused by improper mask fit or pressure Allergic reaction to mask materials From tight straps or poorly fitting masks
Eye	Dry eyes or irritation	Caused by air leaks near the eyes
Oral and dental	Dry mouth Jaw discomfort or bite changes	Common in mouth breathers using nasal masks Prolonged use of full-face masks
Ear	Ear pressure or barotrauma Middle ear pain or infections	Discomfort in the ears due to pressure Rare but possible with prolonged pressure changes
Gastrointestinal	Aerophagia (air swallowing)	Leads to bloating, gas, and abdominal discomfort
Infections	Upper respiratory infections Bacterial or fungal infections	Poor cleaning of CPAP equipment
Cerebrovascular	Cerebrospinal fluid (CSF) leak	Usually occurs in the context of transsphenoidal surgery with reports of pneumocephalus
Psychological discomfort	Claustrophobia Sleep disturbances	Difficulty adapting to wearing a mask Trouble falling asleep due to machine noise or sensation
Equipment-related	Noise from the machine Machine malfunction	May disturb the user or bed partner Interrupted therapy due to equipment failure

12.7 Conclusion

Although CPAP is still the first choice of treatment for OSA, it is not perfect till now and all the discomfort and inconvenience make the acceptance and compliance not satisfactory. Also questions and debate were raised recently for the effectiveness and long-term morbidities and mortalities. There are still efforts needed for a better treatment choice for OSA rather than CPAP.

References

1. Sullivan CE, Issa FG, Berthon-Jones M, Eves L. Reversal of obstructive sleep apnoea by continuous positive airway pressure applied through the nares. Lancet. 1981;1:862–5.
2. Khandpur RS. Continuous positive airway pressure machine. 2019.
3. Gardner K, Douglas J, Johnson C, Johnson K. Treatment of sleep-disordered breathing with positive airway pressure devices: technology update. 2015.
4. What are the different types of CPAP machines. SleepApnea.org. Sleep Foundation; 2023.
5. Cheema FZ, Sasieta HC, Dalai VV, Majid R, Castriotta RJ. 0549 does adaptive servo ventilation help resolve central sleep apnea. Sleep. 2019;42(Supplement_1):A219.
6. Pablo M, Banodkar N, Camacho L, Deas R, Ghamedi S. 0581 evaluating the usability and performance of a PAP mask system in the home environment. Sleep. 2024;47(Supplement_1):A248.
7. Johnson KG, Johnson DC. Treatment of sleep-disordered breathing with positive airway pressure devices: technology update. Medical devices. Auckland; 2015.
8. Lachowicz J, Naughton MT. The CPAP machine, mask and interface. 2022.
9. Chen NH. Advancements of CPAP therapy for obstructive sleep apnea. In: Lin H-C, editor. Sleep-related breathing disorders. Adv Otorhinolaryngol, vol. 80. Basel: Karger; 2017. p. 49–56. https://doi.org/10.1159/000470895.
10. Patil SP, Ayappa IA, Caples SM, Kimoff RJ, Patel SR, Harrod CG. Treatment of adult obstructive sleep apnea with positive airway pressure: an American Academy of Sleep Medicine Clinical Practice Guideline. J Clin Sleep Med. 2019;15(2):335–43.
11. Kribbs NB, Pack AI, Kline LR, et al. Objective measurement of patterns of nasal CPAP use by patients with obstructive sleep apnea. Am Rev Respir Dis. 1993;147(4):887–95.
12. Parthasarathy S, Subramanian S, Quan SF. A multicenter prospective comparative effectiveness study of the effect of physician certification and center accreditation on patient-centered outcomes in obstructive sleep apnea. J Clin Sleep Med. 2014;10(3):243–9.
13. Pamidi S, Knutson KL, Ghods F, Mokhlesi B. The impact of sleep consultation prior to a diagnostic polysomnogram on continuous positive airway pressure adherence. Chest. 2012;141(1):51–7.
14. Zhu X, Wimms AJ, Benjafield AV. Assessment of the performance of nasal pillows at high CPAP pressures. J Clin Sleep Med. 2013;9(9):873–7.
15. Hu XY, Fan JQ, Chen SJ, et al. The role of continuous positive airway pressure in blood pressure control for patients with obstructive sleep apnea and hypertension: a meta-analysis of randomized controlled trials the. J Clin Hypertens. 2015;17(3):215–22.
16. Cruz FC, Drager LF, Queiróz DB, Souza GA, Pedrosa RP, Patriota TL, et al. The effect of continuous positive airway pressure on blood pressure in patients with obstructive sleep apnea and uncontrolled hypertension-study design and challenges during the COVID-19 pandemic. Clinics. 2021;76:e2926.
17. Maier LE, Matenchuk BA, Vucenovic A, Sivak A, Davenport MH, Steinback CD. Influence of obstructive sleep apnea severity on muscle sympathetic nerve activity and blood pressure: a systematic review and meta-analysis. Hypertension. 2022;79(9):2091–104.

18. Qureshi WT, bin Nasir U, Alqalyoobi S, O'Neal WT, Mawri S, Sabbagh S, et al. Meta-analysis of continuous positive airway pressure as a therapy of atrial fibrillation in obstructive sleep apnea. Am J Cardiol. 2015;116:11:1767–73
19. Nalliah C, Wong J, Geoffrey R, et al. Impact of CPAP on the atrial fibrillation substrate in obstructive sleep apnea: the SLEEP-AF study. Clin Electrophysiol. 2022;8(7):869–77.
20. Abuzaid AS, Al Ashry S, Haitham S, et al. Meta-analysis of cardiovascular outcomes with continuous positive airway pressure therapy in patients with obstructive sleep apnea. Am J Cardiol. 2017;120(4):693–9.
21. Cowie MR, Woehrle H, et al. Adaptive servo-ventilation for central sleep apnoea in systolic heart failure: results of the major substudy of SERVE-HF. Eur J Heart Fail. 2017; https://doi.org/10.1002/ejhf.10481048.
22. McEvoy RD, Antic NA, Heeley E, et al. CPAP for prevention of cardiovascular events in obstructive sleep apnea. N Engl J Med. 2016;375:919–31.
23. Arzt M, Floras JS, Logan AG, et al. CANPAP investigators. Suppression of central sleep apnea by continuous positive airway pressure and transplant-free survival in heart failure: A post hoc analysis of the Canadian continuous positive airway pressure for patients with central sleep apnea and heart failure trial. Circulation. 2007;115(25):3173–80.
24. Bryant. Continuous positive airway pressure (CPAP) therapy did not significantly reduce cardiovascular events among patients with obstructive sleep apnoea (OSA) and cardiovascular disease, according to controversial findings from the Sleep Apnea Cardiovascular Endpoints (SAVE) trial. SAVE trial: no cardiovascular benefits for CPAP in OSA Furlow, Lancet Resp Med. 2016;4(11):860.
25. Rosenzweig I, et al. Changes in neurocognitive architecture in patients with obstructive sleep apnea treated with continuous positive airway pressure. EBioMedicine. 2016;7:221–9.
26. Richards KC, et al. Predictors of adherence to continuous positive airway pressure in older adults with apnea and amnestic mild cognitive impairment. J Gerontol. 2023;78(10):1861–70.
27. Wang P-C, Li H-Y, et al. Generic and specific quality-of-life measures in Taiwanese adults with sleep-disordered breathing. Otolaryngology. 2006;135:421–6.
28. Bonsignore MR, et al. Excessive daytime sleepiness in obstructive sleep apnea patients treated with continuous positive airway pressure: data from the European Sleep Apnea Database. Front Neurol. 2021;12:690008.
29. Pépin J-L, et al. Pitolisant for residual excessive daytime sleepiness in OSA patients adhering to CPAP. Chest. 2021;159(4):1598–609.
30. Lin H-J, et al. Continuous positive airway pressure with good adherence can reduce risk of stroke in patients with moderate to severe obstructive sleep apnea: an updated systematic review and meta-analysis. Sleep Med Rev. 2020;54:101354.
31. Parra O, et al. Efficacy of continuous positive airway pressure treatment on 5-year survival in patients with ischaemic stroke and obstructive sleep apnea: a randomized controlled trial. J Sleep Res. 2015;24(1):47–53.
32. Martinez-Ceron E, Barquiel B, et al. Effect of continuous positive airway pressure on glycemic control in patients with obstructive sleep apnea and type 2 diabetes. A randomized clinical trial. Am J Respir Crit Care Med. 2016;194(4):476–85.
33. Loffler KA, Heeley E, Freed R, Meng R, Bittencourt LR, Gonzaga Carvalho CC, et al. Continuous positive airway pressure treatment, glycemia, and diabetes risk in obstructive sleep apnea and comorbid cardiovascular disease. Diabetes Care. 2020;43(8):1859–67.
34. Chou Y-T, Lee P-H, Yang C-T, et al. obstructive sleep apnea: a stand-alone risk factor for chronic kidney disease. Nephrol Dial Transplant. 2011;0:1–6. https://doi.org/10.1093/ndt/gfq821.
35. Lin Y-S, Liu P-H, Lin S-W, et al. Simple obstructive sleep apnea patients without hypertension or diabetes accelerate kidney dysfunction: a population follow-up cohort study from Taiwan. Sleep Breath. 2016; https://doi.org/10.1007/s11325-016-1376-2.
36. Chou YT, Chuang LP, Li HY, Fu JY, Lin SW, Yang CT, Chen NH. Hyperlipidaemia in patients with sleep-related breathing disorders: Prevalence & risk factors. Indian J Med Res. 2010;131:121–5.

37. Ng SS, Wong VW, Wong GL, Chu WC, Chan TO, To KW, Ko FW, Chan KP, Hui DS. Continuous positive airway pressure does not improve nonalcoholic fatty liver disease in patients with obstructive sleep apnea. A randomized clinical trial. Am J Respir Crit Care Med. 2021;203(4):493–501.
38. Patil SP, et al. Treatment of adult obstructive sleep apnea with positive airway pressure: an American Academy of sleep medicine clinical practice guideline. J Clin Sleep Med. 2019;15(02):335–43.
39. Ghadiri M, Grunstein RR. Clinical side effects of continuous positive airway pressure in patients with obstructive sleep apnoea. Respirology. 2020;25(6):593–602.
40. Rotty MC, et al. Mask side-effects in long-term CPAP-patients impact adherence and sleepiness: the InterfaceVent real-life study. Respir Res. 2021;22(1):17.
41. Askland K, et al. Educational, supportive and behavioural interventions to improve usage of continuous positive airway pressure machines in adults with obstructive sleep apnoea. Cochrane Database Syst Rev. 2020;4(4):Cd007736.

13 Mandibular Advancement Devices in Adult Obstructive Sleep Apnea

Yu-Fang Liao

Contents

Abstract

Mandibular advancement devices (MAD) are an effective alternative to continuous positive airway pressure (CPAP) for managing obstructive sleep apnea (OSA), particularly in patients who cannot tolerate CPAP. By repositioning the mandible, MAD enhance airflow and airway patency. Recent guidelines from the American Academy of Sleep Medicine and the European Respiratory Society emphasize the importance of custom-made, titratable devices, and interdisciplinary collaboration in patient care. Treatment success depends on clinical and anatomical factors such as younger age, lower BMI, and specific craniofacial structures. While CPAP is more effective in reducing respiratory events, MAD provide comparable real-world benefits due to higher adherence, though long-term use may cause occlusal changes. Future research should refine predictive markers, optimize device design, integrate digital tools, and assess long-term health outcomes to enhance MAD effectiveness in OSA management.

Y.-F. Liao (✉)
Department of Craniofacial orthodontics, Chang Gung Memorial Hospital, Taoyuan, Taiwan
e-mail: yufang@cgmh.org.tw

H.-Y. Li (ed.), *Snoring and Obstructive Sleep Apnea*,
https://doi.org/10.1007/978-981-95-2648-2_13

Keywords

Obstructive sleep apnea (OSA) · Mandibular advancement device (MAD) Titratable appliance · Treatment predictors · Effectiveness · Side effects Multidisciplinary management

13.1 Introduction

Obstructive sleep apnea (OSA) is a prevalent condition linked to serious health complications, including cardiovascular and metabolic disorders, and places a significant strain on healthcare systems [1]. Although continuous positive airway pressure (CPAP) therapy remains the primary treatment for severe OSA, long-term adherence is a considerable challenge, with nearly half of patients discontinuing its use within three years of initiation [2].

Oral appliances have traditionally served as a first-line therapy for mild-to-moderate OSA or as an alternative for individuals who are unable or unwilling to use CPAP [3]. However, in many countries, their role has expanded to include primary treatment for symptomatic patients across varying OSA severities, particularly in those with a lower burden of comorbidities [4]. This evolving approach highlights their increasing recognition as an effective and flexible option for OSA management.

Oral appliances for OSA are generally classified into two main types: tongue stabilization devices and mandibular advancement devices (MAD), with MAD being the most widely used [5, 6]. These appliances function by repositioning the mandible forward and downward using splints attached to the maxillary and mandibular teeth, thereby enhancing upper airway patency.

MAD can be further categorized into two groups: custom-made (titratable or non-titratable) and prefabricated "boil-and-bite" models (titratable or non-titratable). Custom-made devices are fabricated using a physical impression or digital scan of the patient's dentition, combined with a bite registration. They can be designed as a mono-bloc structure with a fixed mandibular advancement or as a bi-bloc appliance, which consists of separate maxillary and mandibular components that allow for adjustable mandibular advancement.

Prefabricated "boil-and-bite" devices offer a more cost-effective and readily available alternative. However, a recent systematic review and meta-analysis indicates that custom-made MAD demonstrate superior efficacy in OSA treatment. They are also better tolerated and more frequently chosen by patients due to their improved retention and fit [7]. As a result, titratable and custom-made MAD remain the preferred and most effective approach for managing OSA.

13.2 Mechanism of MAD

MAD primarily function by repositioning the mandible forward and downward, effectively enlarging the pharyngeal airway. This movement advances the tongue due to its muscular connections to the mandible, including the genioglossus, while also exerting a secondary effect on the soft palate through muscles, such as the palatoglossus and pharyngeal constrictors. As a result, both the retroglossal and retropalatal airway spaces are expanded [8–10].

The impact of MAD on upper airway dimensions has been extensively documented using advanced imaging modalities, including 3D videofluoroscopy, magnetic resonance imaging (MRI), computed tomography (CT), and drug-induced sleep nasendoscopy (DISE). These techniques consistently show a significant increase in upper airway volume, particularly in the velopharyngeal region [5, 11, 12].

Importantly, the expansion of the airway is not limited to the anteroposterior direction; the most substantial changes are observed in the lateral dimension [10, 13]. This lateral widening is believed to play a crucial role in the effectiveness of MAD therapy.

13.3 Clinical Practice Guidelines for MAD

The American Academy of Sleep Medicine (AASM) and the European Respiratory Society (ERS) have both issued guidelines on the use of MAD for treating OSA, reflecting shared principles and regional distinctions [4, 14].

Both recommend MADs for patients unable to tolerate CPAP or those who prefer alternative treatments. They agree on the importance of titratable and custom-made MAD that allow precise mandibular positioning to improve airway patency and patient comfort. Collaboration between dentists trained in dental sleep medicine and sleep physicians is emphasized in both guidelines to ensure informed decision-making, device optimization, and comprehensive follow-up care. Both also advocate for follow-up studies, such as polysomnography, to monitor treatment efficacy, adherence, and any side effects.

However, the AASM focuses specifically on MAD therapy, stressing long-term monitoring for jaw discomfort or dental changes and recommending dentists with specialized training. MAD are strongly recommended for patients intolerant to CPAP but are considered a second-line option. The ERS guidelines, by contrast, highlight the variability in access to devices and trained professionals across healthcare systems and recognize MAD as a viable alternative, particularly for patients with mild-to-moderate OSA or specific treatment preferences.

13.4 MAD Design and Titration

The design and function of MAD require careful consideration to ensure maximum treatment effectiveness [15].

Despite the variety of available designs, evidence comparing their benefits is limited and inconclusive. Clinical recommendations focus on key features essential for optimal performance:

1. *Retention*: Proper retention is crucial to maintain the forward positioning of the mandible during sleep. Without adequate retention, the device may disengage, allowing the mandible to rotate backward, reducing its effectiveness. While there is no strict requirement for the number of teeth needed, having more healthy teeth improves retention.
2. *Occlusal Coverage*: MAD should cover all occlusal surfaces of the teeth to prevent vertical changes, such as supra-eruption of teeth. This design feature minimizes the risk of occlusal interferences and bite disturbances, ensuring better functional outcomes.
3. *Mandibular Titration*: Given the degree of mandibular advancement increases, increased effectiveness of treatment becomes more likely [16–19]. It is essential for MAD to allow for gradual and adjustable mandibular advancement (titration). The titration starts after a successful acclimatization phase in increments ≤1 mm. This titration process, guided by a trained dentist or performed by the patient under supervision, ensures individualized adjustment to achieve treatment benefits while minimizing discomfort. Personalizing the extent of mandibular advancement reduces the risk of side effects, such as discomfort in the temporomandibular structures and enhances long-term adherence. However, the determination of the optimal level of mandibular advancement is currently not standardized. A potential approach to fine-tuning mandibular advancement involves subjective OSA symptoms during treatment such as sleep quality, daytime sleepiness, and under certain circumstances, home-used sensors monitoring oxygen saturation and heart rate and followed by the objective monitoring the apnea–hypopnea index (AHI) [20]. Remotely controlled mandibular titration devices have the potential to identify treatment responders and the level of advancement required in single night titration polysomnography, which is analogous to a CPAP pressure titration study [19].
4. *Minimal Anterior Vertical Opening*: There is no clear scientific evidence regarding the optimal vertical dimension and its effectiveness on OSA. However, clinical experience has shown that the measured absolute protrusion of the mandible and adherence to MAD decrease with increasing vertical dimension [21–24]. This design feature may help maintain tongue posture and prevent pharyngeal airway narrowing, enhancing the device's overall effectiveness. However, to pre-

vent fracture of the device during treatment, the vertical distance in the posterior region should be large enough to enable adequate dental laboratory work on the respective MAD. The design feature possibly is especially important in specific craniofacial phenotype such as high-angle deformity.

5. *Permissible Lateral Jaw Movement*: Titratable and custom-made MAD allow lateral jaw movement in the coupling mechanisms, which attach the two plates together. Although lateral jaw movement has no effect on upper airway patency, the permissible lateral jaw movement is thought to improve patient comfort and wearability.
6. *Anterior Clearance (Hole)*: It is essential for MAD to have anterior clearance for oral breathing when patients encounter nasal obstruction during sleep.

13.5 Predictors of Treatment Responders

Variability in individual responses to MAD therapy presents a significant clinical challenge, as initiating treatment for patients who ultimately derive little to no benefit is unsatisfactory from both an efficacy and cost perspective. As a result, identifying patient characteristics associated with treatment success and developing reliable predictive methods remain key areas of research.

Several factors have been identified as predictors of a favorable response to MAD therapy in individuals with OSA. Clinical traits such as younger age, lower body mass index (BMI), and a smaller neck circumference have been consistently linked to better outcomes [25, 26].

Certain anatomical characteristics also serve as strong indicators of treatment success. These include a retrusive maxilla or mandible, reduced anterior and posterior facial height, a shorter soft palate, a decreased hyoid-to-third cervical vertebra distance, a shorter airway length, and a smaller minimum airway cross-sectional area [25, 26]. Studies suggest that patients with lower apnea–hypopnea index (AHI) frequently exhibit a uniform forward displacement of the posterior tongue, which enhances airway patency. In contrast, those with a higher AHI display limited posterior tongue advancement and greater alterations in tongue shape, even when BMI differences are accounted for [27]. Responders typically demonstrate greater anterior tongue movement in both the retropalatal and retroglossal regions compared to nonresponders [28].

Polysomnographic parameters further distinguish responders from nonresponders. Patients with a lower AHI and reduced loop gain—an indicator of ventilatory control stability—tend to show better treatment outcomes [29]. Conversely, nonresponders frequently experience lower oxygen saturation levels during sleep [26]. Additionally, responders often require lower optimal CPAP pressures compared to nonresponders [26].

13.6 Patient Preference and Treatment Effectiveness (Versus CPAP)

Patient preference is a key factor influencing adherence to treatment, ultimately affecting health outcomes. Recognizing individual preferences and values enables a more tailored approach to therapy, fostering patient engagement in decision-making and enhancing their sense of control over the treatment process.

Comparative studies examining patient preferences between CPAP and MAD generally indicate a stronger inclination toward MAD. Among six crossover trials, four studies reported a greater preference for MAD [30–33], one slightly favored CPAP [34], and another found no significant difference in preference [35]. The increased preference for MAD is likely due to its greater comfort and ease of use compared to CPAP.

In terms of clinical efficacy, CPAP consistently demonstrates superior effectiveness in normalizing respiratory parameters, reducing AHI by an average of seven to eight events per hour more than oral appliance therapy [4, 14, 36]. Additionally, CPAP is more effective in improving the oxygen desaturation index and minimal oxygen saturation [31]. However, both CPAP and MAD yield comparable improvements in sleep efficiency, daytime sleepiness, and health-related quality of life [31, 37–40]. This comparable real-world effectiveness may be attributed to MAD's higher adherence, which helps to compensate for its slightly lower efficacy [41].

Both CPAP and MAD have a similar impact on nighttime systolic and diastolic blood pressure, regardless of OSA severity [31, 34, 42]. However, CPAP appears to exert a greater effect on nighttime systolic blood pressure in severe OSA patients, with a mean difference of 4.2 mmHg (95% CI: 2.2–6.2 mmHg) [42].

Given that OSA requires lifelong management, adherence plays a crucial role in determining treatment success. To assess long-term effectiveness, indices such as mean disease alleviation (MDA), sleep-adjusted residual AHI (SARAH index), apnea burden, and effective AHI have been introduced [43–46]. These metrics integrate treatment efficacy, adherence, and usage relative to sleep duration, providing a comprehensive evaluation of clinical impact. Findings based on these indices suggest that both CPAP and MAD offer comparable, if not superior, long-term health benefits [43, 45, 47].

However, long-term MAD use has been associated with occlusal changes, with incidence rates reported between 14% and 26% [47, 48]. A recent systematic review and meta-analysis found that prolonged use of oral appliances led to a significant reduction in overjet (mean: −0.99 mm, 95% CI: −1.30 to −0.68 mm) and overbite (mean: −1.0 mm, 95% CI: −1.34 to −0.66 mm), primarily due to lower incisor proclination (mean: 2.07 degrees, 95% CI: 0.37–3.77 degrees) rather than skeletal alterations or mandibular rotation [49].

13.7 Conclusions

MAD are an effective and well-tolerated treatment for selected OSA patients, though their efficacy is more variable and less predictable than CPAP. Future research should focus on optimizing MAD use by identifying reliable predictors of treatment success based on clinical, anatomical, and polysomnographic factors. Evaluating efficacy across different OSA phenotypes and MAD designs will aid in personalized treatment and evidence-based decision-making. Long-term studies on cardiovascular and metabolic outcomes will clarify MAD's broader health impact. Technological advancements, such as remote titration and machine learning, can improve precision and adherence. Additionally, combining MAD with other therapies may enhance effectiveness for diverse patient groups. Advancing these areas will strengthen MAD's role in OSA management, offering personalized and patient-friendly solutions.

References

1. Marin JM, Carrizo SJ, Vicente E, Agusti AG. Long-term cardiovascular outcomes in men with obstructive sleep apnoea-hypopnoea with or without treatment with continuous positive airway pressure: an observational study. Lancet. 2005;365:1046–53.
2. Pepin JL, Bailly S, Rinder P, et al. CPAP therapy termination rates by OSA phenotype: a French nationwide database analysis. J Clin Med. 2021;10:10.
3. Kushida CA, Morgenthaler TI, Littner MR, et al. Practice parameters for the treatment of snoring and obstructive sleep apnea with oral appliances: an Update for 2005. Sleep. 2006;29(2):240–3.
4. Ramar K, Dort LC, Katz SG, Lettieri CJ, Harrod CG, Thomas SM, Chervin RD. Clinical practice guideline for the treatment of obstructive sleep apnea and snoring with oral appliance therapy: an update for 2015. J Clin Sleep Med. 2015;11(7):773–827.
5. Chan ASL, Sutherland K, Schwab RJ, et al. The effect of mandibular advancement on upper airway structure in obstructive sleep apnoea. Thorax. 2010;65:726–32.
6. Marklund M, Verbraecken J, Randerath W. Non-CPAP therapies in obstructive sleep apnoea: mandibular advancement device therapy. Eur Respir J. 2012;39:1241–7.
7. Johal A, Agha B. Ready-made versus custom-made mandibular advancement appliances in obstructive sleep apnea: a systematic review and meta-analysis. J Sleep Res. 2018;27:e12660.
8. Schwab R, Gupta K, Duong D, et al. Upper airway soft tissue structural changes with dental appliances in apneics. Am J Respir Crit Care Med. 1996;153:A719.
9. Ishida M, Inoue Y, Suto Y, et al. Mechanism of action and therapeutic indication of prosthetic mandibular advancement in obstructive sleep apnea syndrome. Psychiatry Clin Neurosci. 1998;52:227–9.
10. Ryan C, Love L, Peat D, et al. Mandibular advancement oral appliance therapy for obstructive sleep apnoea: effect on awake calibre of the velopharynx. Thorax. 1999;54(11):972–7.
11. Dontsos VK, Chatzigianni A, Papadopoulos MA, et al. Upper airway volumetric changes of obstructive sleep apnoea patients treated with oral appliances: a systematic review and meta-analysis. Eur J Orthod. 2021;43:399–407.
12. Van Gaver H, de Beeck SO, Dieltjens M, et al. Functional imaging improves patient selection for mandibular advancement device treatment outcome in sleep-disordered breathing: a prospective study. J Clin Sleep Med. 2022;18:739–50.

13. Sutherland K, Takaya H, Qian J, Petocz P, Ng AT, Cistulli PA. Oral appliance treatment response and polysomnographic phenotypes of obstructive sleep apnea. J Clin Sleep Med. 2015;11(8):861–8.
14. Randerath W, Verbraecken J, de Raaff CAL, et al. European Respiratory Society guideline on non-CPAP therapies for obstructive sleep apnoea. Eur Respir Rev. 2021;30:210200.
15. Johal A, Hamoda MM, Almeida FR, et al. The role of oral appliance therapy in obstructive sleep apnoea. Eur Respir Rev. 2023;32:220257.
16. Kato J, Isono S, Tanaka A, et al. Dose-dependent effects of mandibular advancement on pharyngeal mechanics and nocturnal oxygenation in patients with sleep-disordered breathing. Chest. 2000;117:1065–72.
17. Walker-Engstrom ML, Ringqvist I, Vestling O, Wilhelmsson B, Tegelberg A. A prospective randomized study comparing two different degrees of mandibular advancement with a dental appliance in treatment of severe obstructive sleep apnea. Sleep Breath. 2003;7:119–30.
18. Aarab G, Lobbezoo F, Hamburger HL, Naeije M. Effects of an oral appliance with different mandibular protrusion positions at a constant vertical dimension on obstructive sleep apnea. Clin Oral Investig. 2010;14:339–45.
19. Remmers J, Charkhandeh S, Grosse J, Topor Z, Brant R, Santosham P, Bruehlmann S. Remotely controlled mandibular protrusion during sleep predicts therapeutic success with oral appliances in patients with obstructive sleep apnea. Sleep. 2013;36:1517–25.
20. Bernhardt O, Giannakopoulos NN, Heise M, Meyer A, Norden D, Schlieper J, Kares H. Mandibular advancement device: prescription in adult dental sleep medicine - guideline of the German Society of Dental Sleep Medicine. Sleep Breath. 2023;27(1):389–97.
21. Pitsis AJ, Darendeliler MA, Gotsopoulos H, Petocz P, Cistulli PA. Effect of vertical dimension on efficacy of oral appliance therapy in obstructive sleep apnea. Am J Respir Crit Care Med. 2002;166:860–4.
22. Ahrens A, McGrath C, Hagg U. A systematic review of the efficacy of oral appliance design in the management of obstructive sleep apnoea. Eur J Orthod. 2011;33:318–24.
23. Nikolopoulou M, Naeije M, Aarab G, Hamburger HL, Visscher CM, Lobbezoo F. The effect of raising the bite without mandibular protrusion on obstructive sleep apnoea. J Oral Rehabil. 2011;38:643–7.
24. Vroegop AV, Vanderveken OM, Van de Heyning PH, Braem MJ. Effects of vertical opening on pharyngeal dimensions in patients with obstructive sleep apnoea. Sleep Med. 2012;13:314–6.
25. Chen H, Eckert DJ, van der Stelt PF, et al. Phenotypes of responders to mandibular advancement device therapy in obstructive sleep apnea patients: a systematic review and meta-analysis. Sleep Med Rev. 2020;49:101229.
26. Camañes-Gonzalvo S, Bellot-Arcís C, Marco-Pitarch R, et al. Comparison of the phenotypic characteristics between responders and non-responders to obstructive sleep apnea treatment using mandibular advancement devices in adult patients: systematic review and meta-analysis. Sleep Med Rev. 2022;64:101644.
27. Brown EC, Cheng S, McKenzie DK, et al. Tongue and lateral upper airway movement with mandibular advancement. Sleep. 2013;36:397–404.
28. Jugé L, Yeung J, Knapman FL, et al. Influence of mandibular advancement on tongue dilatory movement during wakefulness and how this is related to oral appliance therapy outcome for obstructive sleep apnea. Sleep. 2021;44:1–10.
29. Edwards BA, Andara C, Landry S, Sands SA, Joosten SA, Owens RL, White DP, Hamilton GS, Wellman A. Upper-airway collapsibility and loop gain predict the response to oral appliance therapy in patients with obstructive sleep apnea. Am J Respir Crit Care Med. 2016;194(11):1413–22.
30. Gagnadoux F, Fleury B, Vielle B, et al. Titrated mandibular advancement versus positive airway pressure for sleep apnoea. Eur Respir J. 2009;34:914–20.
31. Phillips CL, Grunstein RR, Darendeliler MA, et al. Health outcomes of CPAP versus oral appliance treatment for obstructive sleep apnea: a randomized controlled trial. Am J Respir Crit Care Med. 2013;187:879–87.

32. Ferguson KA, Ono T, Lowe AA, al-Majed S, Love LL, Fleetham JA. A shortterm controlled trial of an adjustable oral appliance for the treatment of mild to moderate obstructive sleep apnoea. Thorax. 1997;52:362–8.
33. Ferguson KA, Ono T, Lowe AA, Keenan SP, Fleetham JA. A randomized crossover study of an oral appliance vs nasal-continuous positive airway pressure in the treatment of mild-moderate obstructive sleep apnea. Chest. 1996;109:1269–75.
34. Barnes M, McEvoy RD, Banks S, et al. Efficacy of positive airway pressure and oral appliance in mild to moderate obstructive sleep apnea. Am J Respir Crit Care Med. 2004;170:656–64.
35. Engleman HM, McDonald JP, Graham D, et al. Randomized crossover trial of two treatments for sleep apnea/hypopnea syndrome: continuous positive airway pressure and mandibular repositioning splint. Am J Respir Crit Care Med. 2002;166:855–9.
36. Li P, Ning XH, Lin H, et al. Continuous positive airway pressure versus mandibular advancement device in the treatment of obstructive sleep apnea: a systematic review and meta-analysis. Sleep Med. 2020;72:5–11.
37. Bratton DJ, Gaisl T, Schlatzer C, et al. Comparison of the effects of continuous positive airway pressure and mandibular advancement devices on sleepiness in patients with obstructive sleep apnoea: a network meta-analysis. Lancet Respir Med. 2015;3:869–78.
38. Sharples LD, Clutterbuck-James AL, Glover MJ, et al. Meta-analysis of randomised controlled trials of oral mandibular advancement devices and continuous positive airway pressure for obstructive sleep apnoea–hypopnoea. Sleep Med Rev. 2016;27:108–24.
39. Kuhn E, Schwarz EI, Bratton DJ, et al. Effects of CPAP and mandibular advancement devices on health-related quality of life in OSA: a systematic review and meta-analysis. Chest. 2017;151:786–94.
40. Schwartz M, Acosta L, Hung YL, et al. Effects of CPAP and mandibular advancement device treatment in obstructive sleep apnea patients: a systematic review and meta-analysis. Sleep Breath. 2018;22:555.
41. Sutherland K, Phillips CL, Cistulli PA. Efficacy versus effectiveness in the treatment of obstructive sleep apnea: CPAP and oral appliances. J Sleep Dent Med. 2015;4:175–81.
42. Dal-Fabbro C, Garbuio S, D'Almeida V, et al. Mandibular advancement device and CPAP upon cardiovascular parameters in OSA. Sleep Breath. 2014;18:749–59.
43. Vanderveken OM, Dieltjens M, Wouters K, et al. Objective measurement of compliance during oral appliance therapy for sleep-disordered breathing. Thorax. 2013;68:91–6.
44. Toh ST, Phua CQ, Loh S. Holistic management of obstructive sleep apnea: translating academic research to patient care. Sleep Med Clin. 2019;14:1–11.
45. Ravesloot MJ, de Vries N. Reliable calculation of the efficacy of non-surgical and surgical treatment of obstructive sleep apnea revisited. Sleep. 2011;34:105–110.
46. Dissanayake HU, Sutherland K, Phillips CL, et al. Comparative effects of CPAP and mandibular advancement splint therapy on blood pressure variability in moderate to severe obstructive sleep apnoea. Sleep Med. 2021;80:294–300.
47. Rose EC, Staats R, Virchow C, Jonas IE Jr. Occlusal and skeletal effects of an oral appliance in the treatment of obstructive sleep apnea. Chest. 2002;122(3):871–7.
48. Chen H, Lowe AA, de Almeida FR, Fleetham JA, Wang B. Three-dimensional computer-assisted study model analysis of long-term oral-appliance wear. Part 2. Side effects of oral appliances in obstructive sleep apnea patients. Am J Orthod Dentofacial Orthop. 2008;134(3):408–17.
49. Araie T, Okuno K, Ono Minagi H, Sakai T. Dental and skeletal changes associated with long-term oral appliance use for obstructive sleep apnea: a systematic review and meta-analysis. Sleep Med Rev. 2018;41:161–72.

Nasal Surgery for Sleep-Related Breathing Disorders

14

Ming-Shao Tsai, Hsueh-Yu Li, Li-Ang Lee, Seiichi Nakata, Tae-Bin Won, Chien-Yu Huang, and Geng-He Chang

Contents

M.-S. Tsai (✉) · G.-H. Chang
Department of Otolaryngology – Head and Neck Surgery, Chang Gung Memorial Hospital, Chiayi, Taiwan

Faculty of Medicine, College of Medicine, Chang Gung University, Taoyuan, Taiwan

Graduate Institute of Clinical Medical Sciences, College of Medicine, Chang Gung University, Taoyuan, Taiwan
e-mail: b87401061@cgmh.org.tw

H.-Y. Li · L.-A. Lee
Faculty of Medicine, College of Medicine, Chang Gung University, Taoyuan, Taiwan

Sleep Center, Department of Otolaryngology–Head and Neck Surgery, Chang Gung Memorial Hospital at Linkou, Taoyuan, Taiwan

S. Nakata
Department of Otolaryngology/Head and Neck Surgery, Meitetsu Hospital, Nagoya, Japan

T.-B. Won
Department of Otorhinolaryngology-Head and Neck Surgery, Seoul National University Hospital, Seoul National University College of Medicine, Seoul, South Korea

C.-Y. Huang
Department of Otolaryngology-Head and Neck Surgery, Ditmanson Medical Foundation Chia-Yi Christian Hospital, Chiayi, Taiwan

College of Artificial Intelligence, National Yang-Ming Chiao Tung University, Tainan, Taiwan

H.-Y. Li (ed.), *Snoring and Obstructive Sleep Apnea*,
https://doi.org/10.1007/978-981-95-2648-2_14

Abstract

Nasal surgery, including septoplasty and turbinate reduction, plays a pivotal role in optimizing nasal airflow and improving multimodal treatment outcomes for obstructive sleep apnea (OSA). While nasal surgery alone has a limited impact on reducing the apnea–hypopnea index (AHI), it significantly enhances continuous positive airway pressure (CPAP) adherence by lowering airway resistance and reducing therapeutic pressure requirements. Additionally, it improves nasal breathing, mitigates snoring, and contributes to overall sleep quality. Recent advancements, such as mini-invasive septoturbinoplasty (MIST) and distraction osteogenesis maxillary expansion (DOME), offer structurally targeted and minimally invasive approaches to nasal airway optimization. Adjunctive therapies, including nasal irrigation, further enhance surgical outcomes by promoting mucosal healing and reducing congestion. Despite its benefits, nasal surgery should be integrated within a holistic, patient-specific treatment strategy, combining CPAP, oral appliances, weight management, and multilevel airway surgery for optimal results. Future innovations, including AI-assisted surgical planning and 3D modeling, are expected to refine patient selection and enhance long-term efficacy.

Keywords

Obstructive Sleep Apnea (OSA) · Snoring · Nasal Surgery · Septoplasty · Turbinate Reduction · Mini-Invasive Septoturbinoplasty (MIST) · Distraction Osteogenesis Maxillary Expansion (DOME) · Nasal Obstruction · CPAP Adherence · Multilevel Sleep Surgery · Nasal Airflow · Sleep-disordered Breathing

14.1 Introduction

Nasal surgery, including septoplasty and turbinate reduction, plays a crucial role in improving nasal patency and restoring nasal breathing in patients with obstructive sleep apnea (OSA) [1]. Nasal obstruction exacerbates upper airway collapse, increases airway resistance, and shifts breathing to the oral route during sleep. This worsens pharyngeal collapsibility, increasing the severity of OSA. Additionally, nasal obstruction can significantly impact CPAP compliance, as higher airway resistance results in greater therapeutic pressure requirements, causing discomfort [2, 3]. While nasal surgery alone has a limited effect on apnea–hypopnea index (AHI), its role in multimodal OSA treatment strategies is essential, particularly in enhancing CPAP adherence and facilitating multilevel surgical success [4, 5].

14.2 Structured Discussion

14.2.1 Etiology of OSA

OSA results from repeated partial or complete airway collapse during sleep, often due to a combination of anatomical and neuromuscular factors [6, 7]. Soft tissue hypertrophy (e.g., enlarged tonsils and elongated epiglottis) and pharyngeal hypotonia contribute to airway obstruction. Nasal congestion forces patients to adopt mouth breathing, which destabilizes the pharyngeal airway. Drug-induced sleep endoscopy (DISE) studies confirm that mouth breathing aggravates tongue base and lateral pharyngeal wall collapse, worsening OSA severity. Additionally, risk factors such as aging, male gender, obesity, and craniofacial abnormalities (e.g., retrognathia and high-arched palate) further influence OSA pathogenesis.

14.2.2 The Dilemma in the Treatment of OSA

Although CPAP remains the gold standard for treating OSA, adherence rates are suboptimal, with long-term compliance ranging from 50% to 60% [8]. Common issues include nasal irritation, dryness, claustrophobia, and discomfort due to excessive therapeutic pressures [9]. Additionally, CPAP can worsen upper airway obstruction in cases of epiglottic collapse [10]. While oral appliances provide an alternative therapy, their efficacy is inferior to CPAP, particularly in moderate-to-severe OSA. Surgical options, such as maxillomandibular advancement (MMA) and intrapharyngeal surgery (e.g., UPPP and tongue base reduction), can be effective but are invasive and require careful patient selection. Given these challenges, a multimodal and individualized treatment approach is necessary, with nasal surgery serving as a foundational step in optimizing airway function.

14.2.3 Holistic Care for OSA

OSA management should adopt a holistic approach that incorporates conservative, medical, and surgical therapies. Conservative strategies include weight loss, positional therapy, alcohol avoidance, and myofunctional exercises. CPAP therapy remains first-line, with adjustments such as different mask types and humidification improving tolerance. Oral appliances are effective for mild-to-moderate OSA, but have limitations in severe cases. For patients with craniofacial abnormalities, extrapharyngeal surgeries such as MMA and distraction osteogenesis maxillary expansion (DOME) should be prioritized. Multilevel surgical approaches, including nasal surgery, are essential for patients with multilevel airway obstruction [11].

14.2.4 Role and Benefits of Nasal Surgery in Holistic Care of OSA

Nasal surgery plays a crucial role in the holistic management of OSA by addressing nasal obstruction, which contributes to increased airway resistance and upper airway collapsibility. One of its primary benefits is *improving nasal patency and nasal breathing* (Fig. 14.1), which facilitates smoother airflow and reduces reliance on oral breathing. This stabilization of the upper airway is essential in mitigating OSA severity and improving overall sleep quality [12–14].

As an *important part of holistic care for OSA*, nasal surgery enhances multimodal treatment strategies by optimizing the nasal airway, thereby complementing CPAP therapy, oral appliances, and multilevel surgical interventions. One of its key advantages is *benefiting CPAP compliance*, as studies have shown that reducing nasal resistance lowers the required CPAP pressure, minimizes discomfort, and significantly improves adherence rates among patients who struggle with CPAP intolerance [3, 15, 16].

Beyond its impact on CPAP, nasal surgery contributes to *reducing snoring intensity*, which is not only beneficial for patients but also beneficial for their bed partners [17]. By decreasing airflow turbulence and vibration at the level of the soft tissues, nasal surgery provides significant symptomatic relief, even in cases where AHI reduction may be minimal. Additionally, by improving nocturnal breathing, nasal surgery helps to *mitigate daytime sleepiness*, leading to better daytime alertness, cognitive function, and overall quality of life [18].

Another important benefit is its role in *reducing the respiratory disturbance index (RDI)*, though its effect on the apnea–hypopnea index (AHI) remains controversial [11, 19]. While nasal surgery alone may not provide substantial reductions in AHI, it significantly contributes to overall airway stability and improved sleep architecture, making it a valuable component of a comprehensive treatment approach.

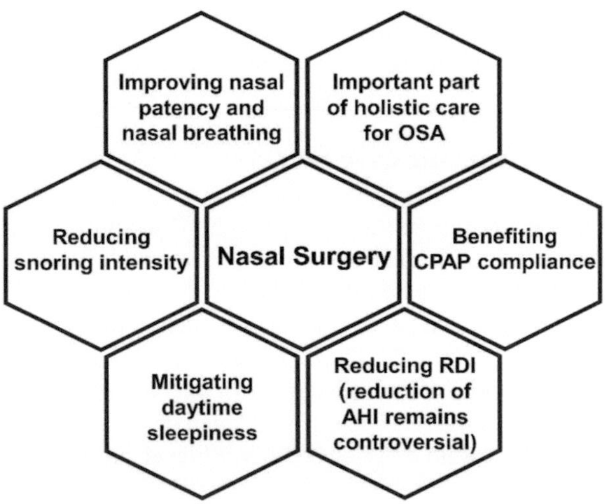

Fig. 14.1 The role and benefits of nasal surgery in OSA management

In summary, nasal surgery is not just a standalone intervention, but an integral part of *holistic OSA management*, enhancing nasal airflow, optimizing CPAP usage, reducing snoring, improving daytime function, and playing a supportive role in multilevel airway surgery. Its incorporation into individualized treatment plans is key to achieving better long-term outcomes for OSA patients.

14.2.5 Mini-Invasive Septoturbinoplasty (MIST): The Shift Toward Minimally Invasive Nasal Surgery.

In recent years, nasal surgery has evolved toward minimally invasive techniques to enhance surgical outcomes while minimizing complications. *Mini-invasive septoturbinoplasty (MIST)* represents a refined approach to traditional septoturbinoplasty, prioritizing tissue preservation and structural stability [4]. Unlike conventional methods, MIST maintains a 1-cm-wide *L-strut* to support the nasal dorsum and tip, ensuring long-term nasal framework integrity. Additionally, the preservation of a *triangular hump in the maxillary crest* reduces intraoperative bleeding and prevents injury to the greater palatine artery (Fig. 14.2).

A key innovation of MIST is the preference for *out-fracture of the inferior turbinate* rather than complete excision, which helps to preserve the nasal mucosa and lowers the risk of *empty nose syndrome (ENS)* [20]. Furthermore, *trans-septal sutures* and incision drainage techniques eliminate the need for nasal packing, reducing patient discomfort and preventing septal hematoma [21]. The incorporation of *endoscopic or loupe-assisted visualization* enhances precision during

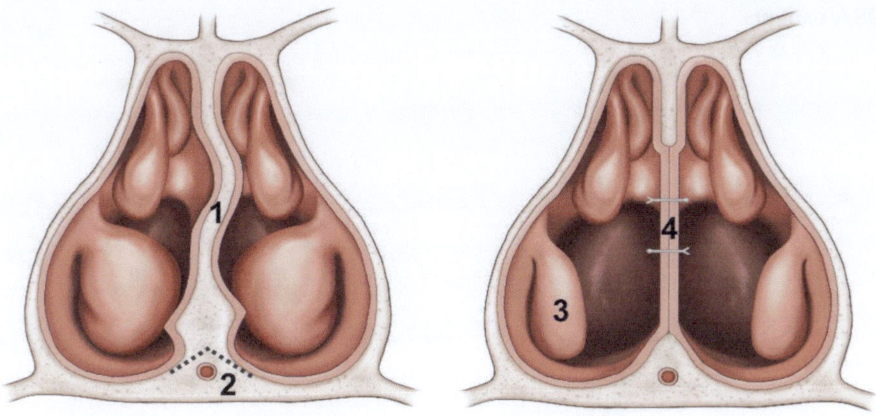

Fig. 14.2 Illustration of the mini-invasive septoturbinoplasty procedure (1). Realignment of the deviated nasal septum to restore nasal patency; (2) preservation of the pyramidal structure of the maxillary crest to minimize the risk of bleeding from the greater palatine artery; (3) out-fracture of the inferior turbinate combined with submucosal cauterization to maintain mucosal integrity and prevent excessive tissue removal; and (4) trans-septal sutures to control bleeding and reduce the likelihood of septal hematoma formation (5)

dissection, minimizing the risk of septal perforation. Collectively, these refinements allow for *a more effective, less invasive approach* to nasal surgery, significantly improving nasal airflow, reducing congestion, and optimizing surgical outcomes in OSA patients.

14.2.6 DOME Surgery and Nasal Patency

For OSA patients with a *high-arched palate or transverse maxillary deficiency*, nasal airway obstruction is often exacerbated by a structurally narrow maxilla. *Distraction osteogenesis maxillary expansion (DOME)* has emerged as an effective surgical intervention to address this issue [22]. The procedure involves *a limited Le Fort I osteotomy* to separate the maxilla, followed by the implantation of an expansion screw that gradually widens the nasal floor and internal nasal valve. This expansion typically begins 5–7 days postoperatively, progressing at a rate of *0.25 mm per day* over several months, ultimately achieving *8–12 mm of transverse maxillary widening* (Fig. 14.3).

DOME significantly improves *nasal breathing by reducing nasal resistance*, particularly at the level of the *internal nasal valve*, which is one of the primary sites of airflow restriction [23]. Beyond its nasal benefits, DOME has been shown to contribute to *reductions in AHI and improvements in rapid eye movement (REM) sleep*, making it a valuable adjunctive therapy for OSA patients with craniofacial abnormalities. Since maxillary expansion may lead to dental misalignment, *orthodontic correction* is typically required during or after the expansion process. With its ability to *enhance nasal airflow, relieve obstruction, and improve sleep parameters*, DOME represents an innovative and structurally targeted intervention for select OSA patients.

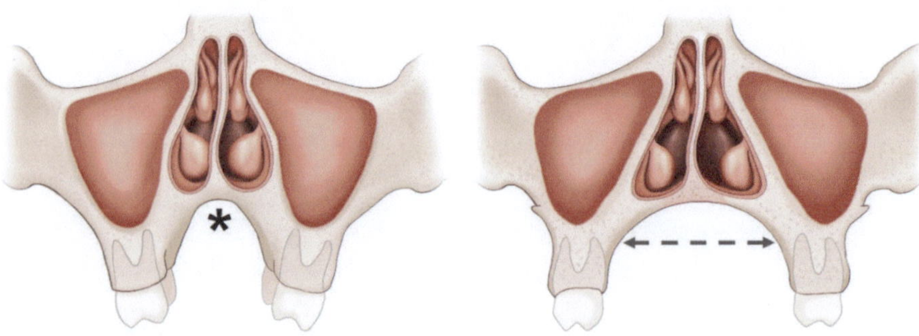

Fig. 14.3 Distraction osteogenesis maxillary expansion (DOME) enhances nasal airflow by expanding the nasal floor. This procedure is designed for patients with a high-arched and narrow palate (asterisk), which contributes to nasal obstruction. Following DOME surgery, the nasal passage broadens, and the space between the tongue and the palate increases (arrows), reducing airway resistance and improving breathing efficiency [5]

14.2.7 Nasal Irrigation Enhances Nasal Breathing and Surgical Outcomes

Nasal irrigation is a widely utilized adjunctive therapy for *improving nasal breathing* by clearing allergens, mucus, and inflammatory mediators from the nasal passage [24]. Regular use of *isotonic (0.9%) or hypertonic (2–3%) saline solutions* restores *mucociliary clearance*, reduces nasal congestion, and alleviates postnasal drip. In postoperative settings, nasal irrigation is particularly beneficial for *removing crusts, minimizing edema, and accelerating mucosal healing*, thereby optimizing surgical outcomes.

Patients with *allergic rhinitis (AR)* or *chronic rhinosinusitis (CRS)* frequently experience turbinate swelling, excessive nasal secretions, and obstructed airflow, which can negatively impact sleep quality. Studies have demonstrated that *herbal irrigation using Glycyrrhiza glabra (licorice extract)* effectively reduces *turbinate congestion, postnasal discharge, and nasal obstruction*. Unlike saline solutions, which may have a bitter or salty taste, *herbal irrigation offers a more pleasant sensory experience*, increasing patient adherence to treatment. Given its *anti-inflammatory, mucolytic, and adherence-promoting properties*, nasal irrigation—particularly with herbal formulations—serves as an essential component of *both surgical recovery and long-term nasal care* for OSA patients.

14.2.8 Challenges and Limitations of Nasal Surgery in OSA Treatment

While nasal surgery offers significant benefits in improving nasal airflow, CPAP adherence, and snoring reduction, its role as a standalone treatment for obstructive sleep apnea (OSA) remains limited. Studies indicate that nasal surgery alone yields *minimal reductions in AHI (16–50%)*, as it primarily addresses nasal resistance rather than directly preventing upper airway collapse. The effectiveness of nasal surgery is highly *patient-dependent*, with the best outcomes observed in individuals with *lower BMI, mild daytime sleepiness, and favorable upper airway anatomy*. In contrast, obese patients or those with severe multilevel airway obstruction may require additional interventions to achieve substantial improvements. Another challenge is the *variability in CPAP benefits* following nasal surgery; while many patients experience *lower therapeutic pressure requirements and improved adherence*, others may continue to struggle with CPAP use due to persistent upper airway obstruction at the level of the *soft palate, tongue base, or epiglottis*. Additionally, nasal surgery carries risks such as *septal perforation, nasal crusting, prolonged congestion, and, in rare cases, ENS,* which can lead to paradoxical nasal obstruction. Given these limitations, nasal surgery should be regarded as *a complementary rather than a curative approach* to OSA treatment, best utilized within a *multimodal strategy that includes palatal or tongue base surgery, maxillomandibular advancement, or positional therapy*. Careful patient selection and a personalized treatment plan are essential to optimizing outcomes and ensuring long-term success.

14.3 Conclusion

Nasal surgery is a crucial component of holistic OSA treatment. While its direct effect on AHI is limited, it significantly enhances *CPAP adherence, snoring reduction,* and *multilevel surgical success.* Personalized treatment approaches, integrating nasal surgery with other modalities, will continue to improve patient outcomes. Emerging technologies, such as *AI-driven surgical planning* and *3D modeling*, will further refine nasal surgery techniques, ensuring *better patient selection and long-term efficacy.*

References

1. Li HY, Lee LA, Wang PC, Chen NH, Lin Y, Fang TJ. Nasal surgery for snoring in patients with obstructive sleep apnea. Laryngoscope. 2008;118(2):354–9.
2. Li H-Y, Lee L-A, Tsai M-S, Chen N-H, Chuang L-P, Fang T-J, et al. How to manage continuous positive airway pressure (CPAP) failure–hybrid surgery and integrated treatment. Auris Nasus Larynx. 2020;47(3):335–42.
3. Poirier J, George C, Rotenberg B. The effect of nasal surgery on nasal continuous positive airway pressure compliance. Laryngoscope. 2014;124(1):317–9.
4. Li HY, Lee LA, Hsin LJ, Fang TJ, Lin WN, Chen HC, et al. Intrapharyngeal surgery with integrated treatment for obstructive sleep apnea. Biom J. 2019;42(2):84–92.
5. Tsai M-S, Chen H-C, Liu SY-C, Lee L-A, Lin C-Y, Chang G-H, et al. Holistic care for obstructive sleep apnea (OSA) with an emphasis on restoring nasal breathing: a review and perspective. J Chin Med Assoc. 2022;85(6):672–8.
6. Tsai MS, Lee LA, Tsai YT, Yang YH, Liu CY, Lin MH, et al. Sleep apnea and risk of vertigo: a nationwide population-based cohort study. Laryngoscope. 2018;128(3):763–8.
7. Tsai MS, Li HY, Huang CG, Wang RYL, Chuang LP, Chen NH, et al. Risk of Alzheimer's disease in obstructive sleep apnea patients with or without treatment: real-world evidence. Laryngoscope. 2020;130(9):2292–8.
8. Wolkove N, Baltzan M, Kamel H, Dabrusin R, Palayew M. Long-term compliance with continuous positive airway pressure in patients with obstructive sleep apnea. Can Respir J. 2008;15(7):365–9.
9. Rakotonanahary D, Pelletier-Fleury N, Gagnadoux F, Fleury B. Predictive factors for the need for additional humidification during nasal continuous positive airway pressure therapy. Chest. 2001;119(2):460–5.
10. Torre C, Camacho M, Liu SYC, Huon LK, Capasso R. Epiglottis collapse in adult obstructive sleep apnea: a systematic review. Laryngoscope. 2016;126(2):515–23.
11. Li H-Y, Wang P-C, Hsu C-Y, Lee S-W, Chen N-H, Liu S-A. Combined nasal–palatopharyngeal surgery for obstructive sleep apnea: simultaneous or staged? Acta Otolaryngol. 2005;125(3):298–303.
12. Li H-Y, Lin Y, Chen N-H, Lee L-A, Fang T-J, Wang P-C. Improvement in quality of life after nasal surgery alone for patients with obstructive sleep apnea and nasal obstruction. Arch Otolaryngol. 2008;134(4):429–33.
13. Nakata S, Noda A, Yasuma F, Morinaga M, Sugiura M, Katayama N, et al. Effects of nasal surgery on sleep quality in obstructive sleep apnea syndrome with nasal obstruction. Am J Rhinol. 2008;22(1):59–63.
14. Won T-B, Park KT, Moon SJ, Moon IJ, Wee JH, Moss T, et al. The effect of septorhinoplasty on quality of life and nasal function in Asians. Ann Plast Surg. 2013;71(1):40–4.

15. Nakata S, Noda A, Yagi H, Yanagi E, Mimura T, Okada T, et al. Nasal resistance for determinant factor of nasal surgery in CPAP failure patients with obstructive sleep apnea syndrome. Rhinology. 2005;43(4):296.
16. Iwata N, Nakata S, Inada H, Kimura A, Hirata M, Yasuma F. Clinical indication of nasal surgery for the CPAP intolerance in obstructive sleep apnea with nasal obstruction. Auris Nasus Larynx. 2020;47(6):1018–22.
17. Li H-Y, Lee L-A, Wang P-C, Fang T-J, Chen N-H. Can nasal surgery improve obstructive sleep apnea: subjective or objective? Am J Rhinol Allergy. 2009;23(6):e51–e5.
18. Li H-Y, Wang P-C, Chen Y-P, Lee L-A, Fang T-J, Lin H-C. Critical appraisal and meta-analysis of nasal surgery for obstructive sleep apnea. Am J Rhinol Allergy. 2011;25(1):45–9.
19. Li H-Y, Wang P-C, Hsu C-Y, Cheng M-l, Liou C-C, Chen N-H. Nasal resistance in patients with obstructive sleep apnea. ORL. 2005;67(2):70–4.
20. Lee DC, Jin S-G, Kim BY, Yoo S, Han S, Lee YJ, et al. Does the effect of inferior turbinate outfracture persist? Plast Reconstr Surg. 2017;139(2):386e–91e.
21. Joo YH, Cho HJ, Jeon YJ, Kim SW, editors. Use of a crossing suture to correct caudal septal deviations. International Forum of Allergy & Rhinology; 2019: Wiley Online Library.
22. Yoon A, Guilleminault C, Zaghi S, Liu SY-C. Distraction osteogenesis maxillary expansion (DOME) for adult obstructive sleep apnea patients with narrow maxilla and nasal floor. Sleep Med. 2020;65:172–6.
23. Abdelwahab M, Yoon A, Okland T, Poomkonsarn S, Gouveia C, Liu SY-C. Impact of distraction osteogenesis maxillary expansion on the internal nasal valve in obstructive sleep apnea. Otolaryngology. 2019;161(2):362–7.
24. Chang G-H, Lin Y-S, Hsu K-H, Cheng Y-C, Yang P-R, Tsai M-S, et al. Nasal irrigation with Glycyrrhiza glabra extract for treatment of allergic rhinitis–a study of in vitro, in vivo and clinical trial. J Ethnopharmacol. 2021;275:114116.

Palatal Hybrid Surgery for Obstructive Sleep Apnea

15

Hsueh-Yu Li, Ming-Shao Tsai, Li-Ang Lee, Shintaro Chiba, and Jing-ying Ye

Contents

H.-Y. Li (✉) · L.-A. Lee
Department of Otolaryngology—Head and Neck Surgery, Sleep Center, Linkou-Chang Gung Memorial Hospital, Taoyuan City, Taiwan

College of Medicine, Chang Gung University, Taoyuan City, Taiwan
e-mail: hyli38@cgmh.org.tw

M.-S. Tsai
College of Medicine, Chang Gung University, Taoyuan City, Taiwan

Departments of Otolaryngology—Head & Neck Surgery, Chang Gung Memorial Hospital, Chayi Branch, Chiayi City, Taiwan

Graduate Institute of Clinical Medical Sciences, College of Medicine, Chang Gung University, Taoyuan, Taiwan

Health Information and Epidemiology Laboratory of Chang Gung Memorial Hospital, Chiayi, Taiwan

S. Chiba
Department of Otorhinolaryngology, Jikei University School of Medicine, Minato, Japan

J.-y. Ye
Departments of Otolaryngology—Head & Neck Surgery, Beijing Tsinghua Changgung Hospital, School of Clinical Medicine, Tsinghua University, Beijing, China

H.-Y. Li (ed.), *Snoring and Obstructive Sleep Apnea*,
https://doi.org/10.1007/978-981-95-2648-2_15

Abstract

The techniques of intrapharyngeal surgery for obstructive sleep apnea (OSA) have evolved from radical excision of the soft tissue to functional reconstruction of the airway. This chapter aims to demonstrate the novel "palatal hybrid surgery" and its outcomes for OSA.

For treating velopharyngeal obstruction in OSA patients, palatal hybrid surgery utilizes an integrated tissue-specific technique that includes mucosa preservation, tonsil excision, fat ablation, and muscle relocation/suspension. Palatal hybrid surgery can be decomposed into 1. pharyngoplasty as stereoscopic reconstruction of the lateral pharyngeal wall after tonsillectomy, 2. palatoplasty as omni-suspension of the soft palate, and 3. uvuloplasty as advancement and shortening of the uvula. The preliminary results showed its safety in no airway compromise or persistent velopharyngeal insufficiency after the surgery; the postoperative bleeding was around 2%, and postoperative pain was mild in the first week. Further, clinical symptoms (snoring and daytime sleepiness) along with polysomnographic data (apnea/hypopnea index, desaturation) improved significantly after surgery. Palatal hybrid surgery reinterprets traditional uvulopalatopharyngoplasty (UPPP) in both concept and technique, and the results showed its low morbidity and high efficacy in treating OSA patients with palatal obstruction.

Keywords

Palatal hybrid surgery · Uvulopalatopharyngoplasty · Omni-suspension of soft palate · Stereoscopic reconstruction of tonsillar fossa · Push-and-pull maneuver · Obstructive sleep apnea

15.1 Introduction

Obstructive sleep apnea (OSA) is characterized by repeated upper airway obstruction during sleep [1]. The importance of OSA emerges in its high prevalence and high morbidities. OSA can cause morbidities through intermittent hypoxia and fragmented sleep. The consequences of OSA have been shown in all ages including attention deficiency, hyperactivity deficiency [2], poor growth [3] (children), cardiovascular disease [4], and metabolic syndrome [5], (adult), Alzheimer's disease [6], and nocturia [7] (elderly). Thus, the treatment of OSA is actually a kind of holistic care.

Continuous positive airway pressure (CPAP) is the gold standard and first-line treatment in adult OSA patients [8], while surgery serves as an alternative or salvage treatment for patients unwilling to or uncompliant with CPAP therapy [9]. Among various *sleep surgeries* for snoring and OSA, uvulopalatopharyngoplasty (UPPP) is the most widely used procedure [10]. That is because soft palate is the most common site of snoring and sleep apnea. Conventional UPPP implements excision of "redundant" soft tissue to enlarge the velopharyngeal airway and, thus, its outcome is closely related to positive effect from tonsillectomy in big tonsil and negative effect from the remaining hypopharyngeal obstruction [11–13]. UPPP has been criticized for its low success (41%) rate and high complications (extreme pain, bleeding, and velopharyngeal insufficiency) [14]. Therefore, modified UPPP is mandatory to enhance safety and improve effectiveness. Among these modified procedures, lateral pharyngoplasty [15] and barbed reposition pharyngoplasty [16] have most significant influence on other palatal surgeries [17–20] in the treatment for snoring and OSA.

From the *pathogenesis*, velopharyngeal narrowing in OSA can be attributed to increased tissue volume with/or decreased muscle tone. Hypertrophic tonsils and fat accumulation contribute to volume-related obstruction, while muscle is tone-related collapse. Therefore, an integrated palatal surgery needs to address both volume-related obstruction and tone-related collapse [21].

In the evolutionary history of palatal surgery, our team proposed various techniques to treat different phenotypes in palatal obstruction: 1. extended uvulopalatal flap (EUPF) to reduce palatal fat volume for enlargement of the velopharynx [22], 2. relocation pharyngoplasty to reconstruct lateral pharyngeal wall for lessening lateral collapse [19], and 3. suspension palatoplasty to suspend the soft palate for alleviating anterior–posterior collapse [20]. This chapter proposes an *integrated technique*—palatal hybrid surgery—that utilizes tissue-specific technique (mucosa preservation, fat ablation, tonsil excision, and muscle relocation/suspension) [23] (Fig. 15.1) to enlarge the velopharyngeal space for respiration during sleep and, in the meanwhile, maintain normal pharyngeal function in swallowing and phonation at daytime.

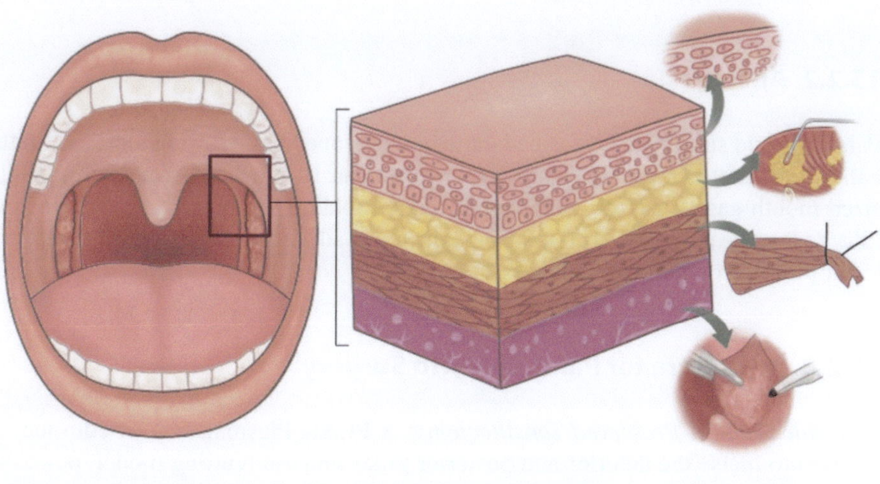

Fig. 15.1 Palatal hybrid surgery integrates tissue-specific techniques including mucosa preservation, fat ablation, tonsil excision, and muscle relocation/suspension

15.2 Discussion

15.2.1 Surgical Candidate, Inclusion/Exclusion Criteria, and Noteworthy Conditions

Surgical intervention is considered for OSA patients who are unsatisfied with conservative treatments, and unwilling for or noncompliant with CPAP therapy. The inclusion criteria are demography (age > 20 years), polysomnography (apnea–hypopnea index>5), physical examination (mouth opening space ≥3 finger breadth), chief complaint (snoring with/without daytime sleepiness), and confirmed velopharyngeal obstruction (Muller's maneuver or drug-induced sleep endoscopy) [24, 25]. Favorable anatomy includes webbing of the posterior pillars, tonsil size≧2, and long uvula. Exclusion criteria include anatomy (significant craniofacial deformity, Friedman tongue position [26] IV) and physiology (severe medical disease, high risk for general anesthesia: ASA [27] > 2).

Three clinical phenotypes are noteworthy in the decision making for palatal hybrid surgery. (1) *Obesity*, a poor predictor for UPPP outcomes, is not a contraindication for palatal hybrid surgery. Our flowchart is to arrange weight loss plans in obese patients (BMI 27–32 Kg/m^2) before or after palatal hybrid surgery, and combined airway and bariatric surgery [28] for morbidly obese patients (BMI >32 Kg/m^2). (2) We did not exclude *retrognathia* in surgical OSA patients. Actually, palatal hybrid surgery is performed concurrently with trans-oral tongue suspension (TOTS) [29] or genioplasty [30] in retrognathia OSA patients to stabilize the collapsing tongue and enlarge the retroglossal space. [3] *High Friedman tongue position* is not a decisive factor for palatal hybrid surgery. Palatal hybrid surgery is commonly performed in conjunction with coblation ablation tongue (CAT) [31] or/and coblation endoscopic lingual lightening (CELL) [32] to improve both palate and tongue obstruction in individual cases.

15.2.2 Anesthesia

Palatal hybrid surgery is performed under general anesthesia through intraoral intubation of endotracheal tube with hyper-extension of the head. An appropriately sized mouth gag is used to well expose the soft palate and tonsils. Local injection (2% Xylocaine merged 1: 100000 Epinephrine) is administered on anterior and posterior pillars around the tonsils.

15.2.3 Procedure for Palatal Hybrid Surgery

1. *Pillar Mucosa-Preserved Tonsillectomy*: A PEAK PlasmaBlade (Medtronic) is used to incise the anterior and posterior pillar mucosa (cutting model, power 4) precisely along its junction with the tonsil [33]. The tonsillar capsule is exposed

at the upper pole by blunt dissection, and tonsils are precisely dissected from the underlying muscles using coagulation model (power 2). The loop dissection starts from the upper pole, lateral pharyngeal wall, and lower pole to medial side (clockwise rotation in left tonsil, and counterclockwise rotation in right tonsil). Pillar retractor is used to pull the mucosa laterally and tonsil forceps is used to grasp and pull the tonsils medially with the purpose of keeping tension to facilitate dissection. The dissection of inferior pole between tonsils and lingual tonsils needs to be above the tongue base to avoid intractable bleeding.

Note: (1) Minimized thermal injury of pharyngeal muscle during tonsillectomy is the prerequisite for integrated pharyngoplasty without dehiscence. (2) Diverse hemostasis is crucial according to the position and amount of bleeding: Oozing is superficially cauterized via PEAK (coagulation model, power 2), pulsatile bleeding especially at the lateral pharyngeal wall is treated by suture ligation (2–0 Vicryl), and active bleeding from inferior pole of tonsillar fossa is treated by electrocoagulation (Bovie, 15 watt). (3) No tonsillectomy palatal surgery is feasible in tiny tonsil (grade I) located at the inferior pole.

2. *Ablation of Supra-Tonsillar Fat*: Using a pillar retractor, the anterior pillar is everted to expose the supra-tonsillar adipose tissue. The PEAK is used to ablate the bulky supra-tonsillar fat (cutting model, power 4) for tissue volume reduction to facilitate further expansion in lateral dimension of the velopharynx [20].

   Note: Ablation of supra-tonsillar fat needs to be in moderation to avoid excessive thermal injury of the palatal mucosa that leads to perforation of anterior pillar.

3. *Relocation of Pharyngeal Muscles (Pharyngoplasty)*: The tonsillar fossa reconstruction involves whole-layer suture of pharyngeal muscle that includes the palatopharyngeus, superior pharyngeal constrictor, and palatoglossus muscles by 2-0 Vicryl starting from the inferior pole (Fig. 15.2a). Keeping the knot at the junction of paratopharyngeus muscle and mucosa of posterior pillar, the tonsillar fossa is closed using a push-and-pull maneuver (simultaneously pulling the string downward and pushing the posterior pillar upward by the needle holder) (Fig. 15.2b), and then tie the knot (paralleling to posterior pillar) tight. It takes 2–3 whole-layer sutures to reconstruct the tonsillar fossa.

Note: (1) During relocation of pharyngeal muscle, whole-layer suture of palatopharyngeus muscle, palatoglossus muscles, and part of superior pharyngeal constrictor muscle is needed to avoid dead space and reduce lateral pharyngeal wall collapse. (2) The reconstruction of tonsillar fossa not only makes the posterior pillar in upright position (three-dimensional structure) to enlarge the oropharyngeal airway and lessen lateral pharyngeal wall collapse, but avoids potentially reduced retroglossal space from pulling the palatoglossus muscle backward during direct suturing of the intra-tonsillar muscles [34]. (3) The whole-layer suture with the knot tied precisely at the junction between paratopharyngeus muscle and posterior pillar mucosa achieves seamless closure of the tonsillar fossa.

4. *Omni-Suspension of Palatopharyngeus Muscle (Palatoplasty)*: Palatal suspension starts with a stay suture of the palatopharyngeus muscle using a 1-0 Vicryl. The needle passes through the palatal mucosa above the pterygomandibular raphe, then presses it down, and penetrates into supra-tonsillar space encompassing whole bundle of the raphe, and finally, the two strings are tied together to accomplish fixing the palatopharyngeus muscle to the pterygomandibular raphe for the suspension of soft palate (Fig. 15.3a). The first suspension is performed at the middle part of the horizontal segment of the palatopharyngeus muscle, and repeated suspensions are implemented at lateral and medial sites from the first suspension (Fig. 15.3b). Mucosa of anterior and posterior pillar is concisely sutured by 3–0 Vicryl.

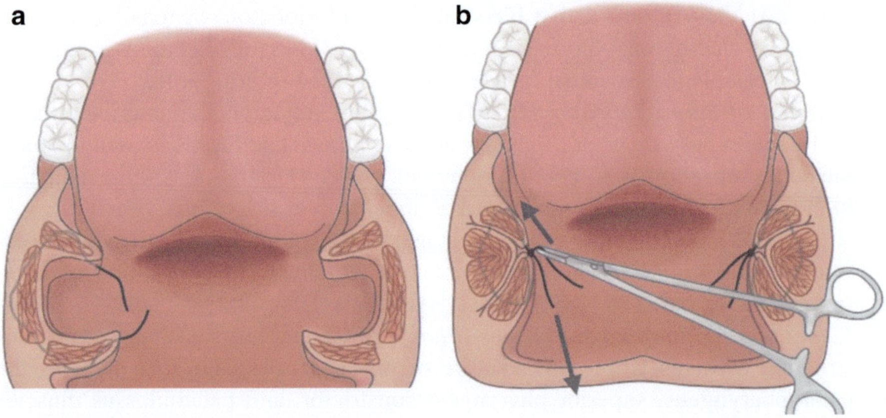

Fig. 15.2 Whole-layer suture of the palatopharyngeus, superior pharyngeal constrictor, and palatoglossus muscles (**a**), push-and-pull maneuver (simultaneously pulling the string downward and pushing the posterior pillar upward by needle holder) (**b**)

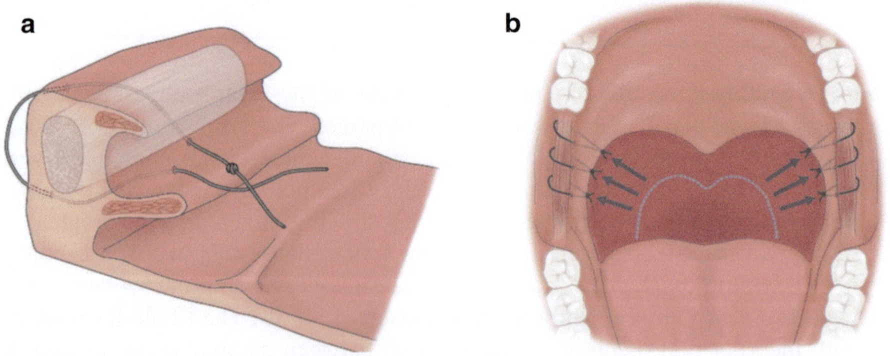

Fig. 15.3 Palatopharyngeus muscle is suspended to pterygomandibular raphe (**a**), omni-suspension (anterior–posterior and lateral) of the soft palate (**b**)

Note: (1) Separating the palatopharyngeus muscle from superior pharyngeal constrictor muscle can be implemented in case of noncompliant palatopharyngeus muscle for further expansion of the velopharynx. (2) At least two suspension sutures toward anterior and lateral directions are implemented to expand the velopharyngeal space in anterior–posterior and lateral dimensions, and that is why we call it omni-suspension. (3) The emergence of uvular bifida usually indicates decreased muscle tone of soft palate and is a poor outcome predictor of palatal suspension.

5. *Ante-Displacement and Shortening of the Uvula (Uvuloplasty)*: The modified uvuloplasty involves a horizontal incision (1 cm) of the mucosa at the uvular root (Fig. 15.4a), oblique ablation of uvular fat in front of uvular muscle via plasma knife (Fig. 15.4b), and mattress sutures of the wound by 3-0 Vicryl with the knot above the incision wound to forward warping the uvula (Fig. 15.4c). Finally, the nonmuscular part of the uvular tip is removed.

Note: (1) The higher position of the knot from the wound is associated with the more forward angle of the uvula. (2) Oblique ablation of uvular fat is helpful to facilitate upward sliding the uvula. (3) Excision of uvula is limited to uvular tip (nonmuscular part) to avoid velopharyngeal insufficiency, and that is important in young age and professional voice user.

15.2.4 Postoperative Care

(1) Intravenous broad-spectrum antibiotic (Cefazolin) is given to avoid wound infection. (2) Pain control is through multi-modal analgesic approach via intravenous NSAID (Ketorolac), [35] topic anesthetics (Benzydamine spray, Kenalog

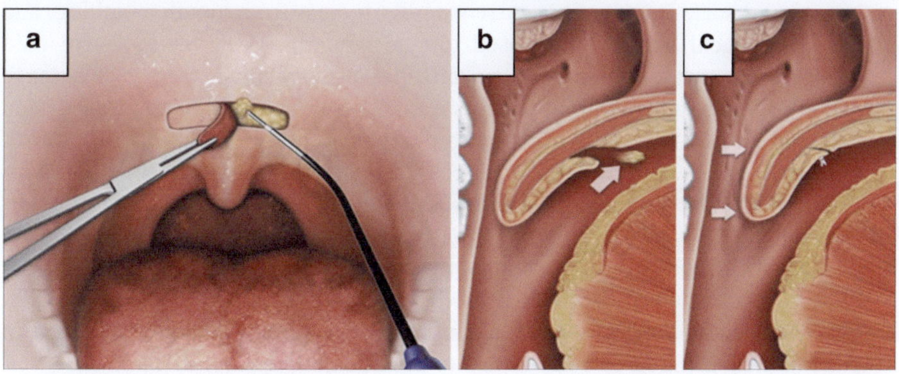

Fig. 15.4 Horizontal incision (1 cm) of the mucosa at the uvular root (**a**), oblique ablation of uvular fat in front of uvular muscle (**b**), forward warping the uvula by a mattress suture of the incision wound (**c**)

application), and ice packing over the chin. In patients experiencing persistent moderate-to-severe pain (visual analog scale of pain ≧ 4) after aforementioned pain management, short-acting analgesic (Dynastat) injection is given and long-acting analgesic (Naldebain) [36] injection is used one day before discharge. (3) Positional therapy via elevation of the head of the bed (45–60°) is routinely performed to prevent choking and tongue obstruction during sleep, and that is imperative in combined tongue surgery [37]. (4) Humidified oxygen is supported with monitor of oxygen saturation by pulse oximeter at night. (5) Proton pump inhibitor is also prescribed for patients with reflux symptom/disease. (6) In diet progress, we prescribe a liquid diet in the first day and soft diet in the next 5–7 days, followed by regular diet afterward. (7) Patients are suggested to ambulate (to decrease edema), talk more (as oropharyngeal excise), and drink more water (to avoid dehydration and improve oral hygiene). Return of physical exercise (weight training, jogging) is suggested three weeks after surgery.

15.2.5 Surgical Outcomes

Subjective improvement was noted in snoring (70% reduction in visual analog scale) and daytime sleepiness (−5.8 in Epworth sleepiness scale). Objective improvement from home sleep apnea test was also significant in apnea/hypopnea index (−21.1 event/hr), lowest oxygen saturation (+9.1%), and snoring index (−134.0 event/hr). The classical success rate in individual Friedman stage was 100% (3/3) in stage I, 63% (12/19) in stage II, and 58% (14/24) in stage III with an overall success rate of 63% [23].

15.2.6 Complications

1. Postoperative *pain* is mild (visual analog scale ≦3) in overwhelming majority, but may increase in case of concurrent nasal (dryness from mouth breathing) and hypopharyngeal surgery (increased tongue wound). Moderate-to-severe pain (visual analog scale >3) during hospitalization usually comes from (1) excessive cautery for active bleeding in patients with hypertrophic tonsils (grade 3/4) and (2) poor surgical plan with excision of pharyngeal muscle or anterior /posterior pillar mucosa, and consequently increased tension of suture in the pharyngoplasty. Persistent pain after discharge is always related to dehiscence of tonsillar wound.

   Note: (1) Cautious use of cautery, precise dissection of tonsils, and firm but not too tight suture of the mucosa are helpful in decreasing postoperative pain. (2) Multi-modal analgesia (short-acting, long-acting, and topical) is needed in patients with difficult-to-control pain to provide good quality of care.
2. Postoperative *bleeding* rate in our surgical cohort was 2% and generally self-limited or ceased following conservative treatment. The timing of bleeding usually occurs between the seventh and 14th day following surgery. Precise

dissection of the tonsils and aforementioned hemostasis models for individual bleeding are complementary to the low bleeding rate. The postoperative bleeding period in our patients is relatively late than ordinary UPPP patients (10th-14th day vs. 1st-7th day) that may be attributable to the use of intravenous NSAID (ketololac) (prolonged bleeding time).

Note: Two-week careful observation with instruction (avoid heavy lifting, big clumps of food, and mouth breathing) is necessary to ride out the danger period of postoperative bleeding.

3. *Stitch abscess* over the raphe may emerge with complaint of pain 1–2 weeks after operation and can be processed properly by open drainage and Betadine throat gargling. The risk factor of stitch abscess is infection from the exposed suture of palatal suspension.

   Note: Reducing the exposed suspension suture is helpful to avoid stitch abscess.

4. *Globus sensation of the throat* in the early postoperative period is mostly due to retained stitches over the palatal edge that can be released dramatically after removing the stitches. There is rare experience of long-term lumping throat because palatal hybrid surgery performs suspension instead of excision of the soft palate. The scarless healing in the edge of soft palate leads to "soft body collision properties" with the tongue during swallowing.

   Note: (1) Pharyngeal hygiene education (avoid mouth breathing and frequent cleaning of the throat) is indispensable to lessen globus sensation of the throat. (2) Transient taste disturbance may occur after palatal surgery, especially in heavy smoking patients due to chronic pharyngitis. Therefore, preoperative notification of potential taste disturbance is essential, especially in taste-related worker.

15.3 Summary

The features of palatal hybrid surgery include preservation of pillar mucosa in tonsillectomy, ablation of supra-tonsillar fat, relocation of tonsillar pharyngeal muscle (pharyngoplasty), omni-suspension of the soft palate (palatoplasty), and ante-displacement and shortening of the uvula (uvuloplasty). Palatal hybrid surgery, utilizing a tissue-specific approach, reinterprets traditional UPPP in both concept and technique.

The low complication rates and improved treatment outcomes support the use of palatal hybrid surgery in surgical OSA patients with velopharyngeal obstruction. Palatal hybrid surgery can be used in multilevel surgery or multidisciplinary surgery for individual phenotype of OSA to improve success rate with high safety. Further research in merging palatal hybrid surgery with skeletal surgery or hypoglossal nerve stimulation and long-term follow-up in diverse outcomes is warranted to verify multidimensional outcomes and the durability of surgical effect over time.

Conflict of Interest The authors have declared that no conflict of interest exists.

References

1. Strollo PJ, Roger RM. Obstructive sleep apnea. N Eng J M. 1996;334:99–104.
2. Li HY, Huang YS, Chen NH, Fang TJ, Lee LA. Impact of adenotonsillectomy on behavior in children with sleep disordered breathing. Laryngoscope. 2006;116:1142–7.
3. Marcus CL, Carroll JL, Koerner CB, Hamer A, Lutz J, Loughlin GM. Determinants of growth in children with the obstructive sleep apnea syndrome. J Pediatr. 1994;125(4):556–62.
4. Bradley TD, Floras JS. Obstructive sleep apnoea and its cardiovascular consequences. Lancet. 2009;373:82–93.
5. Giampá SQC, Lorenzi-Filho G, Drager LF. Obstructive sleep apnea and metabolic syndrome. Obesity (Silver Spring). 2023;31(4):900–11.
6. Tsai MS, Li HY, Huang CG, Wang RYL, Chuang LP, Chen NH, Liu CH, Yang YH, Liu CY, Hsu CM, Cheng WN, Lee LA. Risk of Alzheimer's disease in obstructive sleep apnea patients with or without treatment: real-world evidence. Laryngoscope. 2020;130(9):2292–8.
7. Di Bello F, Napolitano L, Abate M, et al. Nocturia and obstructive sleep apnea syndrome: a systematic review. Sleep Med Rev. 2023;69:101787. https://doi.org/10.1016/j.smrv.2023.101787.
8. Sullivan CE, Issa FG, Berthon-Jones M, Eves L. Reversal of obstructive sleep apnoea by continuous positive airway pressure applied through the nares. Lancet. 1981;1:862–5.
9. Li HY, Lee LA, Tsai MS, Chen NH, Chuang LP, Fang TJ, Shen SC, Cheng WN. How to manage continuous positive airway pressure (CPAP) failure- hybrid surgery and integrated treatment. Auris Naris Larynx. 2020;47:335–42.
10. Fujita S, Conway W, Zorick F, Roth T. Surgical correction of anatomic abnormalities in obstructive sleep apnea syndrome: uvulopalatopharyngoplasty. Otolaryngol Head Neck Surg. 1981;89:923–34.
11. Simmons FB, Guilleminault C, Silvestri R. Snoring and some obstructive sleep apnea can be cured by oropharyngeal surgery: palatopharyngoplasty. Arch Otolaryngol. 1983;109:503–7.
12. Tschopp S, Tschopp K. Tonsil size and outcome of uvulopalatopharyngoplasty with tonsillectomy in obstructive sleep apnea. Laryngoscopy. 2019;129:449–54.
13. Li HY, Wang PC, Li LA, Chen NH, Fang TJ. Prediction of uvulopalatopharyngoplasty outcome: anatomy-based staging system versus severity-based staging system. Sleep. 2006;29:1537–41.
14. Sher AE, Schechtman KB, Piccirillo JF. The efficacy of surgical modifications of the upper airway in adults with obstructive sleep apnea syndrome. Sleep. 1996;19:156–77.
15. Cahali MB. Lateral pharyngoplasty: a new treatment for obstructive sleep apnea hypopnea syndrome. Laryngoscope. 2003;113:1961–8.
16. Vicini C, Hendawy E, Campanini A, et al. Barbed reposition pharyngoplasty (BRP) for OSAHS: a feasibility, safety, efficacy and teachability pilot study. "We are on the giant's shoulders". Eur Arch Otorrinolaringol. 2015;272:3065–70.
17. Friedman M, Ibrahim HZ, Vidyasagar R, et al. Z-palatoplasty (ZPP): a technique for patients without tonsils. Otolaryngol Head Neck Surg. 2004;131:89–100.
18. Pang KP, Woodson BT. Expansion sphincter pharyngoplasty: a new technique for the treatment of obstructive sleep apnea. Otolaryngol Head Neck Surg. 2007;137:110–4.
19. Li HY, Lee LA. Relocation pharyngoplasty for obstructive sleep apnea. Laryngoscope. 2009;119:2472–7.
20. Li HY, Lee LA, Kezirian EJ, Nakayama M. Suspension palatoplasty for obstructive sleep apnea- a preliminary study. Sci Rep. 2018;8:4224.
21. Li HY, Lee LA, Hsin LJ, et al. Intra-pharyngeal surgery with integrated treatment for obstructive sleep apnea. Biom J. 2019;42:84–92.
22. Li HY, Li KK, Chen NH, Wang PC. Modified uvulopalatopharyngoplasty: the extended uvulopalatal flap. Am J Otolaryngol. 2003;24:311–6.
23. Li HY, Tsai MS, Lee LA, et al. Palatal hybrid surgery for obstructive sleep apnea: state-of-the-art annotation of uvulopalatopharyngoplasty. Biom J. 2023;46(3):100568. https://doi.org/10.1016/j.bj.2022.11.001.

24. Amali A, Amirzargar B, Sadeghi M, Saedi B. Muller's maneuver in patients with obstructive sleep apnea. J Sleep Sci. 2016;1:148–50.
25. Kezirian EJ, Hohenhorst W, de Vries N. Drug-induced sleep endoscopy: the VOTE classification. Eur Arch Otorrinolaringol. 2011;268:1233–6.
26. Friedman M, Ibrahim H, Bass L. Clinical staging for sleep-disordered breathing. Otolaryngol Head Neck Surg. 2002;127:13–21.
27. Mayhew D, Mendonca V, Murthy BVS. A review of ASA physical status—historical perspectives and modern developments. Anaesthesia. 2019;74:373–9.
28. Lin CC, Liu KH, Lee LA, et al. Combined airway and bariatric surgery (CABS) for obstructive sleep apnea patients with morbid obesity: a comprehensive alternative preliminary study. J Clin Med. 2022;11(23):7078.
29. Hsin LJ, Lee YC, Lin WN, et al. Transoral tongue suspension for obstructive sleep apnea-a preliminary study. J Clin Med. 2022;11(17):4960.
30. Santos Junior JF, Abrahão M, Gregório LC, Zonato AI, Gumieiro EH. Genioplasty for genioglossus muscle advancement in patients with obstructive sleep apnea-hypopnea syndrome and mandibular retrognathia. Braz J Otorhinolaryngol. 2007;73(4):480–6. https://doi.org/10.1016/s1808-8694(15)30099-9.
31. Lu YA, Wang CJ, Chiang YT, Li HY. Volumetric changes after coblation ablation tongue (CAT) in obstructive sleep apnea patients. J Clin Med. 2022;11(14):4186.
32. Li HY, Lee LA, Kezirian EJ. Coblation endoscopic lingual lightening (CELL) for obstructive sleep apnea. Eur Arch Otorrinolaringol. 2016;273(1):231–6.
33. Tsai MS, Li HY. Comparison of post-operative pain between pulsed electron avalanche knife (PEAK) and electrocautery in uvulopalatopharyngoplasty. JTOLHNS. 2014;49(3):170–6.
34. Li HY, Cheng WN, Chuang LP, Fang TJ, Hsin LJ, Kang CJ, Lee LA. Positional dependency and surgical success of relocation pharyngoplasty among patients with severe obstructive sleep apnea. Otolaryngol Head Neck Surg. 2013;149(3):506–12.
35. Lee LA, Wang PC, Cheng NH, et al. Alleviation of wound pain after surgeries for obstructive sleep apnea. Laryngoscope. 2007;117:1689–94.
36. Tsai HL, Chang TK, Su WC, et al. Comparing efficacy and safety between Naldebain® and intravenous patient-controlled analgesia with fentanyl for pain management post-laparotomy: study protocol for a randomized controlled, non-inferior trial. Trials. 2019;20(1):173. https://doi.org/10.1186/s13063-019-3260-4.
37. Chen WC, Lee LA, Chen NH, et al. Treatment of snoring with positional therapy in patients with positional obstructive sleep apnea syndrome. Sci Rep. 2015;11(5):18188. https://doi.org/10.1038/srep18188.

Hypopharyngeal Surgery for Obstructive Sleep Apnea

16

Hsueh-Yu Li, Li-Jen Hsin, Yi-An Lu, Yi-Chan Lee, Hirotaka Hara, and Sung-Wan Kim

Contents

H.-Y. Li (✉) · L.-J. Hsin · Y.-A. Lu
Departments of Otolaryngology-Head & Neck Surgery, Chang Gung Memorial Hospital, Linkou Main Branch, Taoyuan, Taiwan

School of Medicine, Chang Gung University, Taoyuan, Taiwan
e-mail: hyli38@cgmh.org.tw

Y.-C. Lee
School of Medicine, Chang Gung University, Taoyuan, Taiwan

Departments of Otolaryngology-Head & Neck Surgery, Chang Gung Memorial Hospital, Keelung Branch, Keelung, Taiwan

H. Hara
Department of Otolaryngology-Head and Neck Surgery, Kawasaki Medical School, Kawasaki, Japan

S.-W. Kim
Department of Otorhinolaryngology-Head Neck Surgery, Kyung Hee Medical Center, Kyung Hee University, Seoul, South Korea

H.-Y. Li (ed.), *Snoring and Obstructive Sleep Apnea*,
https://doi.org/10.1007/978-981-95-2648-2_16

Abstract

Hypopharyngeal obstruction in terms of tongue obstruction during sleep has been demonstrated as the second common site of airway obstruction in obstructive sleep apnea (OSA) patients and the commonest cause in the failure of uvulopalatopharyngoplasty for OSA. Assessment of tongue obstruction evolves from oropharyngeal examination, cephalometry, nasopharyngoacopy to drug-induced sleep endoscopy (diagnostic and interventional). Important findings in the dynamic lateral image of drug-induced sleep CT scan showed that obstruction of the tongue not only emerges at tongue base but evenly at tongue base, body with base, and tongue body. Many hypopharyngeal procedures have been used in multilevel surgery to improve success rate that include lingual tonsillectomy (robotic or non-robotic), tongue volume reduction via radiofrequency or coblation, genioglossus advancement/genioplasty, and tongue suspension (transoral or transcervical). The mechanism of action in these procedures is either by volumetric reduction or by stabilization of the tongue but is not in conflict with each other. Volume reduction tongue surgery (targeting on lingual tonsil) is more widely used than stabilization of the tongue (suspension). The use of robotic (transoral robotic surgery) or non-robotic (coblation endoscopic lingual lightening) is the mainstream for volume reduction tongue surgery. Otherwise, radiofrequency or coblation (channeling) has its unique role in reducing intralingual fat volume and tightening lingual muscle for enlargement of retroglossal airspace and lessening tongue collapse. It is noteworthy that stabilization technique via tongue base suspension may increase the risk of epiglottic collapse in OSA patients with long epiglottis.

Keywords

Obstructive sleep apnea · Hypopharyngeal surgery · Coblation endoscopic lingual lightening (CELL) · Coblation ablation tongue (CAT) · Transoral tongue suspension (TOTS) · Coblation- assisted partial epiglottectomy (CAPE)

16.1 Introduction

The hypopharynx traditionally refers to the lower part of the pharynx, which comprises three main subsites: the pyriform sinuses, the posterior pharyngeal wall, and the postcricoid region [1]. However, in the context of hypopharyngeal surgery for obstructive sleep apnea (OSA), this concept is understood more from a functional rather than a strictly anatomical perspective. In sleep medicine, surgeons often categorize the pharynx into two functional segments: an upper and a lower region. The upper segment, termed the retropalatal region, anatomically corresponds to the section of the oropharynx located at the level of the soft palate and tonsils. In contrast, the lower segment, commonly referred to as the hypopharyngeal region, functionally encompasses both the oropharyngeal area at the level of the tongue base and the hypopharynx [2]. Obstruction in the hypopharyngeal region may result from structures such as the base of the tongue, lateral pharyngeal walls, aryepiglottic folds, or epiglottis [3–7]. Consequently, hypopharyngeal surgeries for OSA are specifically designed to address these potential obstruction sites.

To address hypopharyngeal obstruction, various surgical techniques have been developed. Some procedures aim to increase hypopharyngeal airway space by volume reduction, while others focus on modifying airway tissue compliance by ablation or suspension [8]. However, there is currently no consensus on which procedures provide specific benefits for particular patient groups. This chapter presents inclusion/exclusion and assessment of hypopharyngeal obstruction, and current hypopharyngeal surgeries proposed by our sleep team that include coblation endoscopic lingual lightening (CELL) [9, 10], coblation ablation tongue (CAT) [11], transoral tongue suspension (TOTS) [12], and coblation-assisted partial epiglottectomy (CAPE). The full surgical techniques are demonstrated with figures and noteworthy places to enhance safety and effectiveness of hypopharyngeal surgery for OSA.

16.2 Inclusion/Exclusion Criteria, Noteworthy Conditions for Tongue Surgery

The following *inclusion criteria* are suggested: (1) age > 20 years, (2) body mass index <37 kg/m^2, (3) apnea/hypopnea index>5 events/h, (4) symptom: habitual snoring with/without excessive daytime sleepiness, (4) Friedman tongue position II/III, (5) hypopharyngeal obstructions identified by the cephalometry (narrowing of retroglossal space), awake endoscopy (hypertrophic lingual tonsils) or drug-induced sleep endoscopy (total/partial tongue collapse), (6) unwilling or noncompliant with CPAP therapy. *Exclusion criteria* are as follows: (1) high risk in general anesthesia (ASA > 2), (2) poor dentition, trismus (mouth opening <3 fingers' width, 4 cm), (3) bleeding tendency, (4) swallowing, speech, or taste disturbances, and severe extra-esophageal reflux.

It is noteworthy that patients should be instructed to take scheduled medications in the morning of surgery, especially antihypertensive medicine. Discontinuity of anticoagulants is necessary to avoid increased perioperative bleeding. Patients also need to be informed in the risk of postoperative taste disturbance, especially in professional taste user such as chef or sommelier.

16.3 Airway Assessment of Tongue Obstruction

1. *Oropharyngeal Examination and Nasopharyngoscopy:* High modified Mallampati score (III/IV) is likely to have tongue obstruction. Lingual tonsil hypertrophy is indicated for lingual tonsillectomy (robotic or non-robotic [9]). Big tongue without lingual tonsil hypertrophy can be treated by coblation ablation of the whole tongue (base and body) with modest volumetric reduction [11]. Small collapse tongue with low muscle tone (low strength and endurance shown

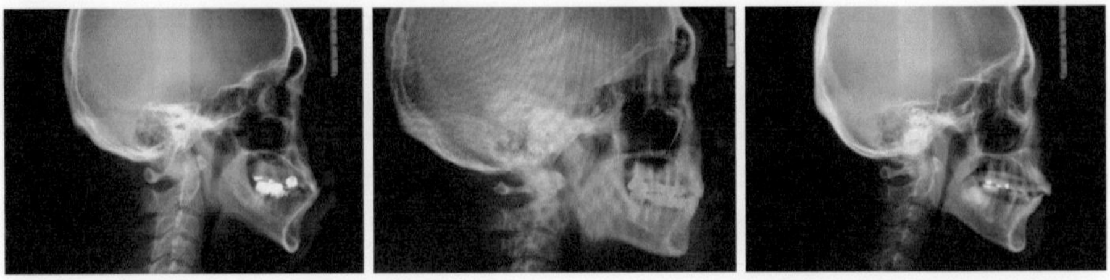

Fig. 16.1 Three shapes of retroglossal airspace: reverse triangular (left), triangular (middle), and long tubular (right)

in Iowa oropharyngeal performance instrument) is suggestive of tongue suspension [12].

2. *Phenotype*: Positive clinical phenotypes (insomnia, daytime sleepiness) and appearance phenotypes (obesity, retrognathia) are relatively poor surgical prognostic factors. In contrast, positive physiological phenotypes (REM stage-, supine position- dependency) are good surgical outcome predictors [13].
3. *Cephalometry*: Narrowest retroglossal airspace (RG) < 1.1 cm is suggestive of tongue obstruction. Mandibular plane to hyoid distance (MP-H) >22 mm is associated with lower success rates of hypopharyngeal surgery [14]. Epiglottic length > 1.66 cm is suggestive of epiglottic collapse [15]. Reverse triangular shape of RG (indicating low tongue base obstruction) is suggestive of lingual tonsillotomy or tonsillectomy. In contrast, the triangular shape of RG (indicating upper tongue base obstruction) is suggestive of tongue ablation by radiofrequency or coblation. Otherwise, the long tubular shape of RG (indicating total tongue obstruction) is a poor outcome predictor for tongue base surgery (Fig. 16.1).
4. *Drug-Induced Sleep Endoscopy (DISE)/ Drug-Induced Sleep CT Scan*: The DISE provides qualitative and semi-quantitative assessment of airway collapse, and that is more precise than awake examinations (such as Muller's maneuver) in assessing nocturnal airway collapse, especially in tongue and epiglottis. This scheme evaluates collapse at 3 segments—the velopharynx (soft palate and tonsil), oropharynx (lateral wall), and hypopharynx (tongue, epiglottis) to better describe obstruction and improve surgical outcomes [16]. Drug-induced sleep CT scan via dynamic changes in the sagittal view further demonstrates 3 types of tongue collapse— tongue base (37%), total tongue (33%, tongue body collapse followed by tongue base collapse) and tongue body (30%), which modify traditional concept of tongue obstruction presenting in tongue base only [17] (Fig. 16.2).

Anesthesia General anesthesia with orotracheal intubation is performed in our surgical patients because of simultaneous multilevel surgery including nasal and palatal surgery. During tongue volume reduction or stabilization surgery, the endotracheal tube is pulled outside (left side usually) by pillar retractor to well expose the tongue

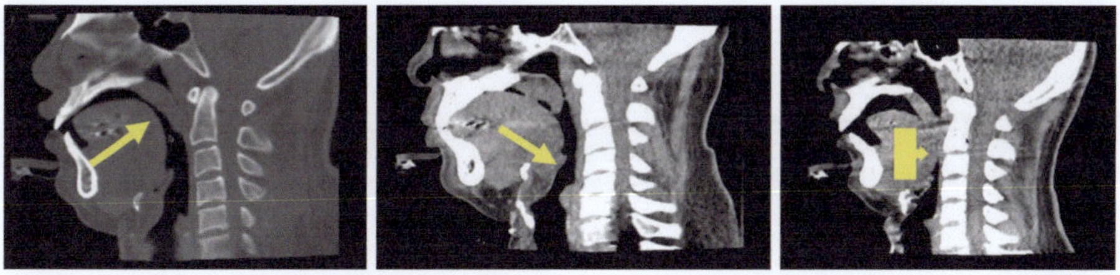

Fig. 16.2 Three types of tongue collapse—tongue body (left), tongue base (middle), and total tongue (right)

base tissue. Patients are routinely extubated after fully recovering from the general anesthesia in the operation room.

16.4 Coblation Endoscopic Lingual Lightening (CELL) [9]

Tongue reduction procedures can be performed by robotic or non-robotic application. A newer technology, Coblation (ArthroCare Corporation, Austin, Texas, USA), excites the electrodes in saline solution to create a plasma field of highly ionized particles so as to break down intercellular bonds in the tissue that enable tissue removal at lower temperatures, thereby offering the potential for tongue tissue resection with decreased risks and morbidity [18–20].

Surgical Technique Under general anesthesia, traction suture via 3-0 silk and mouth gag are used to expose the tongue base. The 30° telescope (Storz) with a mechanical holding system is used to demonstrate the operation field. The operator sits in front of the patient's head and performs the surgery under video-endoscopic guide. The Coblation settings are 7 for ablation mode and 3 for coagulation mode. Tongue base tissue behind circumvallate papillae is ablated by the Coblation wand to develop three vertical trenches: middle (1–1.5 cm in depth) and two paramedian (1 cm lateral to the midline in each side) from circumvallate papillae to the vallecula. The tongue base tissue within the trenches is grasped and excised or scooped out via Coblation until exposing the epiglottis (Fig. 16.3).

Note (1) CELL is more likely to be lingual tonsillotomy (volume reduction) instead of lingual tonsillectomy. (2) CELL uses a mechanical holding system to fix the endoscope and that facilitates bimanual manipulation of tongue base tissue. (3) Currently, a 3-D endoscope (3-D Tipcam 7240BA3D- 30°, Karl Storz, Germany) has been implemented to enhance depth of field for stereo effect during surgery. (4) The manipulation of CELL should stay within the V shaped safety zone (lateral: 1 cm, inferior: 1.5 cm from foramen cecum) and away from the neurovascular bundle to avoid active bleeding. (5) A monopolar electrocauterization is used in case of active bleeding.

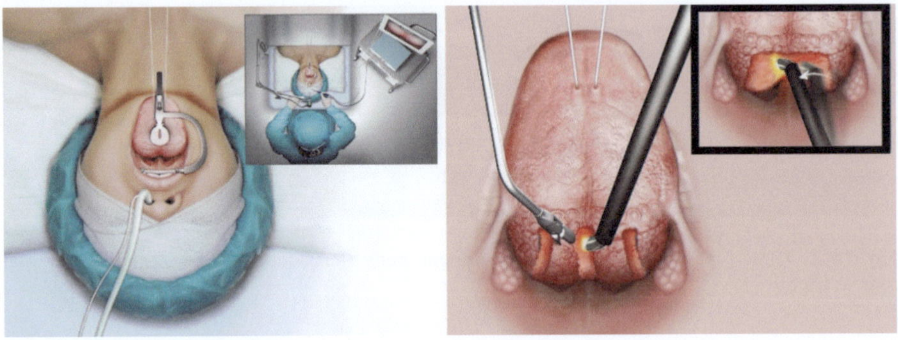

Fig. 16.3 Scenario (left) and surgical techniques of coblation endoscopic lingual lightening (right)

16.5 Coblation Ablation Tongue (CAT) [11]

The use of radiofrequency ablation (RFA) for tongue volume reduction was first demonstrated in 1997 by Powell et al. on an animal model [21]. RFA delivers energy to deep tissues at relatively low temperatures, offering reduced pain and morbidity compared to conventional surgery [22]. However, concerns remain about its limited effectiveness and the lack of standardized criteria for patient selection. Most RFA studies have included patients with mild-to-moderate obstructive sleep apnea (OSA), using fewer than 10 ablation points for tongue volume reduction with little long-term efficacy data for severe OSA cases [23–25].

Surgical Technique Coblation ablation tongue (CAT) is a minimally invasive procedure characterized by mucosal preservation, needle-hole wounds, a hybrid approach (combining muscle scarring/contraction and fat ablation), and full treatment of the tongue body and base. The procedure employs a coblation wand (Reflex 4855, ArthroCare Corp., Austin, TX, USA) in ablation-5 mode for a 15-second energy release per insertion, followed by coagulation-4 mode for hemostasis during wand withdrawal. A 3-0 silk traction suture is placed through the anterior midline tongue, and a reverse T-shaped marking on the dorsal tongue guides the ablation. The wand is inserted fully into the tongue, with the active zone positioned 1 cm below the mucosa to prevent thermal damage to the taste buds. CAT includes 9 ablation points on the tongue body and 9 on the tongue base (Fig. 16.4).

Note (1) CAT is characterized by a mini invasive procedure at the submucosa for taste preservation and by needle-hole tissue ablation instead of excision. (2) CAT is actually whole-tongue hybrid treatment for scarring/contraction of muscle in tongue body (stabilization) and ablation of fat in tongue base (volume reduction). (3) The CAT study provides the data in volumetric reduction of 0.38 cm^3 per ablation that contributes to plan the numbers of ablation needed for optimal tongue volume reduction. (4) CAT is more effective in tightening the tongue to lessen its collapse, but less effective in volumetric reduction of hypertrophic lingual tongue. (5) The

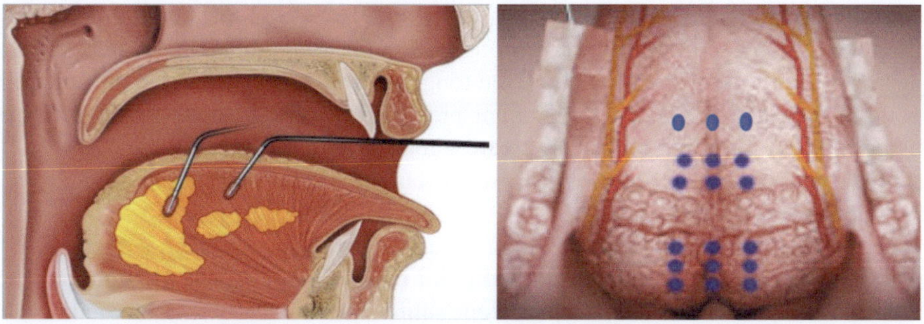

Fig. 16.4 Coblation ablation tongue (CAT) performs ablation in fat of tongue base for volume reduction and muscle of tongue body for stabilization (left). CAT includes 9 ablation points on tongue body and 9 on tongue base (right)

triangular shape of retroglossal airspace in lateral cephalometry (suggestive of upper tongue base obstruction) is more suitable for CAT than reverse triangular shape (suggestive of lower tongue base obstruction) in lessening tongue collapse.

16.6 Transoral Tongue Suspension (TOTS) [12]

Surgical techniques for tongue base suspension (TBS) include TBS with the Repose® (Repose Surgical Kit, CKA AirVance, Medtronic, Inc., Minneapolis, MN, USA) [26] and modified TBS. [27] Repose is more popular than modified TBS and has gradually become a stereotype in performing TBS. However, there was a recent critical discussion of TBS due to its variable outcomes and expensive suspension kit [28]. For some reasons, Repose® was out of production. Due to clinical necessity, our team proposed a new surgical technique of tongue suspension with no need for commercialized kit.

Surgical Techniques Under general anesthesia, the patient was placed in a supine position with full neck extension. The oral cavity was disinfected using Aqua-Better iodine diluted 1:1 with saline. A local anesthetic injection of 1% lidocaine mixed with epinephrine (1:100,000 dilution) was administered at the vestibular area of the inner lower lip. A transverse incision of approximately 3 cm was then made, maintaining a minimum distance of 1.5 cm from the marginal gingiva of the central lower incisors. Blunt dissection was performed over the periosteum of the mandible to expose a 3 × 3 cm^2 rectangle area, with the midline of the central incisors marked for reference. Two parallel tunnels, each 1 cm lateral to the marked midline, were drilled using a 3-mm cutting burr (60,000 RPM, Medtronic, MN, USA). Next, the patient's mouth was opened using a self-retaining retractor, and the tongue was fully extended with 3-0 silk sutures to expose the circumvallate papillae and sulcus terminalis. The middle third of the sulcus terminalis on both sides was identified and marked using electrocautery under a 30° telescope monitor. A suture passer

(Obwegeser mandibular awl, Johnson & Johnson, NJ, USA) loaded with 2-0 polypropylene (Prolene) was inserted through the right mandibular hole, penetrating deep into the floor of the mouth and the tongue before emerging at the pre-marked point at the right tongue base. A similar procedure was performed on the left side using a No. 3 black silk suture to create a loop at the left tongue base. To ensure safety and avoid injury to the neurovascular bundle, the passer was kept within the middle third of the tongue base.

The polypropylene suture was then secured submucosally using a free curved needle and anchored at the left marked point before being passed through the silk loop. The loop and enclosed polypropylene suture were carefully withdrawn through the left mandibular hole. Finally, the distal ends of the polypropylene sutures were tied together within their respective holes, with the knots placed inside to reduce infection risk and minimize foreign body sensation at the chin. After suture tightening, fiberoptic nasopharyngoscopy confirmed an enlarged retro-lingual airway space with a deep dimple at the tongue base. As the final step, multiple interrupted sutures were placed at the pre-mandibular subcutaneous soft tissue and the vestibular mucosa to complete the closure (Fig. 16.5).

Note (1) TOTS implements trans-oral (trans-sublabial) approach instead of trans-cervical approach that leads to aesthetic advantage in scarless exterior. (2) TOTS ties the two ends of suspension string together without the need for an inserted screw and that decreases the potential of infection and foreign body sensation. (3) TOTS implements two interlacing polypropylenes suture to pull the tongue forward that not only provides twofold pulling strength but abates suture migration (muscle-cutting effect) from single string. (4) TOTS is freely available without the need for a commercialized kit that lowers the medical cost for patients. (5) TOTS is more

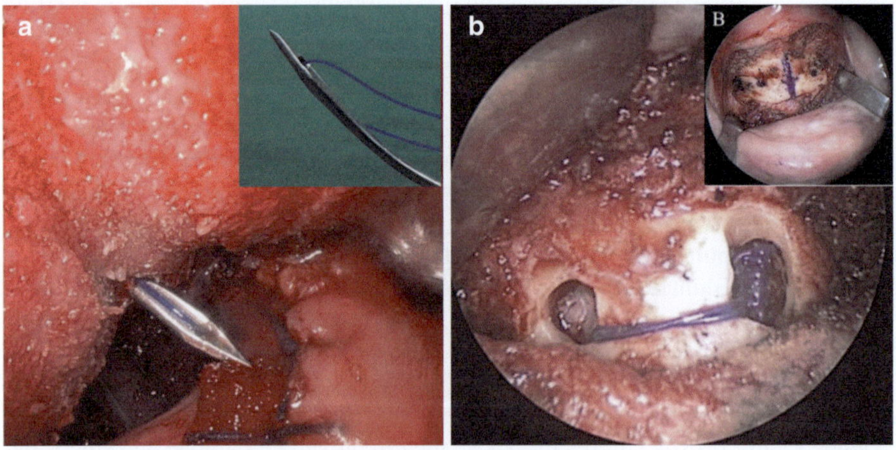

Fig. 16.5 (**a**) Suture passer loaded with 2-0 polypropylene (Prolene) penetrating right tongue base. (**b**) Two parallel tunnels, each 1 cm lateral to the marked midline, the knots placed inside the hole

suitable in small tongue with low endurance shown in Iowa oropharyngeal performance instrument, but less successful in big tongue or lingual tonsillar hypertrophy (grade III/IV). (6) Tongue suspension increases the risk of epiglottic collapse in patients with long epiglottis. Therefore, interventional DISE is necessary to identify epiglottic collapse before TOTS for a comprehensive OSA treatment.

16.7 Coblation-Assisted Partial Epiglottectomy (CAPE)

Epiglottis collapse existed in 25% of moderate-to-severe OSA patients [17] and was deemed to be an potential cause of CPAP failure and UPPP failure [29]. The diagnosis of epiglottis collapse has been widely implemented via drug-induced sleep endoscopy through VOTE classification [30]. A current dynamic image study revealed that the length of epiglottis of more than 1.66 cm achieved 100% sensitivity of epiglottis collapse [15]. Partial epiglottidectomy has been involved in multi-level airway surgery to improve the success rate for OSA. Electrocautery (robotic surgery) has been used to implement partial epiglottidectomy [31]. Side effects from thermal injury in these power instruments include time-consuming wound healing, severe postoperative pain, and a certain degree of hypopharyngeal synechia. To improve the safety of epiglottic surgery, our team used a lower thermal tool-coblation (larynx) and proposed a new technique—coblation-assisted partial epiglottectomy (CAPE).

Surgical Technique Under general anesthesia, direct laryngoscope with built-in 30° telescope (Storz) is used to demonstrate the whole epiglottis (lingual aspect). The operator sits in front of the patient's head and performs the surgery under video-endoscopic guide. A larynx coblator (EIC 7070–01, Smith & Nephew, USA) is used with the fixed model (Ablation:7, Coagulation:4). The middle part of epiglottis is grasped by a micro-forceps and then a vertical ablation is performed by the coblator (Ablation:7) to achieve mid-splitting of the epiglottis (50% in length). The right part of epiglottis is grasped by a micro-forceps again, and excised horizontally from right to left using the same coblator (Ablation:7). Residual left part of epiglottis is removed with the same model. Bleeding from the cutting edge of muco-perichondrium and epiglottic cartilage is controlled by the coblator (Coagulation: 4) (Fig. 16.6).

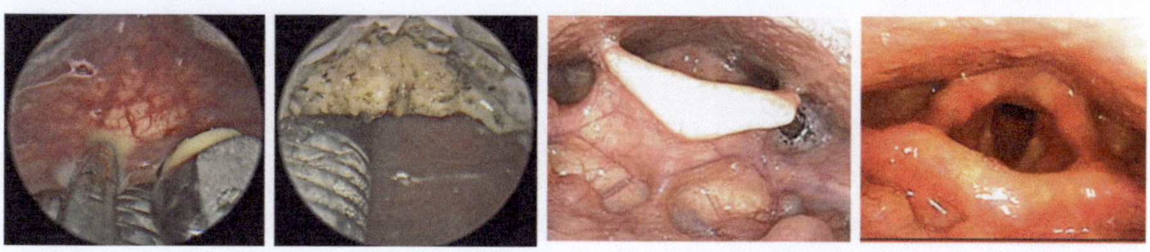

Fig. 16.6 Using coblation (larynx) to ablate epiglottis (right), baseline, and outcome pictures of coblation-assisted partial epiglottectomy (left)

Note (1) Preparatory cautery (Coagulation:4) of the epiglottic muco-perichondrium along the planed cutting route is suggested before ablation of the epiglottic cartilage to avoid bleeding. (2) The junction between epiglottis and aryepiglottic fold needs to be protected from ablation to avoid massive bleeding.

16.8 Summary

Treatment of hypopharyngeal obstruction is almost unavoidable in moderate/severe OSA patients to achieve an optimal outcome. Hypopharyngeal obstruction can be assessed by Friedman stage, cephalometry, or drug-induced sleep endoscopy. Hypopharyngeal surgery is increasingly performed in multilevel surgery for OSA. Hypopharyngeal surgery can be implemented by volumetric reduction or/with stabilization of the tongue. Volume reduction of tongue base (via robotic or coblation) is more widely used than tongue base suspension. There is no rigid guideline of specific technique to be used in particular phenotype group of OSA patients. The determination of hypopharyngeal surgery needs to be based on shared decision making including patient's wish, evidence-based outcomes (safety vs. efficacy), and surgeon's professionalism.

Conflict of Interest The authors have declared that no conflict of interest exists.

References

1. Liao YL, Chi TC, Kuo IC, Yang SW, Lin WN, Hsin LJ, et al. Modified Killian method for flexible nasopharyngoscopic observation of the hypopharynx: a systematic Review and meta-analysis. J Craniofac Surg. 2024;2024 https://doi.org/10.1097/SCS.0000000000010351.
2. Lin HC, Friedman M. Transoral robotic OSA surgery. Auris Nasus Larynx. 2021;48:339–46.
3. Li KK. Hypopharyngeal airway surgery. Otolaryngol Clin North Am. 2007;40:845–53.
4. Schwab RJ, Gefter WB, Hoffman EA, Gupta KB, Pack AI. Dynamic upper airway imaging during awake respiration in normal subjects and patients with sleep disordered breathing. Am Rev Respir Dis. 1993;148:1385–400.
5. Schwab RJ, Pasirstein M, Pierson R, Mackley A, Hachadoorian R, Arens R, et al. Identification of upper airway anatomic risk factors for obstructive sleep apnea with volumetric magnetic resonance imaging. Am J Respir Crit Care Med. 2003;168:522–30.
6. Rojewski TE, Schuller DE, Clark RW, Schmidt HS, Potts RE. Videoendoscopic determination of the mechanism of obstruction in obstructive sleep apnea. Otolaryngol Head Neck Surg. 1984;92:127–31.
7. Ritter CT, Trudo FJ, Goldberg AN, Welch KC, Maislin G, Schwab RJ. Quantitative evaluation of the upper airway during nasopharyngoscopy with the Muller maneuver. Laryngoscope. 1999;109:954–63.
8. Woodson BT. Updated hypopharyngeal surgery for sleep apnea. Adv Otorhinolaryngol. 2017;80:81–9.
9. Li HY, Lee LA, Kezirian EJ. Coblation endoscopic lingual lightening (CELL) for obstructive sleep apnea. Eur Arch Otorhinolaryngol. 2016;273(1):231–6.
10. Li HY, Lee LA, Kezirian EJ. Efficacy of coblation endoscopic lingual lightening in multilevel surgery for obstructive sleep apnea. JAMA Otolaryngol. 2016;142(5):438–43.
11. Lu YA, Wang CJ, Chiang YT, Li HY. Volumetric changes after coblation ablation tongue (CAT) in obstructive sleep apnea patients. J Clin Med. 2022;11(14):4186.

12. Hsin LJ, Lee YC, Lin WN, Lu YA, Lee LA, Tsai MS, Cheng WN, Chiang YT, Li HY. Transoral tongue suspension for obstructive sleep apnea-a preliminary study. J Clin Med. 2022;11(17):4960.
13. Sutherland K, Yee B, Kairaitis K, Wheatley J, Chazal P, Cistulli P. A phenotypic approach for personalized management of obstructive sleep apnoea. Current Otorhinolaryngol Rep. 2021;9 https://doi.org/10.1007/s40136-021-00346-6.
14. Woodson BT. Updated hypopharyngeal surgery for sleep apnea in sleep-related breathing disorders. In: Lin HC, editor. Adv Otorhinolaryngol, vol. 80. Basel: Karger; 2017. p. 81–9.
15. Kuo IC, Hsin LJ, Lee LA, Fang TJ, Tsai MS, Lee YC, Shen SC, Li HY. Prediction of epiglottic collapse in obstructive sleep apnea patients: epiglottic length. Nat Sci Sleep. 2021;13:1985–92. https://doi.org/10.2147/NSS.S336019.
16. Green KK, Kent DT, D'Agostino MA, et al. Drug-induced sleep endoscopy and surgical outcomes: a multicenter cohort study. Laryngoscope. 2019;129(3):761–70.
17. Li HY, Lo YL, Wang CJ, Hsin LJ, Lin WN, Fang TJ, Lee LA. Dynamic drug- induced sleep computed tomography in adults with obstructive sleep apnea. Sci Rep. 2016;6:35849. https://doi.org/10.1038/srep35849.
18. Robinson S, Lewis R, Norton A, McPeake S. Ultrasound guided radiofrequency submucosal tongue-base excision for sleep apnoea: a preliminary report. Clin Otolaryngol Allied Sci. 2003;28:341–5.
19. Woodson BT. Innovative technique for lingual tonsillectomy and midline partial glossectomy for obstructive sleep apnea. Oper Techn Otolaryngol Head Neck Surg. 2007;18:20–8.
20. Babademez MA, Ciftci B, Acar B, Yurekli MF, Karabulut H, Yilmaz A, Karasen RM. Low-temperature bipolar radio frequency ablation (coblation) of the tongue base for supine position-associated obstructive sleep apnea. ORL J Otorhinolaryngol Relat Spec. 2010;72:51–5.
21. Powell NB, Riley RW, Troell RJ, Blumen MB, Guilleminault C. Radiofrequency volumetric reduction of the tongue: a porcine pilot study for the treatment of obstructive sleep apnea syndrome. Chest. 1997;111(5):1348–55.
22. Farrar J, Ryan J, Oliver E, Gillespie MB. Radiofrequency ablation for the treatment of obstructive sleep apnea: a meta-analysis. Laryngoscope. 2008;118(10):1878–83.
23. Atef A, Mosleh M, Hesham M, Fathi A, Hassan M, Fawzy M. Radiofrequency vs laser in the management of mild to moderate obstructive sleep apnoea: does the number of treatment sessions matter? J Laryngol Otol. 2005;119(11):888–93.
24. Blumen MB, Dahan S, Fleury B, Hausser-Hauw C, Chabolle F. Radiofrequency ablation for the treatment of mild to moderate obstructive sleep apnea. Laryngoscope. 2002;112(11):2086–92.
25. Ceylan K, Emir H, Kizilkaya Z, Tastan E, Yavanoglu A, Uzunkulaoğlu H, et al. First-choice treatment in mild to moderate obstructive sleep apnea: single-stage, multilevel, temperature-controlled radiofrequency tissue volume reduction or nasal continuous positive airway pressure. Arch Otolaryngol. 2009;135(9):915–9.
26. Miller FR, Watson D, Malis D. Role of the tongue base suspension suture with The Repose System bone screw in the multilevel surgical management of obstructive sleep apnea. Otolaryngol Head Neck Surg. 2002;126:392–8.
27. Omur M, Ozturan D, Elez F, Unver C, Derman S. Tongue base suspension combined with UPPP in severe OSA patients. Otolaryngol Head Neck Surg. 2005;133:218–23.
28. Tsou YA, Huang CW, Wu TF, Hung LW, Chang WD. The effect of tongue base suspension with uvulopalato-pharyngoplasty on sleep quality in obstructive sleep apnea. Sci Rep. 2018;8:8788.
29. Sung CM, Kim HC, Kim J, Kim JG, Lim SC, Shin MH, Nam K, Lee J, Vena D, White DP, Yang HC, Wellman A. Patients with epiglottic collapse are less adherent to autotitrating positive airway pressure therapy for obstructive sleep apnea. Ann Am Thorac Soc. 2022;19(11):1907–12.
30. Kezirian EJ, Hohenhorst W, de Vries N. Drug-induced sleep endoscopy: the VOTE classification. Eur Arch Otorhinolaryngol. 2011;268(8):1233–6.
31. Arora A, Chaidas K, Garas G, Amlani A, Darzi A, Kotecha B, Tolley NS. Outcome of TORS to tongue base and epiglottis in patients with OSA intolerant of conventional treatment. Sleep Breath. 2016;20(2):739–47.

Maxillomandibular Advancement: Considerations for Asian Patients with OSA

17

Clement Cheng-Hui Lin, Po-fang Wang, Yau Hong Goh, Yu-Chih Wang, Ellen Wen-Ching Ko, and Yu-Ray Chen

Contents

C. C.-H. Lin (✉) · P.-f. Wang · Y.-R. Chen
Craniofacial Center, Department of Plastic and Reconstructive Surgery, Chang Gung Memorial Hospital, Taoyuan, Taiwan

Craniofacial Research Center, Chang Gung Memorial Hospital, Taoyuan, Taiwan

School of Medicine, Chang Gung University, Taoyuan, Taiwan

Y. H. Goh
Ear Nose Throat Head and Neck Surgery, Mount Elizabeth Medical Centre, Singapore, Singapore

Y.-C. Wang · E. W.-C. Ko
Craniofacial Research Center, Chang Gung Memorial Hospital, Taoyuan, Taiwan

Department of Craniofacial Orthodontics, Chang Gung Memorial Hospital, Taipei, Taiwan

Graduate Institute of Dental and Craniofacial Science, Chang Gung University, Taoyuan, Taiwan

H.-Y. Li (ed.), *Snoring and Obstructive Sleep Apnea*,
https://doi.org/10.1007/978-981-95-2648-2_17

Abstract

Obstructive sleep apnea is a multifactorial medical condition with a complex pathophysiology. Surgery becomes a primary treatment option when anatomical factors causing airway narrowing are identified at specific or generalized levels of the pharyngeal airway. Surgical procedures are targeted at the underlying anatomical structures to achieve maximum effectiveness. Key factors in selecting appropriate surgical interventions include body weight, the Mallampati score, and tonsil grade. Surgical weight reduction, maxillomandibular advancement (MMA), and pharyngeal soft tissue surgeries are considered for patients with obesity, maxillomandibular retrognathism, and tonsillar hypertrophy, respectively. A comprehensive craniofacial assessment, including cephalometry and facial aesthetics, is essential for developing an optimal surgical plan for MMA in patients with varying facial patterns. The surgical objective should ideally balance the patient's airway needs, aesthetic considerations, and normal occlusion.

Keywords

Maxillomandibular advancement · OSA · Treatment protocol · Surgical design

17.1 Introduction

Maxillomandibular advancement (MMA) has been the most efficient surgical option to dilate the pharyngeal airway and reduce the apnea–hypopnea index (AHI) [1]. Patients after other surgeries can also benefit from MMA for treating residual or recurrent OSA [2]. MMA's surgical success and cure rates were 86.0% and 43.2%, respectively, according to a meta-analysis [3]. The success rate can reach up to 100% in young (<45 y/o) and thin (BMI < 25) patients with significant jaw retrusion (SNB < 75°) [4].

The three evaluations to select surgical options are the body mass index (BMI), the modified Mallampati score (with the tongue staying in the oral cavity), and the tonsil size grading. Firstly, the *BMI* is an independent factor for OSA—a higher BMI results in higher AHI in patients with the same cephalometries [5]. An increase in BMI is also a primary reason for OSA relapse after all types of surgery. For patients with obesity, bariatric surgery has become one of the treatment options to improve OSA. The *modified Mallampati score* grades the disproportion between the soft tissue volume and the size of the oral cavity. OSA patients with maxillofacial retrognathism commonly have higher Mallampati scores of III or IV. The *tonsil size grading* illustrates the ratio of pharyngeal lateral dimension occupied by hypertrophic tonsils. An OSA patient with a high tonsil grading of III or IV has a better chance of having improvement by tonsillectomy with uvulopalatopharyngoplasty [6]. Ideally, a slim patient with low tonsil grading and a high Mallampati score could be a candidate for MMA as the first-line surgical option (Fig. 17.1). [7]

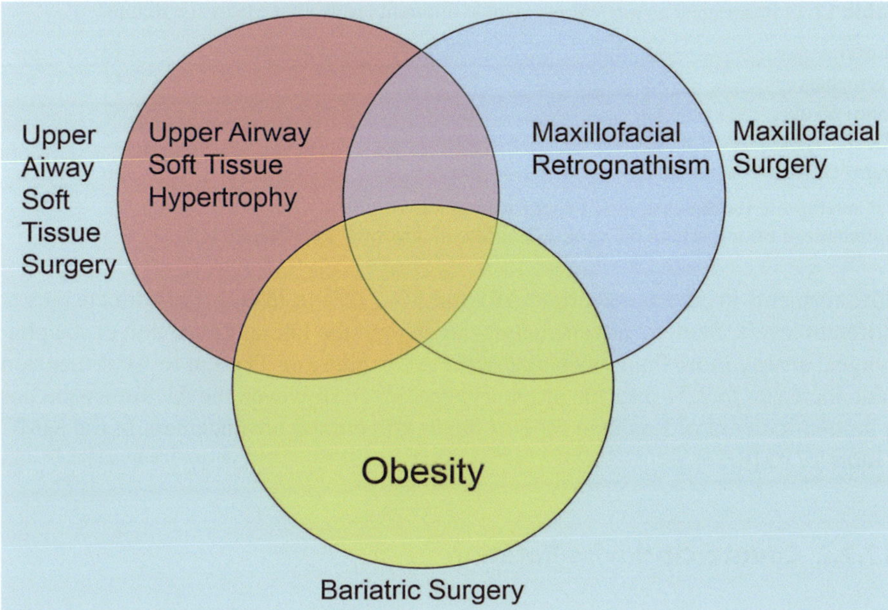

Fig. 17.1 Three types of surgical treatment for OSA are used to correct specific anatomic abnormalities. Multidisciplinary treatments are indicated for patients with multiple types of deformity. (Adopted from Lin et al. Sleep Med Clin 14 (2019) 83–89)

17.2 Discussion

17.2.1 The Advancement and the Airway

Considering the airway, ***every segment of the maxillomandibular skeleton must be advanced as much as possible.***

After osteotomies, the advancement of the upper and lower jaws is limited by the soft tissue attachment, including the *neurovascular bundles* of the greater palatine nerve and vessels to the posterior maxilla, the mandibular nerve, and vessels to the mandible.

The surgical procedure of the conventional MMA includes maxillary and mandibular osteotomies, advancement of the maxillomandibular complex (MMC) by more than 10 mm, and an anterior–inferior mandibular osteotomy with genioglossus advancement as indicated.

MMA enlarges the maxillomandibular skeletal frame, expands the pharyngeal airway, and reinforces the genioglossus muscle tension. The conventional design can be applied to patients with normal facial profiles. With the MMA advancement of 10 mm in a three-dimensional computed tomography (3DCT) upper airway study, the pharyngeal airway increases on average by 45–58% (of 10 mm

Table 17.1 Pharyngeal airway enlargement at different levels after MMA for 10 mm

Level	AP mm	(%)	Lat mm	(%)
Velopharynx	5.6	56	7.1	71
retro-Uvula	4.5	45	5.7	57
Oropharynx	4.9	49	13.2	132
retro-Glossal	5.8	58	6.6	66

AP anteroposterior dimension, *Lat* lateral dimension
Summarized from Fairburn SC et al. J Oral Maxillofac Surg 2007 Jan;65(1):6-12

advancement) in anterioposterior (AP) and 57–132% in lateral (Lat) dimensions at different levels. Skeletal advancement can enlarge the lateral dimension of the pharyngeal airway more than twice that of the AP expansion. The Lat to AP dimension ratio increases to 2.37 over the oropharyngeal level. However, the AP dimension has a limited increase of less than 60% of the unidirectional advancement of the MMC (Table 17.1) [8].

17.2.2 Counterclockwise Rotation

The mandible can be regularly advanced up to a distance of 12–15 mm or more, which is usually beyond the limit of maxillary advancement. For example, to achieve a pharyngeal airway space (PAS) of more than 1 cm for a patient with a PAS of 4 mm, the mandibular advancement could be estimated to be more than 12 mm. Maximizing mandibular advancement can optimize its airway expansion effect. However, a discrepancy between maxillary and mandibular advancement creates a negative overjet and Class III malocclusion. A counterclockwise rotation of MMC becomes crucial to avoid negative overjet while maximizing both maxillary and mandibular advancement. After counterclockwise rotational advancement of the MMC, the pharyngeal airway has 47% and 76% (of the amount of mandibular advancement) increase in AP dimension at the velopharynx and retroglossal areas, respectively [9]. With the MMC rotational forward, pivoted on the facial skeletal landmark N (Nasion), the ANB angle can remain within the normal range, affirming an aesthetically pleasing facial profile (Fig. 17.2).

> *Integrating* ***counterclockwise rotation*** *of the maxilla and mandible, and the occlusal plane angle into the surgical design, the MMC can achieve optimal advancement in both jaws without compromising the occlusion.*

17.2.3 Segmental Osteotomies

For patients with *large overjet* (>6 mm), considering the advancement of the mandible by 10 mm and the normalization of the ANB angle, the maxilla would be advanced around only 3 mm at A point. If the velopharyngeal PAS has inefficient dilatation, a *segmental LeFort I osteotomy* combined with posterior maxillary

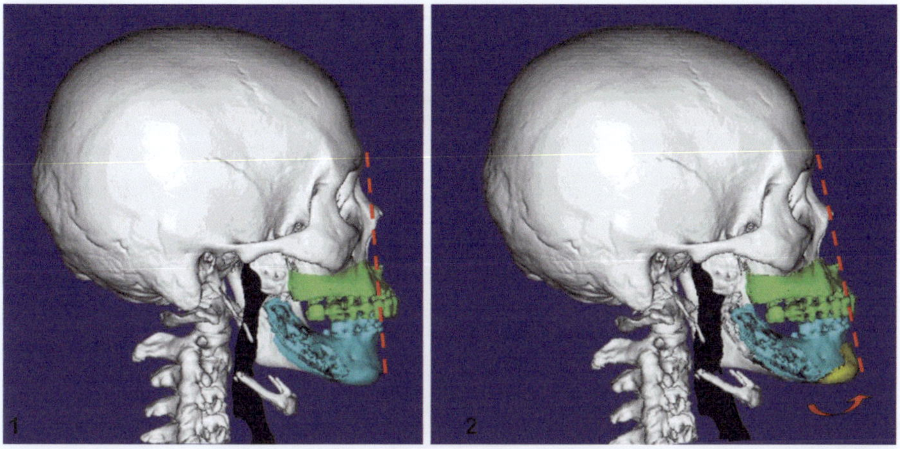

Fig. 17.2 Surgical design of maxillomandibular rotational advancement (MMRA) incorporating maxillomandibular advancement and counterclockwise rotation, pivoted on the skeletal landmark Nasion, maintaining the angular measurement ANB, which represents the relationship between cranium, maxilla, and mandible. (Adopted from Lin et al. Sleep Med Clin 14 (2019) 83–89)

advancement is indicated. Segmental osteotomies are incorporated into the surgical plan of MMA. Combined with counterclockwise rotation, a segmental LeFort I maxillary osteotomy can further advance the posterior nasal spine and palatine bone by 7 mm, facilitated by an alveolar bone reduction after extraction or edentulous spaces. Additional advancement of the posterior border of the maxilla stretches the soft palate and repositions it anteriorly to enlarge the AP dimension of the velopharyngeal airway. By the design of segmental maxillomandibular rotational advancement (*SMMRA*), the pharyngeal airway can be increased in the AP dimension at the velopharynx by 60% and at the oropharynx by 70% of the amount of mandibular advancement (Fig. 17.3) [10].

A modification in surgical design is needed to incorporate and maximize advancement in both jaws and create an optimal pharyngeal airway enlargement.

17.2.4 The Aesthetics

With significant advancements in MMA, facial changes have drawn significant attention from patients and surgeons. By questionnaire evaluation, MMA patients subjectively experienced moderate changes in their facial appearance. However, over 90% of patients still give positive or neutral feedback on the facial changes [11].

Patients in Asia were noted to have more protrusive facial profiles and narrower cranial base angles than Caucasian patients by cephalometric evaluation [12]. For most OSA patients undergoing MMA, despite focusing on the airway and sleep quality during the presurgical consultation, the perioperative emphasis shifts more

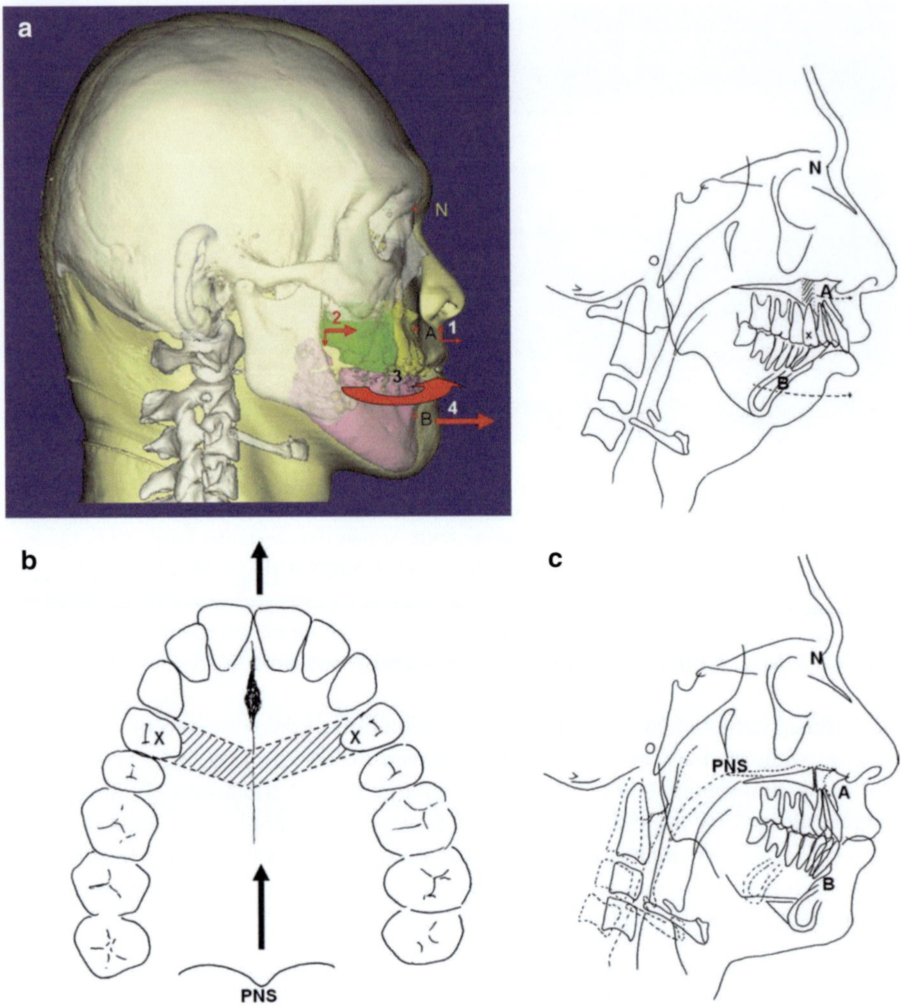

Fig. 17.3 Surgical design of segmental maxillomandibular rotational advancement (SMMRA). SMMRA uses intrusion of the anterior segment (1) and elongation of the posterior segmented maxilla (2) to achieve counterclockwise rotation (3) of the maxillomandibular complex (MMC), pivoted on Nasion (N). (**a**) Cephalometric simulation illustrates the tooth to be extracted (x) and the related bony segment to be removed (slashed area). The direction of rotation after osteotomy on both jaws is demonstrated (curved dash arrows). (**b**) The segmental osteotomy of the maxilla uses an extraction (x) or edentulous space to facilitate further advancement of the posterior segment, including PNS. (**c**) The counterclockwise rotation of MMC reinforces that the mandibular advancement to a distance more than could be achieved in ordinary orthognathic surgery performed for the correction of Class II malocclusion. N: nasion, B: B point, A: A point, PNS: posterior nasal spine. (Adopted from Lin et al., Laryngoscope, 121:1336–1347, 2011)

or less to facial aesthetics. Disturbingly, aesthetic considerations may lead to confusing and paradoxical thoughts with airway considerations.

To clarify, three scenarios need to be addressed:

17.2.4.1 Maxillary/Bimaxillary Protrusion

The protrusive appearance is caused by the strong support provided by the front teeth of the upper and lower jaws to the perioral soft tissue. Frequently used aesthetic techniques are anterior segmental osteotomies of the maxilla (Wassmund procedure) and the mandible (Köle procedure). The protruded front teeth can be altered through pitch rotation of anterior segments, and the protrusive appearance can thus be improved [13]. Modified MMA with segmental osteotomies has been shown to effectively treat OSA while maintaining an acceptable facial profile [14]. Unlike the backward rotation of anterior segments in aesthetic orthognathic surgery, it has been emphasized that all the dentoskeletal segments should be advanced to ensure the enlargement of the maxillofacial frame and oral cavity in OSA patients [10]. Although these key objectives were accomplished, removing premolar teeth and portions of the alveolar ridge during segmental osteotomies significantly reduces the oral cavity's volume. This could impair a certain degree of the forward movement of the tongue and the soft palate, ultimately limiting the expansion of the pharyngeal airway. The resultant impairment in the forward movement of the tongue and soft palate and the limitation imposed on the pharyngeal airway expansion could have accounted for deterioration in the long-term result of the modified MMA despite effective initial postsurgical outcome [15].

17.2.4.2 Counterclockwise Rotation of MMC

Incorporating counterclockwise rotation (CCWR) of the MMC into the surgical planning of MMA helps to balance the facial relationship between the cranium, midface (the maxilla), and lower face (the mandible). Maintaining a normal ANB angle is essential for achieving a harmonic and natural facial relationship. By rotating the MMC forward around the nasion (N) as a pivot point, the ANB angle can be preserved. This approach mainly benefits OSA patients with maxillomandibular retrusion, who often present with deformities such as steep occlusal and mandibular planes. The CCWR of the MMC can enhance these angles while maintaining overall facial harmony (Fig. 17.2) [10].

17.2.4.3 Genioplasty

Chin retrusion combined with a flat labiomental curve is commonly observed in patients with mandibular retrusion. A sliding genioplasty can enhance chin projection and improve the labiomental fold. This involves a horizontal osteotomy across the central portion of the mandible, positioned below the bilateral mental nerves, with the bony segment advanced forward. In OSA patients, the genioplasty technique can be modified to involve a more superior cut below the lower central incisors, including the genioglossus tubercle in the advancing segment. Advancing the genioglossus tubercle increases the tension of the genioglossus and geniohyoid muscles, helping to pull the tongue forward and indirectly suspend the hyoid. Depending on the specific surgical design, the genioplasty can be shaped as a trapezoid, mortise, or triangle, as described in the literature [16].

17.2.5 Phenotypes by the Angle's Classification

Class II malocclusion is commonly observed in patients with OSA; however, individuals with Class I or Class III occlusion are not necessarily free from the disorder. As a result, sleep surgeons must employ diverse strategies to tailor surgical plans for OSA patients with varying facial patterns (Table 17.2). A thorough cephalometric evaluation is essential for effective surgical planning. Cone-beam computed tomography (CBCT) offers three-dimensional (3D) imaging that facilitates cephalometric analysis, airway measurements, and surgical simulations when integrated into a 3D surgical planning program.

17.2.5.1 Class I

Patients with Class I occlusion typically exhibit aesthetic profiles by cephalometric measurements, but have a narrow airway. A suitable surgical approach for these patients is MMA combined with counterclockwise rotation pivoted on the Nasion. This technique preserves the normal ANB angle while increasing the anterior–posterior dimensions of the maxillofacial region without altering the angular relationships between the cranium, maxilla, and mandible. The approach addresses airway and aesthetic concerns while maintaining the Class I occlusal relationship.

17.2.5.2 Class II

OSA patients with Class II malocclusion often present with mandibular retrognathism, a large overjet, and/or a high mandibular plane angle. The mandible can be advanced to correct the abnormal overjet to address Class II malocclusion. The MMC should be advanced to match the mandible's full advancement to maximize pharyngeal airway expansion. For patients with large overjets whose maxillary advances little after MMC repositioning, a segmental osteotomy can help. The maxillary segmental osteotomy can further advance the posterior maxillary border and pull the soft palate forward, enlarging the velopharyngeal space. Genioplasty, including the genioglossus tubercle, enhances tension in the genioglossus and geniohyoid muscles, helping to maintain the retroglossal airway dimensions [10].

Table 17.2 Surgical design of MMA for OSA patients with various facial patterns

Class	Surgical Design
I	MMA +/- CCWR
II	MMRA +/- ASO + G
III	MMRA

MMA maxillomandibular advancement, *CCWR* counterclockwise rotation, *MMRA MMA+CCWR, ASO* anterior segmental osteotomy, *G* genioplasty

17.2.5.3 Class III

OSA patients with Class III malocclusion often have a narrow velopharyngeal airway caused by maxillary hypoplasia and retrusion. They may also exhibit high, narrow palatal vaults, and restricted nasal passages. Correcting Class III malocclusion and enlarging the velopharynx requires maxillary advancement. For patients with high nasal resistance, maxillary expansion or distraction [17] is recommended. If a narrow retroglossal airway is observed, mandibular advancement with counterclockwise rotation can improve the airway and correct the occlusion simultaneously.

17.3 Conclusion

A successful surgical treatment for adult patients with OSA relies on a precise differential diagnosis, which involves clinical history, physical examination of craniofacial structures, and standard overnight polysomnography. To maximize the effectiveness of surgical therapy, procedures must target the anatomical abnormalities underlying the pathophysiology of OSA.

Obesity independently worsens the soft tissue condition of the pharyngeal airway and reduces surgical effectiveness. For OSA patients with a high BMI, combining positive airway pressure therapy with weight reduction, either surgical or nonsurgical, is the preferred treatment approach.

Intrapharyngeal surgeries are highly effective in patients with large tonsils and a low Mallampati score, where soft tissue factors are predominant and maxillofacial retrognathism is minimal. Additionally, extrapharyngeal procedures such as like septoplasty and turbinate reduction can be performed to enhance nasal patency and reduce upper airway resistance.

Maxillomandibular advancement can be the primary treatment option for nonobese patients with a Mallampati score of III or above. Radiological evaluation using 2D lateral cephalograms or 3D CT imaging is recommended. By comparing cephalometric data with normative values of the general population, a tailored surgical plan can be developed to address airway function, aesthetics, and Angle's normal occlusion requirements (Fig. 17.4).

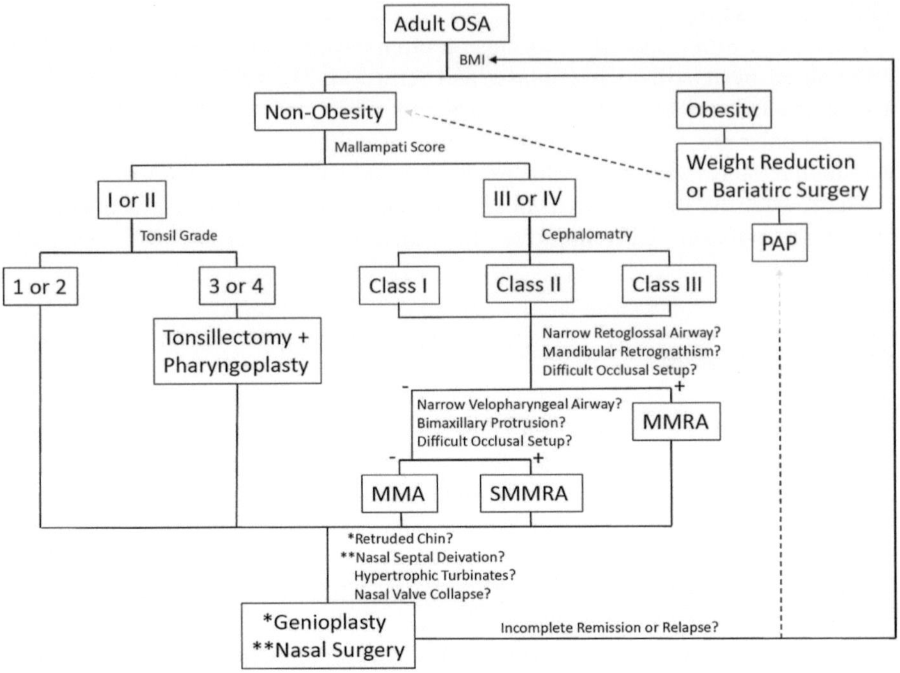

Fig. 17.4 Algorithm of surgical considerations for OSA. (Adopted from Lin et al. Sleep Med Clin 14 (2019) 83–89)

References

1. Prinsell JR. Primary and secondary telegnathic maxillomandibular advancement, with or without adjunctive procedures, for obstructive sleep apnea in adults: a literature review and treatment recommendations. J Oral Maxillofac Surg. 2012;70(7):1659–77.
2. Zagi S, Holty JE, Certal V, Abdullatif J, Guilleminault C, Powee NB, Riley RW, Camacho M. Maxillomandibular advancement for treatment of obstructive sleep apnea: a meta-analysis. JAMA Otolanryngol Head Neck Surg. 2016;142(1):58–66.
3. Holty JE, Guilleminault C. Maxillomandibular advancement for the treatment of obstructive sleep apnea: a systemic review and meta-analysis. Sleep Med Rev. 2010;14(5):287–97. Review.
4. Vigneron A, Tamisier R, Orset E, Pepin JL, Bettega G. Maxillomandibular advancement for obstructive sleep apnea syndrome treatment: long-term results. J Craniomaxillofac Surg. 2017;45(2):183–91.
5. Partinen M, Guilleminault C, Quera-Salva MA, Jamieson A. Obstructive sleep apnea and cephalometric roentgenograms. The role of anatomic upper airway abnormalities in the definition of abnormal breathing during sleep. Chest. 1988;93(6):1199–205.
6. Friedman M, Ibrahim H, Bass L. Clinical staging for sleep-disordered breathing. Otolaryngol Head Neck Surg. 2002;127(1):13–21.
7. Lin C CH, Wang PF, Loh S RH, Lau HT, Hsu S SP. Maxillomandibular rotational advancement: airway, aesthetics, and Angle's considerations. Sleep Med Clin. 2019;14(1):83–9.

8. Fairburn SC, Waite PD, Vilos G, Harding SM, Bernreuter W, Cure J, Charala S. Three-dimensional changes in upper airways of patients with obstructive sleep apnea following maxillomandibular advancement. J Oral Maxillofac Surg. 2007;65(1):6–12.
9. Mehra P, Downie M, Pita MC, Wolford LM. Pharyngeal airway space changes after counterclockwise rotation of the maxillomandibular complex. Am J Orthod Dentofacial Orthop. 2001;120(2):154–9.
10. Lin CH, Liao YF, Chen NH, Lo LJ, Chen YR. Three-dimensional computed tomography in obstructive sleep apneics treated by maxillomandibular advancement. Laryngoscope. 2011;121(6):1336–47.
11. Li KK, Riley RW, Powell NB, Guilleminault C. Patient's perception of the facial appearance after maxillomandibular advancement for obstructive sleep apnea syndrome. J Oral Maxillofac Surg. 2001;59(4):377–80; discussion 380-1.
12. Li KK, Powell NB, Kushida C, Riley RW, Adornato B, Guilleminault C. A comparison of Asian and white patients with obstructive sleep apnea syndrome. Laryngoscope. 1999;109(12):1937–40.
13. Chu YM, Chen P-HR, Morris DE, Ko W-CW, Chen YR. Surgical approach to the patient with bimaxillary protrusion. Clin Plast Surg. 2007;34(3):535–46.
14. Goh YH, Lim KA. Modified maxillomandibular advancement for the treatment of obstructive sleep apnea: a preliminary report. Laryngoscope. 2003;113:1577–82.
15. See A, Loh RHS, Tan DA, Leong ZH, Goh YH, Wong YYC, Toh ST. The long-term results of modified maxillomandibular advancement in Asian OSA patients. Am J Otolaryngol. 2023;44(1):103677.
16. Singhal D, Hsu SS, Lin CH, Chen YC, Chen YR. Trapezoid mortised genioplasty: a further refinement of mortised genioplasty. Laryngoscope. 2013;123(10):2578–82.
17. Li K, Quo S, Guilleminault C. Endoscopically-assisted surgical expansion (EASE) for the treatment of obstructive sleep apnea. Sleep Med. 2019;60:53–9.

Transoral Robotic Surgery (TORS) for Obstructive Sleep Apnea

18

Hsin-Ching Lin, John Ming Ren Loh, and Michael Friedman

Contents

H.-C. Lin (✉)
Department of Otolaryngology, Kaohsiung Chang Gung Memorial Hospital, Kaohsiung, Taiwan

Department of Otolaryngology, Kaohsiung Municipal Ta-Tung Hospital, Kaohsiung, Taiwan

College of Medicine, Chang Gung University, Taoyuan and Kaohsiung, Taiwan

Sleep Center and Robotic Surgery Center, Kaohsiung Chang Gung Memorial Hospital, Kaohsiung, Taiwan

Department of Business Management, Institute of Healthcare Management, Institute of Biomedical Sciences, National Sun Yat-sen University, Kaohsiung, Taiwan
e-mail: hclin@adm.cgmh.org.tw

J. M. R. Loh
Department of Otorhinolaryngology, Head & Neck Surgery, Changi General Hospital, Simei, Singapore

Department of Sleep Medicine, Surgery & Science, Changi General Hospital, Simei, Singapore

M. Friedman
Department of Otolaryngology, Division of Sleep Surgery, Rush University Medical Center, Chicago, IL, USA

Department of Otolaryngology, Advanced Center for Specialty Care, Advocate Illinois Masonic Medical Center, Chicago, IL, USA

H.-Y. Li (ed.), *Snoring and Obstructive Sleep Apnea*,
https://doi.org/10.1007/978-981-95-2648-2_18

Abstract

Obstructive sleep apnea/hypopnea syndrome (OSA) is characterized by repetitive obstruction of the upper airway, often resulting in intermittent oxygen desaturation and arousal during sleep. Untreated OSA is strongly associated with the increase of cardiovascular and cerebrovascular disorders. Continuous positive airway pressure (CPAP) is the primary treatment of obstructive sleep apnea/hypopnea syndrome (OSA). Clinically, many OSA patients cannot or will not use CPAP or fail their conservative therapies.

As experts of upper airway diseases, we often examine the airway of patients with OSA clearly and help OSA patients understand the importance of assessment and treatment for OSA. Surgery for OSA is not a substitute for CPAP, but is an alternative treatment for those who failed conservative OSA therapies. Transoral robotic surgery (TORS) is a relatively innovated technique and a valid surgical modality with minor postoperative morbidities for selected OSA patients.

In this chapter, we provide an updated overview of the role and evidence of TORS for the treatment of OSA.

Keywords

Snoring · Obstructive sleep apnea · Transoral robotic surgery · Tongue base reduction/resection

18.1 Introduction

In sleep surgery literature, obstruction below the levels of the tonsils and palate is historically referred to as "hypopharyngeal obstruction." Fujita reported that unaddressed hypopharyngeal obstruction could be a common cause of uvulopalatopharyngoplasty (UPPP) failure [1]. From an anatomical perspective, the tongue base is within the oropharynx and the epiglottis is part of the larynx. We prefer the term tongue base or retroglossal obstruction and classify epiglottic collapse separately rather than purely using the term "hypopharyngeal obstruction." Tongue base obstruction occurs in more than 50% of sleep apnea patients [2], and plays a key role in obstructive sleep apnea (OSA).

The development of tongue base procedures for OSA has progressed over the years through an interplay of innovative surgical techniques and new technology. Fujita in 1991 described midline glossectomy and linguoplasty using carbon dioxide laser for tissue vaporization [3]. Riley and Powell described their 2-stage protocol in 1993: Phase 1 consisted of mandibular osteotomy with genioglossus advancement—hyoid myotomy and suspension, followed by a phase 2 maxillary-mandibular advancement in the event of surgical failure [4]. In 1999, Powell et al. subsequently pioneered the use of temperature-controlled radiofrequency, which is designed to maintain the tissue temperature at a therapeutic level for the treatment site without causing rapid tissue desiccation or destruction for tissue volume reduction of the tongue base as a minimally invasive approach [5]. In the same year, Chabolle described a transcervical approach for tongue base reduction combined with hyoepiglottoplasty [6].

Transoral robotic surgery (TORS) was first adopted in the head and neck region by Weinstein and O'Malley for tongue base neoplasms in 2004 [7]. The use of the da Vinci robotic surgical system (Intuitive Surgical, Sunnyvale, CA) allowed for excellent 3-dimensional stereoscopic visualization. Angled scopes together with articulated surgical arms allowed for visualization and instrument access around the curve of the tongue base.

In 2009, Vicini et al. pioneered the use of TORS tongue base resection for patients with OSA. A prophylactic tracheostomy was routinely performed in all cases [8]. Subsequently, Friedman reported 27 cases of TORS partial glossectomy that were performed safely without a tracheostomy [9].

The daVinci S and Si models gained the United States Food and Drug Administration (FDA) approval for transoral resection of early-stage benign or malignant tumors in 2009, and subsequently for benign base of tongue resection in 2014 [10].

18.2 Clinical Indications and Patient Selection

We define TORS-OSA surgery as a combination of posterior midline glossectomy and lingual tonsillectomy, with or without epiglottidectomy. In patients without prior pharyngeal surgery, it is usually done in conjunction with multilevel soft tissue surgery of the palate and/or nose.

Some authors may restrict TORS-TBR for patients with moderate-to-severe OSA / apnea–hypopnea index (AHI) $\geq$ 15 [11]. However, we would still consider patients who have mild OSA if they have significant symptoms and/or have moderate-to-severe OSA during REM sleep as well as have prominently lingual tonsil hypertrophy (Friedman Type 3 or 4) [12], epiglottic collapse, and/or obstructive muscular bulk at the base of tongue. Nonlymphoid tongue base obstruction can be determined clinically through transoral and nasoendoscopic examination with Friedman tongue position classification [13, 14]. Drug-induced sleep endoscopy (DISE) may be used to guide treatment planning, selection, and alternatives. Patients with lateral wall collapse of the oropharynx may respond less well to TORS; however, DISE has not been shown to improve overall TORS outcome [15–18].

Some exceptions to these can been made if patients have exhausted all other treatment options, and the surgeon has the appropriate expertise. We initially proposed a minimum of 2-cm inter-incisor distance for mouth opening [19], but with experience, we now have been able to operate on patients with smaller mouth openings.

Before TORS-OSA surgery, the patient must be counseled on alternative options to tongue base reduction, which include endoscopic coblation open tongue base resection (ECO-TBR) [20], radiofrequency reduction base of tongue reduction (RF-BOT) [21], and its variants [22], and/or upper airway stimulation (UAS).

Intraoperative ultrasound may be used to evaluate the course of the lingual artery [23], but we find that unnecessary once the surgeon is familiar with the anatomical limits of resection.

The extent of resection is thus kept posterior to the foramen cecum, and within 1.5–2 cm from the midline, and to a maximum intramuscular depth of around 1.5–2 cm. We leave at least 5 mm of tissue laterally from the glossopharyngeal fold and lateral pharyngeal wall to avoid the lingual artery, glossopharyngeal nerve, and to avoid communication with the tonsillar fossa, which may cause circumferential scarring. The exact volume of tissue excised does not significantly influence clinical outcomes [24, 25]; thus, it is more important to tailor the resection margins to each patient's anatomy rather than to aim for a specific volume.

18.3 Procedure and Tips

We usually describe the technique for TORS tongue base excision for OSA based on the da Vinci Xi platform. TORS for tongue base and epiglottis resection in OSA patients can be performed with the da Vinci Xi or Si system. The technique with Xi platform is similar to what we previously described for the Si system [19]. Advantages of the Xi system include easier patient cart docking and the availability of bipolar cautery with the Maryland forceps. One potential downside is that the smallest Xi endowrist instruments are 8 mm compared to the 5-mm instruments available for the Si. In our experience, this does not hinder transoral surgery.

The patient is positioned with neck extension and slight flexion of the atlanto-occipital joint. We prefer intubation with a south Ring–Adair–Elwyn (RAE) endotracheal tube. In our experience, prophylactic tracheostomy is not required. In patients who require multilevel surgery, our preferred sequence is to perform nasal surgery first, followed by tonsillectomy, modified uvulopalatoplasty, and then TORS is performed last.

Posterior midline glossectomy is started with an initial vertical midline incision that is made with monopolar cautery, followed by the superior incision perpendicular to the midline, forming a T shape. The lateral limits of the dissection are measured with a ruler and marked with cautery 1.5–2 cm from the midline. Sequential dissection is performed from a superior–inferior and lateral-to-medial direction. With experience, we now resect the entire specimen enbloc rather than in two halves.

After resection of the tongue base, the epiglottis comes into view. A partial epiglottidectomy is performed, if indicated, by excising the superior third with a horizontal incision (Figs. 18.1 and 18.2). The robot is then undocked and the gag is removed.

In the last decade, we added an additional step to check for hemostasis again after removing the robot and gag. The pharynx is re-examined with the Jennings gag, which does not have tongue blade, and a 70-degree endoscope. Any bleeding points are coagulated with the coblator wand.

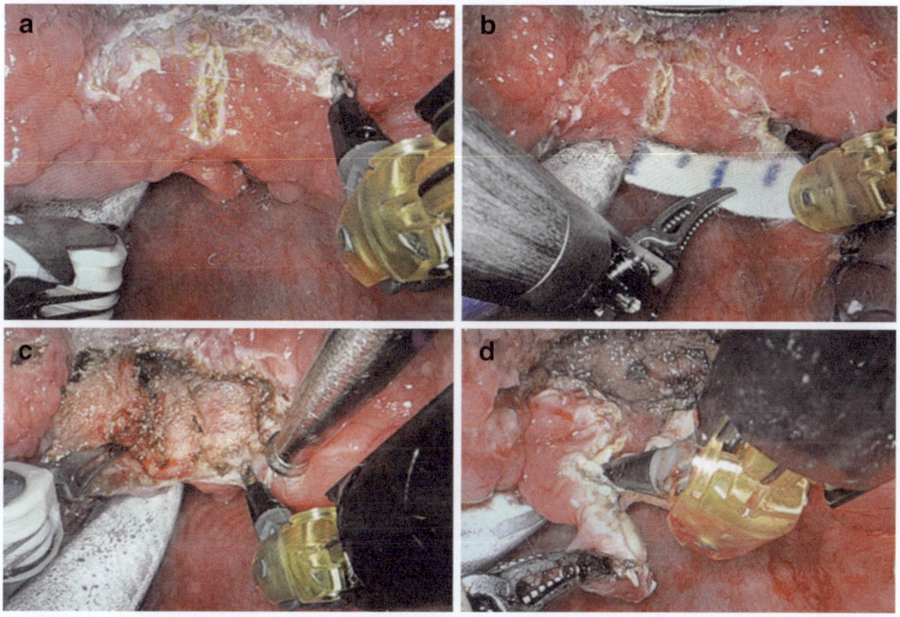

Fig. 18.1 Operative steps. (**a**) Initial T-shaped incision. (**b**) Lateral extent of resection measured around 1.5 cm from midline. (**c**) Sequential dissection from lateral to medial with suction providing counter-traction. (**d**) Superior third of epiglottis excised

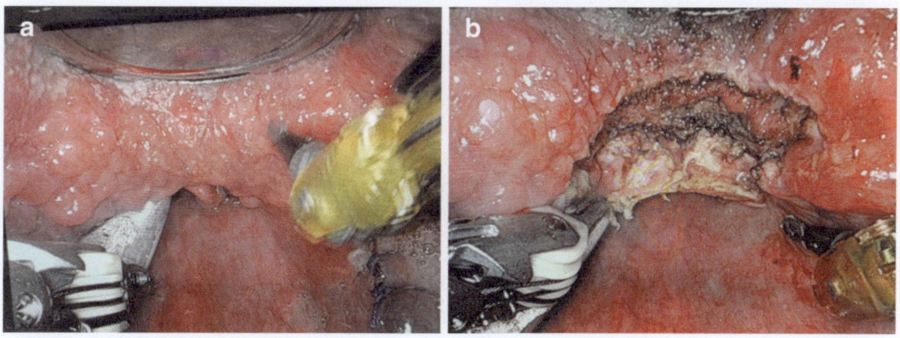

Fig. 18.2 Before (**a**) and after (**b**) TORS-OSA surgery

18.4 Postoperative Care

Patients are extubated at the end of the surgery when the patient is completely awake in the operating room. In our institution, patients are sent to the general ward with continuous oxygen saturation monitoring and sleep in a 30–45 degree head up position overnight. In some centers where nursing care and monitoring are limited, we suggest that high dependency or intensive care monitoring may be prudent.

The patient is prescribed oral cephalexin, paracetamol, and topical benzydamine spray. Intravenous dexamethasone is given to reduce airway edema. Opioids are avoided to minimize the risk of respiratory depression. We allow the patient to eat and drink as tolerated, with a patient-led escalation from fluids to diet. Intravenous fluids are given in the first two days and weaned off as oral intake improves. Patients can be discharged home on postoperative day 2–5, depending on their pain score and oral intake.

18.5 Complications

TORS has a similar complication rate to that of other soft tissue surgery for OSA [25]. Minor short-term complications include dehydration and pain, which can be mitigated with careful postoperative care.

The most feared acute complications for TORS surgery are airway obstruction, bleeding, and death. Emergency airway equipment, tracheostomy set, and tubes are prepared on stand-by in the operating room. Patients undergoing TORS are consented for potential emergency tracheostomy, but this has not been required in our institution thus far.

Prompt identification and management of bleeding in TORS for OSA must be done to prevent asphyxiation. In severe bleeding, the airway must be secured by intubation or tracheostomy.

The reported incidence of postoperative bleeding for TORS OSA cases was an average of 3.3% [25]. In our experience of over 656 cases (unpublished data), none of our patients have required hemostasis again under general anesthesia in the operating room after TORS-OSA surgery.

Functional complications of taste (13.8%) and swallowing (6%) after tongue base surgery tend to improve within the first 3 months, but can persist in a small number of patients [26–28]. Patients must be made aware of these risks.

18.6 Surgical Outcomes

Recent meta-analyses showed a surgical success rate of 68–69% in TORS-OSA surgery, reduction of AHI by 24.25 events/hour, reduction of Epworth sleepiness scale by 7.92 and improvement in lowest oxygen saturation by 6.04% for TORS tongue base reduction when done alone or as part of multilevel surgery [29, 30].

In addition to traditional metrics of OSA surgical success, multilevel surgery with TORS has also been shown to reduce homocysteine levels [31], and overactive bladder (OAB) and lower urinary tract symptoms (LUTS) in men [32].

18.7 Conclusion

TORS for tongue base and/or epiglottic resection in OSA patients presents a viable, effective option for the surgical management of OSA. Thorough preoperative planning, meticulous surgical technique, and vigilant postoperative care are essential to minimize adverse outcomes and maximize patient benefits.

Disclosures

Competing Interests **Dr. Hsin-Ching Lin, Dr. Michael Friedman, and Dr. John Ming Ren Loh** declared no potential conflict of interest.

References

1. Fujita S. Obstructive sleep apnea syndrome: pathophysiology, upper airway evaluation and surgical treatment. Ear Nose Throat J. 1993;72(1):67–76. https://doi.org/10.1177/014556139307200115.
2. Lee EJ, Cho JH. Meta-analysis of obstruction site observed with drug-induced sleep endoscopy in patients with obstructive sleep apnea. Laryngoscope. 2019;129(5):1234–43. https://doi.org/10.1002/lary.27320.
3. Fujita S, Woodson BT, Clark JL, Wittig R. Laser midline glossectomy as a treatment for obstructive sleep apnea. Laryngoscope. 1991;101(8):805–9. https://doi.org/10.1288/00005537-199108000-00001.
4. Riley RW, Powell NB, Guilleminault C. Obstructive sleep apnea syndrome: a surgical protocol for dynamic upper airway reconstruction. J Oral Maxillofac Surg. 1993;51(7):742–7. https://doi.org/10.1016/S0278-2391(10)80412-4.
5. Powell NB, Riley RW, Guilleminault C. Radiofrequency tongue base reduction in sleep-disordered breathing: a pilot study. Otolaryngol Head Neck Surg. 1999;120(5):656–64. https://doi.org/10.1053/hn.1999.v120.a96956.
6. Chabolle F, Wagner I, Blumen MB, Séquert C, Fleury B, De Dieuleveult T. Tongue base reduction with hyoepiglottoplasty: a treatment for severe obstructive sleep apnea. Laryngoscope. 1999;109(8):1273–80. https://doi.org/10.1097/00005537-199908000-00017.
7. O'Malley BW, Weinstein GS, Snyder W, Hockstein NG. Transoral robotic surgery (TORS) for base of tongue neoplasms. Laryngoscope. 2006;116(8):1465–72. https://doi.org/10.1097/01.mlg.0000227184.90514.1a.
8. Vicini C, Dallan I, Canzi P, Frassineti S, La Pietra MG, Montevecchi F. Transoral robotic tongue base resection in obstructive sleep apnoea-hypopnoea syndrome: a preliminary report. ORL. 2010;72(1):22–7. https://doi.org/10.1159/000284352.
9. Friedman M, Hamilton C, Samuelson CG, et al. Transoral robotic glossectomy for the treatment of obstructive sleep apnea-hypopnea syndrome. Otolaryngol Head Neck Surg. 2012;146(5):854–62. https://doi.org/10.1177/0194599811434262.
10. Intuitive Surgical. FDA clears da Vinci(R) system use in benign base of tongue resection procedures. September 18, 2014. Accessed November 24, 2024. https://isrg.intuitive.com/news-releases/news-release-details/fda-clears-da-vincir-system-use-benign-base-tongue-resection
11. Baptista PM, Zufiaurre ND, Garaycochea O, et al. TORS as part of multilevel surgery in OSA: the importance of careful patient selection and outcomes. J Clin Med. 2022;11(4):990. https://doi.org/10.3390/jcm11040990.
12. Friedman M, Yalamanchali S, Gorelick G, Joseph NJ, Hwang MS. A standardized lingual tonsil grading system: Interexaminer agreement. Otolaryngol Head Neck Surg. 2015;152(4):667–72. https://doi.org/10.1177/0194599815568970.
13. Robinson S, Krishnan S, Hodge JC, Foreman A. Conventional tongue base volumetric reduction for obstructive sleep apnea. Oper Tech Otolaryngol Head Neck Surg. 2012;23(1):36–44. https://doi.org/10.1016/j.otot.2011.10.002.
14. Friedman M, Salapatas AM, Bonzelaar LB. Updated Friedman staging system for obstructive sleep apnea. Adv Otorhinolaryngol. 2017;80:41–8. https://doi.org/10.1159/000470859.
15. Lin HS, Rowley JA, Folbe AJ, Yoo GH, Badr MS, Chen W. Transoral robotic surgery for treatment of obstructive sleep apnea: factors predicting surgical response. Laryngoscope. 2015;125(4):1013–20. https://doi.org/10.1002/lary.24970.

16. Meraj TS, Muenz DG, Glazer TA, Harvey RS, Spector ME, Hoff PT. Does drug-induced sleep endoscopy predict surgical success in transoral robotic multilevel surgery in obstructive sleep apnea? Laryngoscope. 2017;127(4):971–6. https://doi.org/10.1002/lary.26255.
17. Hoff PT, Glazer TA, Spector ME. Body mass index predicts success in patients undergoing transoral robotic surgery for obstructive sleep apnea. ORL. 2014;76(5):266–72. https://doi.org/10.1159/000368415.
18. Garas G, Kythreotou A, Georgalas C, et al. Is transoral robotic surgery a safe and effective multilevel treatment for obstructive sleep apnoea in obese patients following failure of conventional treatment(s)? Ann Med Surg. 2017:19. https://doi.org/10.1016/j.amsu.2017.06.014.
19. Lin HC, Friedman M. Transoral robotic OSA surgery. Auris Nasus Larynx. 2021;48(3):339–46. https://doi.org/10.1016/j.anl.2020.08.025.
20. Lin HC, Friedman M, Chang HW, Yalamanchali S. Z-palatopharyngoplasty combined with endoscopic coblator open tongue base resection for severe obstructive sleep apnea/hypopnea syndrome. Otolaryngol Head Neck Surg. 2014;150(6):1078–85. https://doi.org/10.1177/0194599814524722.
21. Stuck BA, Köpke J, Hörmann K, et al. Volumetric tissue reduction in radiofrequency surgery of the tongue base. Otolaryngol Head Neck Surg. 2005;132(1):132–5. https://doi.org/10.1016/j.otohns.2004.09.009.
22. Lee JA, Byun YJ, Nguyen SA, Lentsch EJ, Gillespie MB. Transoral robotic surgery versus plasma ablation for tongue base reduction in obstructive sleep apnea: meta-analysis. Otolaryngol Head Neck Surg. 2020;162(6):839–52. https://doi.org/10.1177/0194599820913533.
23. Chang CC, Wu JL, Hsiao JR, Lin CY. Real-time, intraoperative, ultrasound-assisted transoral robotic surgery for obstructive sleep apnea. Laryngoscope. 2021;131(4):1383–90. https://doi.org/10.1002/lary.29135.
24. Cammaroto G, Meccariello G, Costantini M, et al. Trans-oral robotic tongue reduction for OSA: does lingual anatomy influence the surgical outcome? J Clin Sleep Med. 2018;14(8):1347–51. https://doi.org/10.5664/jcsm.7270.
25. Tsou YA, Chang WD. Comparison of transoral robotic surgery with other surgeries for obstructive sleep apnea. Sci Rep. 2020;10(1):18163. https://doi.org/10.1038/s41598-020-75215-1.
26. Paker M, Duek I, Awwad F, et al. Long-term swallowing performance following transoral robotic surgery for obstructive sleep apnea. Laryngoscope. 2019;129(2):422–8. https://doi.org/10.1002/lary.27364.
27. Eesa M, Montevecchi F, Hendawy E, D'Agostino G, Meccariello G, Vicini C. Swallowing outcome after TORS for sleep apnea: short- and long-term evaluation. Eur Arch Otorrinolaringol. 2015;272(6):1537–41. https://doi.org/10.1007/s00405-014-3480-x.
28. Lin HC, Hwang MS, Liao CC, Friedman M. Taste disturbance following tongue base resection for OSA. Laryngoscope. 2016;126(4):1537–41. https://doi.org/10.1002/lary.25580.
29. Lechien JR, Chiesa-Estomba CM, Fakhry N, et al. Surgical, clinical, and functional outcomes of transoral robotic surgery used in sleep surgery for obstructive sleep apnea syndrome: a systematic review and meta-analysis. Head Neck. 2021;43(7):2216–39. https://doi.org/10.1002/hed.26702.
30. Miller SC, Nguyen SA, Ong AA, Gillespie MB. Transoral robotic base of tongue reduction for obstructive sleep apnea: a systematic review and meta-analysis. Laryngoscope. 2017;127(1):258–65. https://doi.org/10.1002/lary.26060.
31. Chiu LW, Lin CW, Lin PW, et al. Homocysteine levels in severe OSA patients before and after TORS-OSA surgery. Otolaryngol Head Neck Surg. 2023;168(5):1238–44. https://doi.org/10.1002/ohn.218.
32. Chuang YC, Lin PW, Lin HC, et al. Effects of TORS-OSA surgery on lower urinary tract symptoms, overactive bladder symptoms, and nocturia in male patients with obstructive sleep apnea/hypopnea syndrome. Nat Sci Sleep. 2022;14:546–56. https://doi.org/10.2147/NSS.S349807.

Hypoglossal Nerve Stimulation Implant 19

S. T. Toh, A. C. W. Ng, and T. J. Fang

Contents

Abstract

Hypoglossal nerve neurostimulation implant is a surgically implanted device that stimulates protrusor branches of the hypoglossal nerve to effect upper airway opening during sleep. It has emerged as a treatment for patients with moderate-to-severe obstructive sleep apnea (OSA) who are intolerant to continuous posi-

S. T. Toh (✉) · A. C. W. Ng
Singapore General Hospital, SingHealth Duke – NUS Sleep Centre, Department of Otorhinolaryngology – Head and Neck Surgery, Singapore, Singapore

T. J. Fang
Linkou Chang Gung Memorial Hospital, Department of Otorhinolaryngology – Head and Neck Surgery, Taoyuan City, Taiwan

H.-Y. Li (ed.), *Snoring and Obstructive Sleep Apnea*,
https://doi.org/10.1007/978-981-95-2648-2_19

tive airway pressure therapy. This chapter reviews the current evidence on indications, procedure, safety, efficacy, and outcomes associated with HGNS implant in the treatment of OSA.

Keywords

Obstructive Sleep Apnea · Hypoglossal Nerve Neurostimulation · Upper Airway neurostimulation · Selective Upper Airway neurostimulation

19.1 Introduction

An estimated 425 million adults aged 30–69 years globally suffer from moderate-to-severe obstructive sleep apnea (OSA), with estimates reaching as high as 30% in some countries [1]. Traditional treatments for OSA, such as continuous positive airway pressure (CPAP), have been highly effective but are met with poor compliance [2]. As a result, there has been growing interest in alternative therapeutic approaches, such as hypoglossal nerve neurostimulation (HGNS). Instead of altering upper airway anatomy, the HGNS implant stimulates protrusor muscles innervated by the hypoglossal nerve (CN-XII) to effect upper airway opening and prevent upper airway collapse from loss of muscle tone during sleep. It has emerged as an alternative treatment for patients with moderate-to-severe OSA who are intolerant of CPAP, and a contender as the treatment of choice for some.

19.2 Indications

Since April 2014, the Inspire HGNS implant (Inspire Medical Systems, Maple Grove, MN) has been approved by the United States Food and Drug Administration (FDA). This was largely due to promising patient outcomes and safety profile demonstrated by the STAR trial [3]. Initially, FDA guidelines limited HGNS to patients with a maximum apnea–hypopnea index (AHI) of 65 events/hour and body mass index (BMI) 32 kg/m^2. However, recent studies demonstrated that higher AHI [4, 5] and BMI [4, 6] did not diminish HGNS therapy benefits. Equivalent improvements in AHI and Epworth sleepiness scale (ESS) were seen in patients with AHI 65–100 and BMI 32–40, with no significant difference when compared to groups with lower AHI or BMI.

Currently, this procedure is indicated for adults ˃18 years of age with:

- Moderate-to-severe OSA (AHI 15–100 events/hour).
- CPAP failure or intolerance.
  - CPAP failure is defined as the inability to eliminate OSA with CPAP, with residual AHI >15 events/hour.
  - CPAP intolerance is defined as poor compliance (<5 nights/week or <4 h/night) or unwillingness.

- BMI <40 kg/m^2.
- Inpatient polysomnogram <25% mixed and central apneas.
- Drug-induced sleep endoscopy (DISE) shows no soft palate complete concentric collapse.

Since March 2023, Inspire HGNS also received FDA approval for pediatric patients aged 13–18 years with Down syndrome, and:

- Severe OSA (AHI 10–50 events/hour).
- Unable to use or get consistent benefit from CPAP.
- Not effectively treated by adenotonsillectomy.
- Drug-induced sleep endoscopy shows no soft palate complete concentric collapse.

19.3 Surgical Anatomy and Mechanism of Action

The surgical anatomy of the distal CN-XII is complex. A clear understanding is paramount for a proper identification of the functional breakpoint between inclusion and exclusion branches for accurate stimulation cuff placement, as this ultimately affects the tongue movement and surgical success.

CN-XII innervates 3 main muscle groups of the tongue—extrinsic retractors (styloglossus and hyoglossus), extrinsic protrusor (genioglossus), and intrinsic stiffeners (transverse and vertical intrinsic muscles) [7–9]. The first cervical spinal nerve (C1) innervates the geniohyoid muscle, which causes anterosuperior hyoid displacement and acts as a dynamic hyoid suspension. This C1 branch and CN-XII branches to the extrinsic protrusor and intrinsic stiffeners need to be included, whereas those to the extrinsic retractors should be excluded, to achieve an unhindered protrusion of a stiffened tongue when stimulated (Fig. 19.1) [9]. This results in an increase in retroglossal and velopharyngeal (via palatoglossus coupling) airway dimensions [10–13]. Additionally, C1 stimulation opens the hypopharyngeal airway and causes anterior displacement of the epiglottis [9].

The anterior border of the hyoglossus muscle is a useful landmark as most exclusion CN-XII branches are found posterior to this landmark. Inclusion CN-XII branches extend anterior to this landmark and lie superficial on the hyoglossus muscle. However, it is important to note that a late hyoglossus branch which branches off the main CN-XII distal to the anterior border of the hyoglossus may sometimes be present and needs to be identified and excluded for optimal tongue movement and surgical outcomes. This branch can be confirmed using a bipolar nerve integrity monitoring (NIM) probe, which shows mixed activation signal. Mistakenly including this branch in the cuff leads to poor tongue protrusion or even retraction.

The C1 branch usually runs inferior to the CN-XII main trunk, parallel to the overlying Ranine vein. In some cases, C1 takes off very proximally at an acute angle from the CN-XII, making it difficult to include in the cuff. When the C1 branch is not included, the epiglottis may be floppier and hypopharynx more collapsible [9].

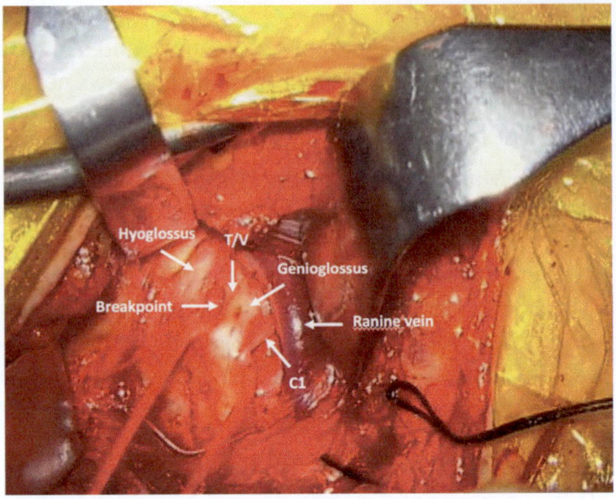

Fig. 19.1 Neuroanatomy of the distal hypoglossal nerve. Red vessel loop placed around inclusion branches. T/V: Branches supplying transverse and vertical intrinsic muscles

The sensing lead detects respiratory effort, which directs the implantable pulse generator (IPG) to deliver a mild electrical stimulus to the inclusion muscles at the beginning of inspiration to maintain upper airway patency.

19.4 Procedures and Tips

19.4.1 Placement of NIM Electrodes

Single-use 18-mm Prass Paired Electromyography Electrodes (Medtronic Xomed, Ref 82227304) are used for surgical nerve monitoring. The inclusion channel (blue electrode) is placed lateral to the frenulum of the tongue just posterior to the lingual surface of the mandible, deep into the floor of mouth targeting the genioglossus muscle. The exclusion channel (red electrode) is placed 4 cm from the tongue tip, just inferior to the dorsum edge of the tongue superficially to target the hyoglossus (Fig. 19.2).

19.4.2 Neck Incision

The HGNS implant is usually performed on the right side. The neck incision is placed midway between the hyoid and mandible, 1 cm lateral to midline, 3–4 cm in length (Fig. 19.3). The incision is deepened past the platysma muscle, until the anterior belly of the digastric is encountered. The digastric tendon is identified and retracted downwards. The submandibular gland may be encountered postero-superiorly and should be retracted upwards. The posterior border of the mylohyoid muscle is identified and retracted anteriorly to expose the CN-XII main branch. Surgical loupes or microscope can be used to carefully trace the CN-XII

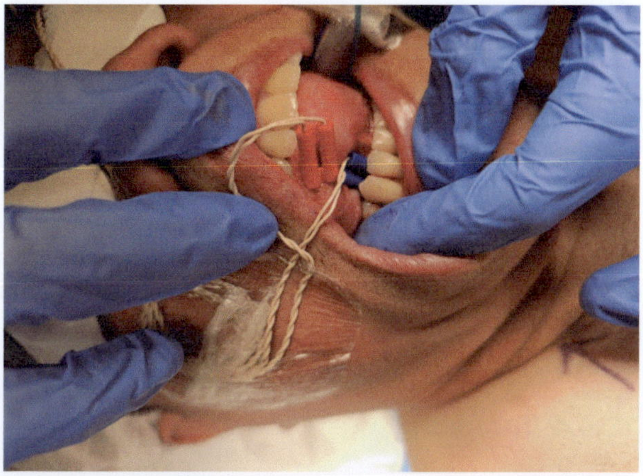

Fig. 19.2 Nerve integrity monitoring electrode placement

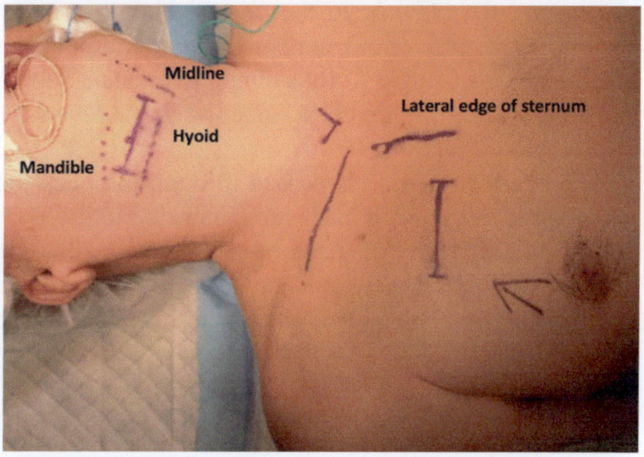

Fig. 19.3 Neck and chest incision markings

anteriorly to identify the functional break point. The C1 branch should be identified and included whenever possible. Any overlying Ranine vein can be mobilized and retracted inferiorly, or ligated if necessary. NIM testing (NIM 3.0, Medtronic Xomed, Jacksonville, Florida, USA) combined with familiar understanding of neuroanatomy helps to accurately identify inclusion and exclusion branches. It is important to ensure that a late hyoglossus branch is identified and excluded. A vessel loop is placed around all inclusion branches and subsequently replaced by the stimulation cuff. The stimulation lead is then anchored to the digastric tendon with silk sutures.

19.4.3 Chest Incision

The chest incision is placed midway between the second and third ribs, 3 cm lateral to the lateral edge of the sternum, 4–5 cm in length (Fig. 19.3). The incision is deepened until the pectoralis fascia is encountered. A soft tissue pocket (3-finger breadth width) is created inferiorly for subsequent IPG placement. The pectoralis major and minor muscles are bluntly dissected with retractors until a fat pad is encountered. This fat pad is carefully dissected to expose the underlying external and internal intercostal muscles (fibers forming a "V" outline). A small pocket extending laterally from the medial edge of the external intercostal muscle is carefully dissected for sensing lead placement. There should not be any resistance as the sensing lead is placed in this pocket between the external and internal intercostal muscles. The sensing lead is then secured to the medial external intercostal muscle fibers with silk sutures.

19.4.4 Tunneling

The tunneler is bent 90° and passed from the neck incision in the subplatysmal plane, over the clavicle to the chest incision in the supra-pectoralis plane. Care should be taken to avoid puncturing the external jugular vein. Tunneling can be made easier by using a Kelly's clamp to create a superior tunnel from the neck and inferior tunnel from the chest, meeting at the level of the clavicle before inserting the tunneler device.

The stimulation lead is passed through the neck tunnel to the IPG pocket. The stimulation and sensing leads are connected to the IPG, which is placed in the IPG pocket and secured to the pectoralis fascia with silk sutures. It is important to ensure that the stimulation and sensing wires are placed underneath the IPG to prevent these wires from being accidentally cut in future during IPG change.

19.4.5 Test Device

The device is tested at various amplitudes and configurations to ensure good tongue protrusion and absence of retraction throughout. Proper sensing lead placement is confirmed with good respiratory waveforms (Fig. 19.4). If the sensing lead is placed too near a rib, the respiratory waveform will have a sawtooth appearance and the sensing lead needs to be adjusted. The incisions are closed in layers with pressure dressings placed over.

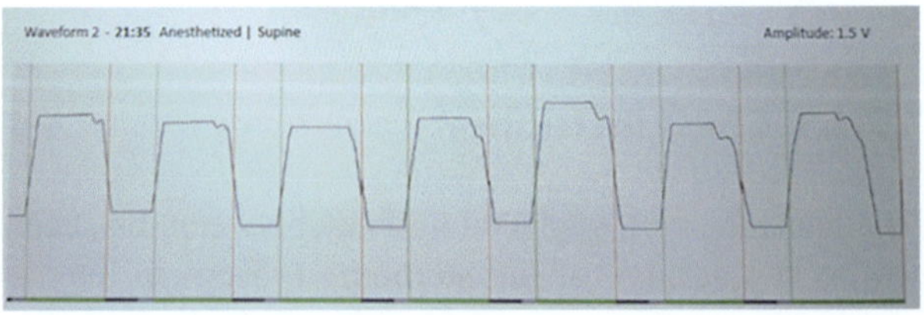

Fig. 19.4 Good respiratory waveforms. Orange line indicates stimulation off, and green line indicates stimulation on at the beginning of inspiration

19.4.6 Special Considerations

Special considerations which may affect surgery or follow-up care include the following:

- Previous neck or thorax surgery or irradiation
- Previous mastectomy, breast irradiation, or augmentation
- Frequent magnetic resonance imaging (MRI)
- Other implanted devices (pacemaker, defibrillator, spinal cord, or deep brain stimulator)
- Contact sports

19.5 Postprocedural Care

Postoperative X-rays (lateral neck and chest anteroposterior) are taken to visualize IPG, stimulation and sensing leads. The chest X-ray also aids in excluding pneumothorax. The patient can be discharged on the same day and advised to avoid strenuous activities and heavy lifting for 4–6 weeks.

Skin sutures are removed 1 week postoperatively and the device is activated 1 month postoperatively. In patients with CN-XII neuropraxia, device activation is delayed until neuropraxia is fully recovered. Upon activation, the patient is kept at the functional threshold (lowest amplitude whereby the tongue protrudes past the lower incisors) for a week and advised to increase amplitude by 1 level every week until it becomes uncomfortable. At 8–12 weeks postoperatively, the patient is scheduled for a compliance check to review usage and comfort. A fine-tuning HGNS titration inpatient polysomnogram is scheduled 12–16 weeks postoperatively to check efficacy.

IPG change is required every 10–11 years and Inspire HGNS is MRI compatible up to 1.5 Tesla.

19.6 Outcomes

Several clinical trials and observational studies had demonstrated significant objective and subjective improvements in OSA with Inspire HGNS. The stimulation therapy for apnea reduction (STAR) trial was a pivotal prospective multicenter randomized controlled trial with 126 patients, which showed sustained improvements in median AHI from baseline 29.3 events/hour to 9, 6, and 6.2 events/hour at 1-year [3], 3-year, [14] and 5-year [15] follow-ups, respectively. ESS normalized from a baseline of 11 to 6 at 1-year and remained constant at 3 and 5 years. This translated into a 78.8% reduction in AHI, 65.7% reduction in oxygen desaturation index (ODI), and 40.5% reduction in ESS at 5 years [15]. Surgical success as defined by Sher's criteria (ε50% reduction in AHI and absolute AHI <20 events/hour) was 75% at 5 years [15]. Similar results were seen in the German Post Market Study [16] and

ADHERE registry [17]. Consistent use of HGNS reduces diastolic [18] and mean arterial blood pressure (BP), the number of BP medications required [19] and atrial fibrillation burden [20].

Compared to CPAP, HGNS patients had higher nightly device usage (6.5 vs. 5.6 hours) [21] and significantly greater ESS reduction (8 vs. 3.9) [22]. Matched populations comparing HGNS to CPAP showed greater mean disease alleviation (AHI and usage combined) in the HGNS group (59% vs. 55%) [22]. ADHERE registry patient impression questionnaires revealed that 92% of patients felt that Inspire HGNS was better than CPAP and would choose Inspire again, 93% would recommend Inspire to someone, and 90% were satisfied with Inspire [5].

There is a need to show that treatment of OSA with HGNS can reduce long-term medical comorbidities and mortality, compared to no treatment or CPAP treatment.

Compared to traditional upper airway soft tissue surgery, HGNS has been shown to be more effective [23, 24] and safer [24–26]. A meta-analysis conducted in 2021 demonstrated that patients who underwent HGNS had a significantly greater AHI reduction (23.9 vs. 15.5), mean success rates (86% vs. 51%). and cure rates (63% vs. 22%) compared to those who had soft tissue surgery [24].

19.7 Complications and Safety

HGNS has a good safety profile, with low intraoperative (<1%) and postoperative (<2%) serious adverse events [6, 25], such as hematoma, infection, failed implant, pneumothorax, system explant, or revision. Common adverse events are mostly nonserious, such as stimulation-related discomfort, tongue abrasion, and temporary tongue weakness.

HGNS is also safer compared to the traditional upper airway soft tissue surgery. HGNS patients had significantly shorter hospital stays [24, 26], lower readmission rates (4% vs. 15–20%), and less overall adverse events (2% vs. 15–20%) [26, 27].

19.8 Future Directions

In August 2024, Inspire obtained FDA approval for the Next Generation Inspire V System, whereby the respiratory sensing lead is replaced by an accelerometer within the IPG. This potentially reduces surgical time, eliminates the risk of pneumothorax and revision surgeries related to sensing lead issues.

The Genio device (Nyxoah SA, Mont-Saint-Guibert, Belgium) is another HGNS implant that stimulates bilateral protrusor branches of CN-XII. The neurostimulator is implanted through a single submental incision and is activated by an adhesive external activation chip. As it produces bilateral stimulation, it is suitable for patients with complete concentric collapse and a preoperative DISE is not required. It does not require repeated surgeries for battery change and is MRI-compatible up to 3 Tesla. It is currently being evaluated in an FDA-supervised international multicenter prospective single-arm pivotal trial (DREAM study).

The concept of ansa cervicalis stimulation (ACS) was introduced in 2021, whereby the resultant sternothyroid muscle contraction significantly increased retropalatal cross-sectional area by 211% in OSA patients. Supplementing HGNS with ACS significantly led to a 180% increase in retropalatal cross-sectional area compared to HGNS alone, indicating that the addition of ACS further augments HGNS efficacy by reducing pharyngeal collapsibility [28].

19.9 Conclusion

HGNS is efficacious and safe in appropriately selected OSA patients. With emerging new technologies and innovations, upper airway stimulation continues to evolve and holds exciting future prospects for the treatment of OSA.

References

1. Benjafield AV, Ayas NT, Eastwood PR, Heinzer R, Ip MS, Morrell MJ, Nunez CM, Patel SR, Penzel T, Pépin JL, Peppard PE. Estimation of the global prevalence and burden of obstructive sleep apnoea: a literature-based analysis. Lancet Respir Med. 2019;7(8):687–98.
2. Weaver TE, Grunstein RR. Adherence to continuous positive airway pressure therapy: the challenge to effective treatment. Proc Am Thorac Soc. 2008;5(2):173–8.
3. Strollo PJ Jr, Soose RJ, Maurer JT, et al. Upper-airway stimulation for obstructive sleep apnea. N Engl J Med. 2014;370(2):139–49.
4. US Food and Drug Administration. FDA summary of safety and effectiveness data. PMA P130017. 2014.
5. Bosschieter PF, De Vries N, Mehra R, Manchanda S, Padhya TA, Vanderveken OM, Ravesloot MJ, ADHERE Registry Investigators. Similar effect of hypoglossal nerve stimulation for obstructive sleep apnea in 5 disease severity categories. J Clin Sleep Med. 2022;18(6):1657–65.
6. Suurna MV, Steffen A, Boon M, Chio E, Copper M, Patil RD, Green K, Hanson R, Heiser C, Huntley C, Kent D. Impact of body mass index and discomfort on upper airway stimulation: ADHERE registry 2020 update. Laryngoscope. 2021;131(11):2616–24.
7. Sanders I, Mu L. A three-dimensional atlas of human tongue muscles. Anat Rec (Hoboken). 2013;296(7):1102–14.
8. Mu L, Sanders I. Human tongue neuroanatomy: nerve supply and motor endplates. Clin Anat. 2010;23(7):777–91.
9. Heiser C, Knopf A, Hofauer B. Surgical anatomy of the hypoglossal nerve: a new classification system for selective upper airway stimulation. Head Neck. 2017;39(12):2371–80.
10. Bishara H, Odeh M, Schnall RP, Gavriely N, Oliven A. Electrically-activated dilator muscles reduce pharyngeal resistance in anaesthetized dogs with upper airway obstruction. Eur Respir J. 1995;8(9):1537–42.
11. Schwartz AR, Eisele DW, Hari A, Testerman R, Erickson D, Smith PL. Electrical stimulation of the lingual musculature in obstructive sleep apnea. J Appl Physiol (1985). 1996;81(2):643–52.
12. Safiruddin F, Vanderveken OM, de Vries N, et al. Effect of upper-airway stimulation for obstructive sleep apnoea on airway dimensions. Eur Respir J. 2015;45(1):129–38.
13. Heiser C, Edenharter G, Bas M, Wirth M, Hofauer B. Palatoglossus coupling in selective upper airway stimulation. Laryngoscope. 2017;127(10):E378–83.
14. Woodson BT, Soose RJ, Gillespie MB, Strohl KP, Maurer JT, de Vries N, Steward DL, Baskin JZ, Badr MS, Lin HS, Padhya TA. Three-year outcomes of cranial nerve stimulation for obstructive sleep apnea: the STAR trial. Otolaryngol Head Neck Surg. 2016;154(1):181–8.

15. Woodson BT, Strohl KP, Soose RJ, Gillespie MB, Maurer JT, De Vries N, Padhya TA, Badr MS, Lin HS, Vanderveken OM, Mickelson S. Upper airway stimulation for obstructive sleep apnea: 5-year outcomes. Otolaryngol Head Neck Surg. 2018;159(1):194–202.
16. Steffen A, Sommer UJ, Maurer JT, Abrams N, Hofauer B, Heiser C. Long-term follow-up of the German post-market study for upper airway stimulation for obstructive sleep apnea. Sleep Breath. 2020;24:979–84.
17. Thaler E, Schwab R, Maurer J, Soose R, Larsen C, Stevens S, Stevens D, Boon M, Huntley C, Doghramji K, Waters T. Results of the ADHERE upper airway stimulation registry and predictors of therapy efficacy. Laryngoscope. 2020;130(5):1333–8.
18. McKinlay AJ, Walters BK, Aden JK, Scalzitti NJ. Upper airway stimulation therapy effect on blood pressure. Otolaryngol Head Neck Surg. 2023;168(6):1551–6.
19. Freeman CG, Durgham R, Ren E, Conti KR, Yeakel H, Fan T, Brown DM, Ohlstein JF, Keeler JA. The impact of hypoglossal nerve stimulation on secondary health outcomes. Ear Nose Throat J. 2024:01455613241235538.
20. Bakhshaei S, Vu D, Nguyen D, Boone K, Nguyen E, Reddy KT, Ho AT. Spontaneous resolution of atrial fibrillation following hypoglossal nerve stimulation procedure for obstructive sleep apnea, a case report. J Am Coll Cardiol. 2023;81(8_Supplement):3680.
21. Walia HK, Thompson NR, Strohl KP, Faulx MD, Waters T, Kominsky A, Foldvary-Schaefer N, Mehra R. Upper airway stimulation vs positive airway pressure impact on BP and sleepiness symptoms in OSA. Chest. 2020;157(1):173–83.
22. Heiser C, Steffen A, Strollo PJ Jr, Giaie-Miniet C, Vanderveken OM, Hofauer B. Hypoglossal nerve stimulation versus positive airway pressure therapy for obstructive sleep apnea. Sleep Breath. 2023;27(2):693–701.
23. Huntley C, Boon M, Tschopp S, Tschopp K, Jenks CM, Thaler E, Baptista Jardin P, Shah J, Kominsky A, Kezirian EJ, Heiser C. Comparison of traditional upper airway surgery and upper airway stimulation for obstructive sleep apnea. Ann Otol Rhinol Laryngol. 2021;130(4):370–6.
24. Neruntarat C, Wanichakorntrakul P, Khuancharee K, Saengthong P, Tangngekkee M. Upper airway stimulation vs other upper airway surgical procedures for OSA: a meta-analysis. Sleep Breath. 2022;26(1):407–18.
25. Heiser C, Steffen A, Boon M, Hofauer B, Doghramji K, Maurer JT, Sommer JU, Soose R, Strollo PJ Jr, Schwab R, Thaler E. Post-approval upper airway stimulation predictors of treatment effectiveness in the ADHERE registry. Eur Respir J. 2019;53(1):1801405.
26. Van Daele DJ, Cromwell JW, Hsia JK, Nord RS. Post-operative complication rate comparison between airway surgery and upper airway stimulation using NSQIP and ADHERE. OTO Open. 2021;5(4):2473974X211051313.
27. Nord R, Fitzpatrick T, DeShazo JP, Reiter ER. Comparison of readmission and complication rates between traditional sleep surgery and hypoglossal nerve stimulation. Laryngoscope Investig Otolaryngol. 2022;7(5):1659–66.
28. Kent DT, Scott WC, Zealear D, Schwartz AR. Ansa cervicalis stimulation increases pharyngeal patency in patients with obstructive sleep apnea. J Appl Physiol. 2021;131(2):487–95.

Multilevel Sleep Surgery for Obstructive Sleep Apnea

20

Yung-An Tsou and Pei-Shao Liao

Contents

Abstract

Obstructive sleep apnea (OSA) often involves multilevel airway collapse, making multilevel sleep surgery more effective than single-level procedures. Preoperative assessments like drug-induced sleep endoscopy (DISE) are crucial

Y.-A. Tsou (✉)
Department of Otorhinolaryngology-Head and Neck Surgery, China Medical University Hospital, Taichung, Taiwan

Department of Otorhinolaryngology-Head and Neck Surgery, Asia University Hospital, Taichung, Taiwan

School of Medicine, China Medical University, Taichung, Taiwan

P.-S. Liao
Department of Otorhinolaryngology-Head and Neck Surgery, China Medical University Hospital, Taichung, Taiwan

School of Medicine, National Taiwan University, Taipei, Taiwan

H.-Y. Li (ed.), *Snoring and Obstructive Sleep Apnea*,
https://doi.org/10.1007/978-981-95-2648-2_20

for identifying obstruction sites and guiding surgical strategies. Common techniques include uvulopalatopharyngoplasty (UPPP), maxillomandibular advancement (MMA), and tongue base reduction.

While multilevel surgery improves the apnea–hypopnea index (AHI) and patient outcomes, it carries higher risks of postoperative bleeding and airway obstruction, particularly in patients with narrow facial skeletal structures or challenging intubation. Close monitoring and extended intubation may be necessary to manage complications. Advances, such as ultrasound-assisted transoral robotic surgery (TORS), further enhance precision and safety.

This review highlights the importance of tailored surgical approaches and meticulous technique to optimize outcomes and minimize risks in OSA management.

Keywords

Multilevel sleep surgery · Obstructive Sleep Apnea · Drug-induced sleep endoscopy

20.1 Introduction

Obstructive sleep apnea (OSA) is a complex disorder characterized by recurrent episodes of upper airway obstruction during sleep, leading to fragmented sleep and various health consequences. Effective surgical management often requires addressing multiple anatomical levels simultaneously. This chapter explores the rationale behind multilevel sleep surgery, details the anatomy of the upper airway, and examines surgical approaches targeting various anatomical structures. The discussion includes combinations of procedures such as uvulopalatopharyngoplasty (UPPP) with tongue base surgeries, maxillomandibular advancement (MMA), genioglossus advancement (GGA), and epiglottis surgery.

20.2 Structured Discussion

20.2.1 Why Multilevel Surgery?

OSA arises from airway obstructions that can occur at one or multiple anatomical levels, including the nasal cavity, soft palate, tongue base, and epiglottis.

Drug-induced sleep endoscopy (DISE) evaluates upper airway obstruction under pharmacologic sedation mimicking natural sleep. Introduced in 1991 and widely adopted globally, DISE employs fiberoptic endoscopy for assessment [1]. The VOTE classification, developed by Kezirian et al. in 2011, identifies obstructive structures, including velum, oropharynx-lateral walls, tongue base, and epiglottis [2]. Capasso et al. compared VOTE findings using propofol versus dexmedetomidine in obstructive sleep apnea (OSA) patients [3]. A meta-analysis of eight studies (535 patients) revealed by Certal et al. showed that surgical plans changed in 50.24% of cases post-DISE, predominantly addressing hypopharyngeal or laryngeal obstructions [4].

Overall, single-level surgeries often fail to adequately resolve OSA in patients with multilevel obstruction [5]:

A. *Nasal Region*

The nasal region contributes to about half of the total airway resistance in the upper airway. Notably, the internal nasal valve—comprising the septum, upper lateral cartilage, inferior turbinates, and nasal floor—plays a critical role. In the context of OSA, increased nasal airway resistance and nasal obstruction lead to a shift from nasal to mouth breathing. This transition exacerbates airway issues by narrowing the pharyngeal lumen, reducing the retroglossal diameter due to the tongue's posterior displacement, elongating the soft palate, and causing oscillations in redundant pharyngeal tissues [6].

In fact, the nasal cycle change exacerbated flow limitation during change of sleep architecture, REM-related change nasal cycle worsened flow limitation in marked deviated nasal septal side, and it affected sleep efficiency and sleep quality with change of sleep position. Even though the nasal surgery did not significantly improve AHI, the nasal surgery improved sleep snores and nCPAP compliance [7].

B. *Soft Palate*

Anatomical features of the soft palate in individuals with OSA typically include greater thickness, increased length, and heightened dimensions, all contributing to reduced airway space. In obese patients, excessive fat deposition in the parapharyngeal region extends to the submandibular space, located outside the maxillomandibular bony enclosure. This accumulation of fat intensifies concentric narrowing of the retropalatal pharynx, thereby heightening the risk of OSA. The soft palatal surgery really contributes to the reduction of sleep apnea severity, and patients with multilevel obstruction should consider the modified palatal surgery for not only curative intention treatment for OSA but also improved nCPAP use by preservation of uvula with multiple palatal surgery, as suspension pharyngoplasty, relocation palatoplasty, and holistic palatal surgery not only increase AP but also lateral space of retropharyngeal airway [8].

C. *Tongue Base*

The tongue's physiological influence on OSA is closely linked to tongue fat and the activity of the genioglossus muscle. At the onset of sleep, the genioglossus muscle's activity diminishes in both OSA patients and healthy individuals. However, due to anatomical alterations associated with OSA, this decline further aggravates upper airway collapsibility. The tongue base collapse is frequently noted in 33% of moderate-to-severe OSA [9]. Among them, one-third is post tongue collapse, one-third is tongue body collapse, and one-third is whole-tongue collapse, and diagnosis of the precise collapse leads to precise treatment strategy for treating tongue collapse [10].

D. *Lateral Pharyngeal Wall*

Thickening of the lateral pharyngeal wall, along with an increased volume of the parapharyngeal fat pad, contributes to oropharyngeal lateral collapse. A clinical review by Soares et al. indicated that approximately 53% of OSA patients experienced notable lateral oropharyngeal collapse at the retrolingual level. This condition was predominantly observed in males, individuals with higher BMI, and those with an elevated apnea–hypopnea index (AHI). The lateral collapse is difficult to treat and as it is prominent in obesity patients, proper treatment strategy should be made accordingly [11].

Hsueh-Yu Li proposed the raphe-based omni-suspension technique for the soft palate, utilizing three suspension sutures directed anteriorly, medially, and laterally to enlarge the velopharyngeal space and reduce collapse [12].

E. *Maxilla and Mandible*

A reduced size of the maxilla and mandible can result in a posteriorly positioned maxilla and a steep mandibular plane, contributing to increased upper and lower facial heights. The positioning of the mandible significantly affects the tongue's position; thus, a steep projection of the maxilla and mandible influences both the tongue and the soft palate [13].

Bony deformity-associated airway narrowing is frequently noted in Asian OSA, and it is also related to the central OSA. Proper treatment might correct some of the mild central OSA [14].

F. *Hyoid Bone*

In OSA patients, the hyoid bone is often located more inferiorly and posteriorly. An increased distance between the hyoid and the mandibular plane is critical in determining OSA severity. This positioning, due to the attachment of the lingual musculature, pulls the tongue backward, leading to pharyngeal airway narrowing [15].

G. *Epiglottis*

Epiglottic "trapdoor" behavior or floppiness has been documented in adults with sleep-related breathing disorders and can be assessed using drug-induced sleep endoscopy (DISE). The collapse of the epiglottis is aggravated by continuous positive airway pressure (CPAP). Interestingly, patients exhibiting epiglottic collapse tend to have less severe OSA and a lower BMI [16].

20.2.2 Indications and Contraindications (Table 20.1)

Table 20.1 General indications and relative contraindications from Mandavia and Holty [17, 18]

General indications
Moderate-to-severe obstructive sleep apnea (OSA)
Significant excessive daytime sleepiness
Coexisting with other medical conditions, such as arrhythmias or hypertension
Presence of anatomical abnormalities within the airway structure
Failure of nonsurgical or medical management
Relative contraindications
Severe obesity, except in cases where bariatric surgery or tracheostomy is performed
Severe or unstable cardiopulmonary diseases
Active misuse of substances, including alcohol and illicit drugs
Advanced age, which may affect surgical outcomes
Unstable psychological conditions or mental health issues
Unrealistic expectations regarding the effectiveness of surgical treatment

20.2.3 Anatomy of the Upper Airway and Targeted Surgeries (Table 20.2)

Table 20.2 Obstruction sites and targeted surgical intervention from Tan et al. [5]

Airway level	Obstruction	Surgical intervention
Nasal cavity/ Nasopharynx	Deviated nasal septum, Hypertrophied turbinates, Nasal polyps	Septoplasty, Turbinate reduction, Functional endoscopic sinus surgery (FESS)
Soft palate/ Uvula	Collapse of the soft palate, Excessive uvular tissue	Uvulopalatopharyngoplasty (UPPP), Palatal implants, Expansion sphincter pharyngoplasty
Tongue base	Prolapse of the tongue base, Bulky lingual tonsil, Excessive adipose tissue	Genioglossus advancement (GGA), Tongue base reduction, Hypoglossal nerve stimulation
Lateral pharyngeal wall	Collapse during inspiration	Barbed reposition pharyngoplasty, Lateral pharyngoplasty
Maxilla/ Mandible	Posterior displacement of the tongue and soft palate	Maxillomandibular advancement
Hyoid bone	Backward pull of the tongue	Hyoid myotomy and suspension
Epiglottis	Retroflexion, Excessive epiglottic tissue	Epiglottectomy, Epiglottoplasty

20.2.4 Comparison of Surgical Efficacy for OSA

Holty et al.'s review (Table 20.3) highlights that maxillomandibular advancement (MMA) demonstrates the highest surgical success rate at 86% and a cure rate of 43%, based on case series data. Bariatric surgery can offer benefits to morbidly obese patients, though less than half achieve complete resolution of obstructive sleep apnea (OSA) after weight loss. In contrast, soft palate surgeries, such as uvulopalatopharyngoplasty (UPPP), yield lower efficacy, with success and cure rates of 50% and 16%, respectively. A multimodality surgical strategy may provide additional advantages. Ultimately, for patients unable to tolerate continuous positive airway pressure (CPAP), tailored surgical interventions targeting specific airway obstructions can be a viable alternative [18].

20.2.5 Different Approaches and Effectiveness of Multilevel Surgery (Table 20.4)

Lin et al. divided the published data on multilevel treatment into four groups: [19]

A. *Group 1: Basic Multilevel Approach*

The most commonly performed multilevel approach involves uvulopalatopharyngoplasty (UPPP) as a foundational technique, combined with additional

Table 20.3 Comparison of surgical efficacy for obstructive sleep apnea, from Holty et al. [18]

Surgery	N	Studies	AHI before surgery	AHI after surgery	% change	Cure (%)	Success (%)
Bariatric surgery	437	16	53.3 ± 38.2	15.3 ± 18.7	−72.6 ± 60.6	44	–
GA	91	4	53.9	17.3	−67.8	–	62
HS	101	4	38.7	25.0	−33.0	–	50
HS and GA	328	7	33.5	15.2	−58.0	–	55
LAUP	72	3	–	–	–	7	49
MMA	627	22	63.9 ± 26.7	9.5 ± 10.7	−85.0 ± 18.2	43	86
Midline glossectomy	74	5	53.0	24.2	−54.4	–	50
Radiofrequency ablation (tongue)	394	11	37.0	23.4	−35.7	–	36
Tonsillectomy	1079	23	18.6	4.9	−73.7	60	–
Tongue base suspension	77	6	29.0	16.3	−32.9	–	35
Tracheostomy	33	2	98.9 ± 36.0	26.2 ± 29.2	−79.2 ± 25.8	–	73
UPPP	992	37	60.0	–	−38.2	16	52
Multimodality surgery	1978	58	48.0	–	−60.3	–	66

Abbreviations: *AHI* apnea–hypopnea index, *GA* genioglossus advancement, *HS* hyoid suspension, *LAUP* laser-assisted uvulopalatoplasty, *MMA* maxillomandibular advancement, *N* number, *UPPP* uvulopalatopharyngoplasty

Table 20.4 Classification of multilevel surgery from Lin et al. [19]

Group 1	Group 2	Group 3	Group 4
Basic multilevel approach	Radical hypopharyngeal surgery	Bimaxillary advancement	Minimally invasive techniques
UPPP combined with additional procedures for hypopharynx	More invasive and radical surgeries, such as tongue base resection	Staged surgeries with bimaxillary advancement	Usually radiofrequency surgery
Success rate: 20–100%	Success rate: 44–100%.	Success rate: 65.2–97.5%.	Success rate: 47.5–59%

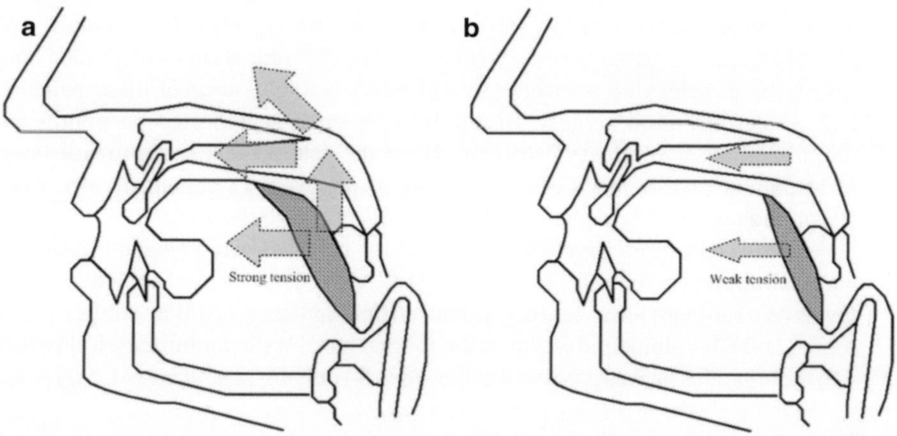

Fig. 20.1 The palatal suture mechanism of BRP plus TORSTBR (**a**) and UPPP plus TORSTBR (**b**) techniques from Tsou et al. [20]

procedures to enhance the hypopharyngeal airway. These supplementary procedures often include genioglossus advancement, thyrohyoid advancement, radiofrequency (RF) tissue volume reduction at the tongue base, and occasionally, tongue base suspension. The reported success rates for these treatments range widely from 20% to 100%.

In the realm of palatal surgery, research examining the combination of transoral robotic tongue base procedures and palatal interventions suggests that barbed reposition pharyngoplasty may provide a more effective anterior–posterior suspension vector for addressing obstructions at the palatal level compared to the modified uvulopalatopharyngoplasty technique [20] (Fig. 20.1).

B. *Group 2: Radical Hypopharyngeal Surgery*

The second category includes patients who have undergone more invasive and radical hypopharyngeal surgeries, such as open tongue base resection. Due to the aggressive nature of these interventions, most patients required a temporary tracheostomy and significant hospitalization. Postoperative complications were notable, with the success rates varying between 44% and 100%.

A review by Tsou et al. highlighted that the application of transoral robotic surgery (TORS) in managing obstructive sleep apnea achieves comparable success and failure rates to alternative surgical methods, with no statistically significant variation in operative durations [21].

C. *Group 3: Bimaxillary Advancement*

The third group consists of patients undergoing bimaxillary advancement as part of their multilevel treatment program. These patients often had staged surgeries, typically starting with UPPP and genioglossus advancement as initial procedures. Success rates in this group ranged from 65.2% to 97.5%.

D. *Group 4: Minimally Invasive Techniques*

The fourth category includes patients treated with minimally invasive multilevel techniques, particularly for mild-to-moderate OSAHS. In one study by Steward et al., 22 patients underwent combined RF reduction of the palate and tongue base, achieving a success rate of 59%. Notably, none of these patients had concurrent nasal surgery. Similarly, Fischer et al. reported on multilevel minimally invasive surgery involving RF on the palate, tonsils, and tongue base in 15 patients, while Stuck et al. published results from 18 patients treated with RF on the palate and tongue base for mild-to-moderate disease.

In 2017, Lin et al. presented a minimally invasive, single-stage multilevel surgical approach for mild-to-moderate OSAHS. This method included three levels of treatment: nasal surgery, palatal stiffening using the pillar implant technique, and RF volume reduction at the tongue base. With a minimum follow-up of 6 months, classical success was achieved in 54 out of 122 patients (47.5%) [22].

The severity of the condition and the sites of airway obstruction vary across these four groups, necessitating careful patient selection in clinical practice. Preoperative assessments, such as drug-induced sleep endoscopy (DISE), are crucial for tailoring surgical combinations to the specific needs of each patient.

20.2.6 Risks of Multilevel Surgery

Multilevel surgery carries inherent risks due to the complexity of addressing multiple airway structures:

A. *Risk of Airway Obstruction*
   1. Swelling and inflammation postsurgery can lead to temporary airway narrowing.
   2. Patients with narrow facial skeletal structures and challenging intubation are at an elevated risk of postoperative airway obstruction, particularly within the first 48 hours following surgery.

B. *Need for Temporary Airway Support*
   1. Endotracheal Intubation: Obese patients and those who may be experiencing airway compromise following sleep surgery should necessitate prolonged intubation to ensure airway protection.

2. Tracheostomy: Routine tracheostomy for TORS patients is now uncommon, as evidence shows that it reduces operative time, hospital stays, and aspiration risk. However, it remains necessary when combined with extensive procedures like MMA, which increase airway edema risk [23].

C. *Bleeding and Infection*
1. Postoperative bleeding often occurs following infection, highlighting the importance of maintaining oral hygiene after surgery. The use of mouthwash for cleaning is recommended.
2. With the assistance of intraoperative ultrasound, surgeons can confidently determine position of lingual artery. Chang et al. introduced a real-time intraoperative technique, utilizing ultrasound to aid transoral robotic surgery (TORS) for identifying the lingual artery during procedures [24].

D. *Postoperative Dysphagia and Pain*

Swallowing difficulties and pain are common in the immediate recovery phase, but generally resolve within weeks.

A comprehensive cohort study revealed a significantly elevated risk of severe complications when uvulopalatopharyngoplasty (UPPP) was conducted in conjunction with non-nasal procedures, compared to performing UPPP alone. The adjusted risk ratio for these combined procedures was calculated as 4.8, with a 95% confidence interval ranging from 2.5 to 9.1. Conversely, when UPPP was performed alongside nasal procedures, no notable increase in complication risk was identified [25].

20.2.7 Cost Efficiency of Single-Stage Multilevel Surgery

Single-stage multilevel surgical strategies have demonstrated a potential for lowering total hospitalization expenses when compared to procedures performed in multiple stages [26]. Research by Kieff and Busaba [27] assessed the feasibility of same-day or overnight discharge. They concluded that, for carefully selected patients without major comorbid conditions, discharging on the same day after undergoing combined nasal and palatal surgeries for obstructive sleep apnea (OSA) can be considered a reasonably safe practice.

20.2.8 Postoperative Care

Reviewing the previously mentioned surgical risks, patients undergoing multilevel sleep surgery face a higher likelihood of bleeding and airway obstruction compared to those receiving single-level sleep surgery. Consequently, the duration of endotracheal tube placement may need to be extended, and initial postoperative observation in the intensive care unit may be warranted. Additionally, meticulous hemostasis during the procedure is crucial to minimize the risk of postoperative bleeding.

20.3 Conclusion

Obstructive sleep apnea often results from multisite airway collapse, making multilevel sleep surgery more effective than single-level procedures. Various surgical techniques are available to address specific sites of airway obstruction, and thorough preoperative assessment combined with careful patient selection can optimize postoperative outcomes. However, it is crucial to recognize the increased risks of airway obstruction and bleeding following multilevel sleep surgery, necessitating close postoperative monitoring to ensure patient safety.

References

1. Croft CB, Pringle M. Sleep nasendoscopy: a technique of assessment in snoring and obstructive sleep apnoea. Clin Otolaryngol Allied Sci. 1991;16(5):504–9.
2. Kezirian EJ, Hohenhorst W, de Vries N. Drug-induced sleep endoscopy: the VOTE classification. Eur Arch Otorrinolaringol. 2011;268:1233–6.
3. Capasso R, Rosa T, Tsou DYA, Nekhendzy V, Drover D, Collins J, et al. Variable findings for drug-induced sleep endoscopy in obstructive sleep apnea with propofol versus dexmedetomidine. Otolaryngol Head Neck Surg. 2016;154(4):765–70.
4. Certal VF, Pratas R, Guimaraes L, Lugo R, Tsou Y, Camacho M, et al. Awake examination versus DISE for surgical decision making in patients with OSA: a systematic review. Laryngoscope. 2016;126(3):768–74.
5. Tan SN, Yang HC, Lim SC. Anatomy and pathophysiology of upper airway obstructive sleep apnoea: review of the current literature. Sleep Med Res. 2021;12(1):1–8.
6. Georgalas C. The role of the nose in snoring and obstructive sleep apnoea: an update. Eur Arch Otorrinolaringol. 2011;268(9):1365–73.
7. Scapuccin M, Schneider L, Rashid N, Zaghi S, Rosa T, Tsou YA, et al. Integrating the divided nasal cannula into routine polysomnography to assess nasal cycle: feasibility and effect on outcomes. J Clin Sleep Med. 2018;14(4):641–50.
8. de Souza Pinto GN, Iwaki Filho L, dos Santos Previdelli IT, Ramos AL, Yamashita AL, Stabile GAV, et al. Three-dimensional alterations in pharyngeal airspace, soft palate, and hyoid bone of class II and class III patients submitted to bimaxillary orthognathic surgery: a retrospective study. J Craniomaxillofac Surg. 2019;47(6):883–94.
9. Ono T. Tongue and upper airway function in subjects with and without obstructive sleep apnea. Jpn Dent Sci Rev. 2012;48(2):71–80.
10. Li HY, Lo YL, Wang CJ, Hsin LJ, Lin WN, Fang TJ, et al. Dynamic drug-induced sleep computed tomography in adults with obstructive sleep apnea. Sci Rep. 2016;6(1):35849.
11. Soares D, Sinawe H, Folbe AJ, Yoo G, Badr S, Rowley JA, et al. Lateral oropharyngeal wall and supraglottic airway collapse associated with failure in sleep apnea surgery. Laryngoscope. 2012;122(2):473–9.
12. Li HY, Tsai MS, Lee LA, Hsin LJ, Lee YC, Lin WN, et al. Palatal hybrid surgery for obstructive sleep apnea-state-of-the-art annotation of uvulopalatopharyngoplasty. Biom J. 2023;46(3):100568.
13. Johal AMA, Patel SI, Battagel JM. The relationship between craniofacial anatomy and obstructive sleep apnoea: a case-controlled study. J Sleep Res. 2007;16(3):319–26.
14. Neelapu BC, Kharbanda OP, Sardana HK, Balachandran R, Sardana V, Kapoor P, et al. Craniofacial and upper airway morphology in adult obstructive sleep apnea patients: a systematic review and meta-analysis of cephalometric studies. Sleep Med Rev. 2017;31:79–90.

15. Watanabe T, Isono S, Tanaka A, Tanzawa H, Nishino T. Contribution of body habitus and craniofacial characteristics to segmental closing pressures of the passive pharynx in patients with sleep-disordered breathing. Am J Respir Crit Care Med. 2002;165(2):260–5.
16. Torre C, Camacho M, Liu SYC, Huon LK, Capasso R. Epiglottis collapse in adult obstructive sleep apnea: a systematic review. Laryngoscope. 2016;126(2):515–23.
17. Mandavia R, Mehta N, Veer V. Guidelines on the surgical management of sleep disorders: a systematic review. Laryngoscope. 2020;130(4):1070–84.
18. Holty JEC, Guilleminault C. Surgical options for the treatment of obstructive sleep apnea. Med Clin. 2010;94(3):479–515.
19. Lin HC, Friedman M, Chang HW, Gurpinar B. The efficacy of multilevel surgery of the upper airway in adults with obstructive sleep apnea/hypopnea syndrome. Laryngoscope. 2008;118(5):902–8.
20. Tsou YA, Hsu CC, Shih LC, Lin TC, Chiu CJ, Tien VHC, et al. Combined transoral robotic tongue base surgery and palate surgery in obstructive sleep apnea syndrome: modified uvulopalatopharyngoplasty versus barbed reposition pharyngoplasty. J Clin Med. 2021;10(14):3169.
21. Tsou YA, Chang WD. Comparison of transoral robotic surgery with other surgeries for obstructive sleep apnea. Sci Rep. 2020;10(1):18163.
22. Lin HC, Friedman M, Chang HW, Bonzelaar L, Salapatas AM, Lin MC, et al. Minimally invasive, single-stage, multilevel surgery for obstructive sleep apnea in Asian patients. JAMA Otolaryngol Head Neck Surg. 2017;143(2):147–54.
23. Mouratidou S, Chaidas K. Transoral robotic surgery for patients with obstructive sleep apnoea: a systematic literature review of current practices. Life. 2024;14(12):1700.
24. Chang CC, Wu JL, Hsiao JR, Lin CY. Real-time, intraoperative, ultrasound-assisted transoral robotic surgery for obstructive sleep apnea. Laryngoscope. 2021;131(4):E1383–90.
25. Kezirian EJ, Weaver EM, Yueh B, Khuri SF, Daley J, Henderson WG. Risk factors for serious complication after uvulopalatopharyngoplasty. Arch Otolaryngol Head Neck Surg. 2006;132:1091–8.
26. Li HY, Wang PC, Hsu CY, Lee SW, Chen NH, Liu SA. Combined nasal-palatopharyngeal surgery for obstructive sleep apnea: simultaneous or staged? Acta Otolaryngol. 2005;125:298–303.
27. Kieff DA, Busaba NY. Same-day discharge for selected patients undergoing combined nasal and palatal surgery for obstructive sleep apnea. Ann Otol Rhinol Laryngol. 2004;113:128–31.

Pediatric Obstructive Sleep Apnea Syndrome

21

Kun-Tai Kang and Wei-Chung Hsu

Contents

K.-T. Kang
Department of Otolaryngology, National Taiwan University Hospital, Taipei, Taiwan

Department of Otolaryngology and Sleep Center, Taipei Hospital, Ministry of Health and Welfare, New Taipei City, Taiwan

W.-C. Hsu (✉)
Department of Otolaryngology, National Taiwan University Hospital, Taipei, Taiwan

Sleep Center, National Taiwan University Hospital, Taipei, Taiwan

Department of Otolaryngology, College of Medicine, National Taiwan University, Taipei, Taiwan
e-mail: hsuwc@ntu.edu.tw

H.-Y. Li (ed.), *Snoring and Obstructive Sleep Apnea*,
https://doi.org/10.1007/978-981-95-2648-2_21

Abstract

Pediatric obstructive sleep apnea (OSA) is a serious disease characterized by repetitive narrowing of the upper airway and cessation of airflow during sleep. The prevalence of pediatric OSA in general population ranged from 0.7% to 13.0%. Children with OSA have symptoms such as snoring, observed apnea, mouth breathing, hyperactivity, excessive daytime sleepiness, and learning problems. Overnight attended polysomnography is the gold standard for diagnosing pediatric OSA. Adenotonsillectomy is an effective first-line treatment for pediatric OSA. Pediatric OSA-related morbidities include increased cardiometabolic risk, impact of neurocognition and behavior, and impaired somatic growth that may be mitigated by proper managements. Recent advancements in imaging modality, home sleep apnea test, therapeutic approach, telemedicine, and artificial intelligence facilitate diagnosis and treatment of OSA in children.

Keywords

Adenotonsillectomy · Child · Polysomnography · Obstructive sleep apnea

21.1 Introduction

Pediatric obstructive sleep apnea (OSA) is a serious disease condition, characterized by repetitive narrowing of the upper airway during sleep, resulting in increased respiratory effort, sleep fragmentation, and impaired ventilation [1–3]. Untreated OSA in children is associated with adverse cardiometabolic, neurobehavior, and somatic morbidities [1, 2]. In pediatric population, the gold standard for diagnosing OSA is polysomnography (PSG) [1, 2], and the first-line therapy is adenotonsillectomy because hypertrophy of the tonsils and adenoids is the main cause of upper airway occlusion [3]. After firstly described by Professor Guilleminault in 1976 [4], pediatric OSA data showed great progress in annual number of publications in recent years. This chapter provides an overview of epidemiology, clinical manifestation, imaging, and recent advancements in diagnosing and treating OSA and OSA-related morbidity in pediatric population.

21.2 Epidemiology: Prevalence and Risk Factor

The prevalence of pediatric OSA has been estimated [5, 6]. In 2008, a systematic review by Lumeng et al. reported the prevalence of pediatric OSA to be 1–4% [5]. More recently, a 2024 systematic review by Magnusdottir et al. announced that the prevalence of pediatric OSA was in the range of 0.7–13.0%, and the prevalence in studies published during 2016–2023 was higher than that before 2014, suggesting that the prevalence of pediatric OSA may possibly have been increasing over the past decade [6].

Adenotonsillar hypertrophy and obesity increase risk of pediatric OSA [7, 8]. The main anatomic factor contributing to pediatric OSA is enlarged tonsils and adenoids [1–3]. There are ample data confirming that OSA associated with obesity is highly prevalent in children and adolescents [7]. Recent evidence indicated that children with tonsillar hypertrophy, adenoid hypertrophy, and obesity have increased 3.5-fold, 1.6-fold, and 1.7-fold risk, respectively, of OSA [8].

21.3 Clinical Symptoms

Children with OSA experience nocturnal symptoms such as snoring, observed apnea, and difficult breathing, as well as daytime symptoms such as mouth breathing, excessive daytime sleepiness, hyperactivity, and learning problems [1, 2]. However, classical daytime symptoms in adult OSA, excessive daytime sleepiness, is less commonly reported in children [2]. Moreover, children with OSA can present with behavioral problems including hyperactivity and poor school performance [2].

For diagnostic accuracy in detecting pediatric OSA, snoring reported by parents or caregivers had high sensitivity but low specificity, while excessive daytime sleep apnea, observed apnea, and difficult breathing during sleep had high specificity but low sensitivity [9]. However, neither single nor combined symptoms have satisfactory performance in predicting pediatric OSA, and warrant the need for subsequent sleep study to confirm the diagnosis [9].

21.4 Questionnaire

The two most common used questionnaires in pediatric OSA are "Pediatric sleep questionnaire (PSQ)" and the OSA-18 [10, 11]. The PSQ is a parent-reported questionnaire and contains 22 questions, comprising three symptom complexes: snoring, excessive daytime sleepiness, and inattentive or hyperactive behavior [10]. The OSA-18 is a caregiver-administered quality-of-life survey containing 18 items divided into five subscales: sleep disturbance, physical symptoms, emotional distress, daytime function, and caregiver concerns [11]. The PSQ exhibits high sensitivity (74%) for detecting pediatric OSA and may serve as alternatives when PSG is not available [12]. The OSA-18 is commonly used for quantitative quality-of-life (QoL) measures demonstrating the impact of pediatric OSA on QoL and short-term and long-term QoL improvement after adenotonsillectomy [13].

21.5 Imaging Study

In view of the major role of structural narrowing, upper airway imaging could be a useful tool for investigating OSA and in establishing the sites of obstruction [14–17].

21.5.1 2-Dimensional Imaging: Skull Lateral Radiograph and Cephalometry

Two-dimensional imaging is obtained from lateral skull radiographs or cephalometry [14, 15]. Fujioka proposed a measure of adenoidal–nasopharyngeal ratio, a simple linear measures from lateral skull radiographs for standardization of adenoid size across ages [14]. Besides, several craniofacial disharmonies, including mandibular retrognathia, decreased pharyngeal diameters at the levels of the adenoids, longer facial profile, and a narrower intercanine width, have been identified from cephalometry analysis among children with OSA [15].

21.5.2 3-Dimensional Imaging

Several 3-dimensioanl radiological techniques, such as magnetic resonance imaging (MRI) or cone-beam computed tomography, and subsequent processing of these images using computational fluid dynamics, have been prescribed [16, 17]. Children with OSA presented a low total airway volume, small minimal cross section, and short antero-posterior dimension that is dramatically improved after OSA treatment [17].

21.5.3 Sleep Endoscopy

Sleep endoscopy evaluates sites of multilevel obstruction in upper airway using a flexible endoscope while patient in a pharmacologically induced sleep-like state [18]. Sleep endoscopy is typically indicated for children with persistent OSA after surgery, those with OSA but without tonsillar hypertrophy, predisposing multiple sites of obstruction or laryngomalacia. However, there is currently no universally accepted anesthetic protocol or scoring system for sleep endoscopy in pediatric population [18].

21.6 Sleep Study

The American Academy of Sleep Medicine classified sleep studies into four levels, of which types 2 through 4 are applicable to home testing [19, 20].

Level 1 sleep study is overnight-attended PSG with seven or more channels that records cardiorespiratory and neurophysiological parameters including electroencephalography, electrooculography, electromyography, airflow, respiratory effort, oxygen saturation, and electrocardiography in a laboratory setting [19]. Overnight PSG is the gold standard for diagnosing OSA, and the apnea–hypopnea index (AHI) in PSG is widely used to define disease severity and surgical success [1, 2]. Guidelines from the American Academy of Sleep Medicine [19], the American Academy of Pediatrics [1], and the European Respiratory Society [2] address clinical applications of PSG studies in pediatric population. In general, pediatric OSA is diagnosed when the PSG reports an AHI ≥ 1 event/h. The severity of pediatric OSA is determined by the AHI and mild, moderate, or severe OSA corresponds to the thresholds of 1–5, 5–9, and >10 events/h, respectively [1, 2, 19].

The home sleep apnea test is considered an alternative to PSG, as it is potentially more cost-effective, convenient, and accessible [20]. Level 2 sleep study is unattended PSG (seven or more channels). Level 3 sleep study consists of studies using four to seven channels. Level 4 sleep study includes only one or two channels, with pulse oximetry traditionally. Limited data comparing level 2 sleep study with level 1 attended PSG indicate no significant differences in respiratory parameters [20]. To date, home sleep apnea tests have not been validated to be suitable replacements for PSG in children, but offers a viable alternative as screening tools for pediatric OSA with high specificity in detecting moderate-to-severe OSA in otherwise healthy children [12, 20].

21.7 Treatment

21.7.1 Adenotonsillectomy

Adenotonsillectomy is the first-line and most effective treatment for pediatric OSA. There is compelling evidence that confirms the beneficial effects of adenotonsillectomy in alleviating severity of disease and improving QoL in pediatric OSA [1–3, 21–23].

In 2013, the Childhood Adenotonsillectomy Trial (CHAT) study was one of the most famous and outstanding randomized controlled trial in pediatric OSA research, which was a single-blinded, multicenter study from 7 academic sleep centers in the United States [22]. After randomly assigned 464 OSA children aged 5–9 years, adenotonsillectomy group had a significantly greater improvement in the AHI and QoL, and a higher rate of normalization of the PSG findings (79% vs. 46%) than those in the watchful-waiting group at the 7-month follow-up [22].

Another randomized controlled trial, the Pediatric Adenotonsillectomy Trial for Snoring (PATS), showed that children with mild OSA who underwent adenotonsillectomy, compared with watchful waiting, had improved symptoms, behavior, QoL, and blood pressure at 12-month follow-up [23].

According to the latest meta-analysis for PSG outcome after adenotonsillectomy, in which 51 studies with 3413 children were analyzed, adenotonsillectomy results in a significant reduction of AHI of 12.4 event/h in AHI, along with an increase in nocturnal oxygen saturation, sleep efficiency, and deep sleep [21]. However, the overall success rate defined as postoperative AHI <1 was only 51%, indicating roughly half of the children had persistent OSA after adenotonsillectomy, especially those with severe disease and obesity, making additional treatment strategies and/or long-term follow-up highly desirable [21].

21.7.2 Persistent OSA After Adenotonsillectomy

The American Thoracic Society announced clinical practice guideline in 2024 [24]. For children with persistent mild OSA (AHI 1–5 events/h), the use of anti-inflammatory medications is recommended, whereas continuous positive airway pressure (CPAP) or surgery is reserved for children with persistent AHI ⩾5 events/h. Drug-induced sleep endoscopy or MRI is considered in children with suspected surgically modifiable sites of airway obstruction. Orthodontic treatment is recommended for children with specific craniofacial features, and weight loss intervention is suggested in overweight or obese children with persistent OSA. These recommendations highlight the need for multidisciplinary management for persistent OSA after adenotonsillectomy [24].

21.7.3 Positive Airway Pressure

CPAP is another effective therapy for pediatric OSA [1, 2, 25]. However, the CPAP adherence rate in children with OSA was 46.56% [25]. Young age and high AHI were associated with acceptable CPAP adherence in children with OSA [25].

21.7.4 Dental Therapy

Orthodontic treatment is suggested to correct transverse maxillary deficiency and mandibular retrusion [26]. Pertinent meta-analysis revealed that rapid maxillary expansion and mandibular advancement produced a significant reduction in AHI and increase in nocturnal oxygen saturation in children with OSA at 6 and 12 months from baseline [26]. However, orthodontic indication is needed to support this therapy and a careful monitoring is required to ensure positive improvement in OSA parameters.

21.7.5 Pharmacotherapy

Anti-inflammatory drugs have been proposed as a potential nonsurgical treatment option in children with OSA [27]. According to the 2020 Cochrane review, intranasal corticosteroids may have short-term beneficial effects on oxygen saturation in children with mild-to-moderate OSA, and montelukast has short-term beneficial treatment effects by significantly reducing the number of apneas and hypopnoeas [27]. However, long-term efficacy and safety data on the use of anti-inflammatory medications for the treatment of OSA in childhood are still not available.

21.7.6 Myofunctional Therapy

Myofunctional therapy include different types of soft palate, tongue, and facial muscle exercises that promote the sensitivity, proprioception, mobility, coordination, and strength of orofacial structures. Myofunctional therapy in pediatric OSA showed positive effects on decreased AHI and increased oxygen saturations in children with mild-to-moderate OSA but with heterogeneity in exercises and protocols [28].

21.8 Pediatric OSA Evolution from Childhood to Adolescence

Longitudinal data elaborated how OSA evolves from childhood to adolescence [29, 30]. Bixler et al. reported that the remission of childhood OSA in adolescence was 52.9% [29]. A 10-year follow-up study by Li et al. showed that the remission rate of childhood OSA was 30% [30]. Children with mild OSA tend to resolve naturally [29], while male sex and higher degree of obesity were associated with incident OSA in adolescent [30].

21.9 OSA-Related Morbidity

21.9.1 Cardiometabolic Risk

Recent research highlights the impact of OSA on cardiovascular risk in children and adolescents [31]. Pediatric OSA is associated with elevated BP [32]. A meta-analysis by Ai et al. found that the mean systolic BP was significantly higher in children with OSA compared to the healthy controls [32]. In prospective studies, moderate-to-severe childhood OSA was associated with a risk of elevated systolic BP in adulthood with a mean difference of 4.02 mm Hg [32]. Relevant studies also investigated the impacts of pediatric OSA on arrhythmia, abnormal ventricular morphology, impaired ventricular contractility, and elevated right heart pressure [31].

Longitudinal studies assessing cardiovascular outcome after treatment are limited [33, 34]. The available literature indicates positive effects on autonomic function, heart function, and C-reactive protein following treatment for pediatric OSA [33]. OSA Children with hypertension experience a significant BP reduction after adenotonsillectomy [34].

21.9.2 Neurocognitive and Behavioral Impact

Several studies enlightened the detrimental effects of pediatric OSA on neurocognitive functions such as learning, memory, executive functions, and school performance, and behavioral abnormalities such as aggressive behavior, hyperactivity and attention-deficit hyperactivity disorder (ADHD) [35].

Neurocognitive dysfunction raised by pediatric OSA may be reversed by treatment [35, 36]. A meta-analysis by Song et al. found improvement in neurocognitive function after pediatric adenotonsillectomy [36]. Children with ADHD symptoms related to OSA also improved after adenotonsillectomy [35].

21.9.3 Impaired Somatic Growth

Many studies reported growth failure in children with OSA [1, 2]. Possible mechanisms for growth impairment include increased energy expenditure during sleep and altered growth hormone production. Fortunately, growth and biomarker increased significantly after pediatric adenotonsillectomy [37].

21.10 Advances in Recent Years

21.10.1 Telemedicine

Telemedicine is applicable at every stage of pediatric OSA management, from diagnosis to treatment monitoring [38]. In pediatric OSA, telemedicine offers convenience in physical examination, access to diagnostic tools, education, and counseling, but also accompany some disadvantages, such as technology barriers and privacy concerns [38].

21.10.2 Artificial Intelligence

The application of artificial intelligence through machine-learning approach has good reliability to diagnose pediatric OSA [39]. However, the capacity of personalizing treatment and predicting treatment outcomes for children with OSA remain unclear [39].

21.11 Conclusion

Pediatric OSA is a serious disease worthy of great attention. Future work should focus on advancement for diagnosis and treatment of OSA in conjunction with understanding and alleviation of OSA-related morbidities.

References

1. Marcus CL, Brooks LJ, Draper KA, Gozal D, Halbower AC, Jones J, Schechter MS, Sheldon SH, Spruyt K, Ward SD, Lehmann C, Shiffman RN, American Academy of Pediatrics. Diagnosis and management of childhood obstructive sleep apnea syndrome. Pediatrics. 2012;130(3):576–84.
2. Kaditis AG, Alonso Alvarez ML, Boudewyns A, Alexopoulos EI, Ersu R, Joosten K, Larramona H, Miano S, Narang I, Trang H, Tsaoussoglou M, Vandenbussche N, Villa MP, Van Waardenburg D, Weber S, Verhulst S. Obstructive sleep disordered breathing in 2- to 18-year-old children: diagnosis and management. Eur Respir J. 2016;47(1):69–94.
3. Kang KT, Hsu WC. Efficacy of adenotonsillectomy on pediatric obstructive sleep apnea and related outcomes: a narrative review of current evidence. J Formos Med Assoc. 2024;123(5):540–50.
4. Guilleminault C, Eldridge FL, Simmons FB, Dement WC. Sleep apnea in eight children. Pediatrics. 1976;58(1):23–30.
5. Lumeng JC, Chervin RD. Epidemiology of pediatric obstructive sleep apnea. Proc Am Thorac Soc. 2008;5(2):242–52.
6. Magnusdottir S, Hill EA. Prevalence of obstructive sleep apnea (OSA) among preschool aged children in the general population: a systematic review. Sleep Med Rev. 2024;73:101871.
7. Narang I, Mathew JL. Childhood obesity and obstructive sleep apnea. J Nutr Metab. 2012;2012:134202.
8. Hasuneh MM, Toubasi AA, Khraisat B, Aldabbas H, Al-Iede M. Risk factors of obstructive sleep apnea (OSA) in pediatric patients: a systematic review and meta-analysis. J Pediatr Health Care. 2024;38(5):717–26.
9. Certal V, Catumbela E, Winck JC, Azevedo I, Teixeira-Pinto A, Costa-Pereira A. Clinical assessment of pediatric obstructive sleep apnea: a systematic review and meta-analysis. Laryngoscope. 2012;122(9):2105–14.
10. Chervin RD, Hedger K, Dillon JE, Pituch KJ. Pediatric sleep questionnaire (PSQ): validity and reliability of scales for sleep-disordered breathing, snoring, sleepiness, and behavioral problems. Sleep Med. 2000;1(1):21–32.
11. Franco RA Jr, Rosenfeld RM, Rao M. First place-resident clinical science award 1999. Quality of life for children with obstructive sleep apnea. Otolaryngol Head Neck Surg. 2000;123(1 Pt 1):9–16.
12. Wu CR, Tu YK, Chuang LP, Gordon C, Chen NH, Chen PY, Hasan F, Kurniasari MD, Susanty S, Chiu HY. Diagnostic meta-analysis of the Pediatric Sleep Questionnaire, OSA-18, and pulse oximetry in detecting pediatric obstructive sleep apnea syndrome. Sleep Med Rev. 2020;54:101355.
13. Todd CA, Bareiss AK, McCoul ED, Rodriguez KH. Adenotonsillectomy for obstructive sleep apnea and quality of life: systematic review and meta-analysis. Otolaryngol Head Neck Surg. 2017;157(5):767–73.
14. Fujioka M, Young LW, Girdany BR. Radiographic evaluation of adenoidal size in children: adenoidal-nasopharyngeal ratio. AJR Am J Roentgenol. 1979;133(3):401–4.
15. Fagundes NCF, Gianoni-Capenakas S, Heo G, Flores-Mir C. Craniofacial features in children with obstructive sleep apnea: a systematic review and meta-analysis. J Clin Sleep Med. 2022;18(7):1865–75.

16. Slaats MA, Van Hoorenbeeck K, Van Eyck A, Vos WG, De Backer JW, Boudewyns A, De Backer W, Verhulst SL. Upper airway imaging in pediatric obstructive sleep apnea syndrome. Sleep Med Rev. 2015;21:59–71.
17. Liu CN, Kang KT, Yao CJ, Chen YJ, Lee PL, Weng WC, Hsu WC. Changes in cone-beam computed tomography pediatric airway measurements after adenotonsillectomy in patients with OSA. JAMA Otolaryngol Head Neck Surg. 2022;148(7):621–9.
18. Wilcox LJ, Bergeron M, Reghunathan S, Ishman SL. An updated review of pediatric drug-induced sleep endoscopy. Laryngoscope Investig Otolaryngol. 2017;2(6):423–31.
19. Aurora RN, Zak RS, Karippot A, Lamm CI, Morgenthaler TI, Auerbach SH, Bista SR, Casey KR, Chowdhuri S, Kristo DA, Ramar K, American Academy of Sleep Medicine. Practice parameters for the respiratory indications for polysomnography in children. Sleep. 2011;34(3):379–88.
20. Tan HL, Kheirandish-Gozal L, Gozal D. Pediatric home sleep apnea testing: slowly getting there! Chest. 2015;148(6):1382–95.
21. Lee CH, Hsu WC, Chang WH, Lin MT, Kang KT. Polysomnographic findings after adenotonsillectomy for obstructive sleep apnoea in obese and non-obese children: a systematic review and meta-analysis. Clin Otolaryngol. 2016;41(5):498–510.
22. Marcus CL, Moore RH, Rosen CL, Giordani B, Garetz SL, Taylor HG, Mitchell RB, Amin R, Katz ES, Arens R, Paruthi S, Muzumdar H, Gozal D, Thomas NH, Ware J, Beebe D, Snyder K, Elden L, Sprecher RC, Willging P, Jones D, Bent JP, Hoban T, Chervin RD, Ellenberg SS, Redline S, Childhood Adenotonsillectomy Trial (CHAT). A randomized trial of adenotonsillectomy for childhood sleep apnea. N Engl J Med. 2013;368(25):2366–76.
23. Redline S, Cook K, Chervin RD, Ishman S, Baldassari CM, Mitchell RB, Tapia IE, Amin R, Hassan F, Ibrahim S, Ross K, Elden LM, Kirkham EM, Zopf D, Shah J, Otteson T, Naqvi K, Owens J, Young L, Furth S, Connolly H, Clark CAC, Bakker JP, Garetz S, Radcliffe J, Taylor HG, Rosen CL, Wang R, Pediatric Adenotonsillectomy Trial for Snoring (PATS) Study Team. Adenotonsillectomy for snoring and mild sleep apnea in children: a randomized clinical trial. JAMA. 2023;330(21):2084–95.
24. Ehsan Z, Ishman SL, Soghier I, Almeida FR, Boudewyns A, Camacho M, Carno MA, Coppelson K, Ersu RH, Ho ATN, Kaditis AG, Machado AJ Jr, Mitchell RB, Resnick CM, Swaggart K, Verhulst S. Management of persistent, post-adenotonsillectomy obstructive sleep apnea in children: an Official American Thoracic Society Clinical Practice Guideline. Am J Respir Crit Care Med. 2024;209(3):248–61.
25. Sawunyavisuth B, Ngamjarus C, Sawanyawisuth K. Adherence to continuous positive airway pressure therapy in pediatric patients with obstructive sleep apnea: a meta-analysis. Ther Clin Risk Manag. 2023;19:143–62.
26. Alessandri-Bonetti G, Incerti-Parenti S, D'Antò V. Effect of orthopedic and functional orthodontic treatment in children with obstructive sleep apnea: a systematic review and meta-analysis. Sleep Med Rev. 2023;67:101730.
27. Kuhle S, Hoffmann DU, Mitra S, Urschitz MS. Anti-inflammatory medications for obstructive sleep apnoea in children. Cochrane Database Syst Rev. 2020;1(1):CD007074.
28. Bandyopadhyay A, Kaneshiro K, Camacho M. Effect of myofunctional therapy on children with obstructive sleep apnea: a meta-analysis. Sleep Med. 2020 Nov;75:210–7.
29. Bixler EO, Fernandez-Mendoza J, Liao D, Calhoun S, Rodriguez-Colon SM, Gaines J, He F, Vgontzas AN. Natural history of sleep disordered breathing in prepubertal children transitioning to adolescence. Eur Respir J. 2016 May;47(5):1402–9.
30. Chan KC, Au CT, Hui LL, Ng SK, Wing YK, Li AM. How OSA evolves from childhood to young adulthood: natural history from a 10-year follow-up study. Chest. 2019;156(1):120–30.
31. Baker-Smith CM, Isaiah A, Melendres MC, Mahgerefteh J, Lasso-Pirot A, Mayo S, Gooding H, Zachariah J, American Heart Association Athero, Hypertension and Obesity in the Young Committee of the Council on Lifelong Congenital Heart Disease and Heart Health in the Young. Sleep-disordered breathing and cardiovascular disease in children and adolescents: a scientific statement from the American Heart Association. J Am Heart Assoc. 2021;10(18):e022427.

32. Ai S, Li Z, Wang S, Chen S, Chan JW, Au CT, Bao Y, Li AM, Zhang J, Chan KC, Wing YK. Blood pressure and childhood obstructive sleep apnea: a systematic review and meta-analysis. Sleep Med Rev. 2022;65:101663.
33. Ehsan Z, Ishman SL, Kimball TR, Zhang N, Zou Y, Amin RS. Longitudinal cardiovascular outcomes of sleep disordered breathing in children: a meta-analysis and systematic review. Sleep. 2017;40(3):zsx015.
34. Kang KT, Chiu SN, Lee CH, Lin MT, Hsu WC. Effect of adenotonsillectomy on blood pressure in children with obstructive sleep apnea: a meta-analysis. Sleep Med. 2021;84:334–42.
35. Trosman I, Trosman SJ. Cognitive and behavioral consequences of sleep disordered breathing in children. Med Sci (Basel). 2017;5(4):30.
36. Song SA, Tolisano AM, Cable BB, Camacho M. Neurocognitive outcomes after pediatric adenotonsillectomy for obstructive sleep apnea: a systematic review and meta-analysis. Int J Pediatr Otorhinolaryngol. 2016;83:205–10.
37. Bonuck KA, Freeman K, Henderson J. Growth and growth biomarker changes after adenotonsillectomy: systematic review and meta-analysis. Arch Dis Child. 2009;94(2):83–91.
38. Rizzo L, Barbetta E, Ruberti F, Petz M, Tornesello M, Deolmi M, Fainardi V, Esposito S. The role of telemedicine in children with obstructive sleep apnea syndrome (OSAS): a review of the literature. J Clin Med. 2024;13(7):2108.
39. Gutiérrez-Tobal GC, Álvarez D, Kheirandish-Gozal L, Del Campo F, Gozal D, Hornero R. Reliability of machine learning to diagnose pediatric obstructive sleep apnea: systematic review and meta-analysis. Pediatr Pulmonol. 2022;57(8):1931–43.

Obstructive Sleep Apnea in Women: Evolution Across Life Span

22

Pei-Lin Lee, Naricha Chirakalwasan, and Wan-Ju Cheng

Contents

P.-L. Lee (✉)
Center of Sleep Disorder, National Taiwan University Hospital, National Taiwan University, Taipei, Taiwan

Department of Internal Medicine, College of Medicine, National Taiwan University, Taipei, Taiwan
e-mail: leepeilin@ntu.edu.tw

N. Chirakalwasan
Division of Pulmonary and Critical Care Medicine, Department of Medicine, Faculty of Medicine, Chulalongkorn University, Bangkok, Thailand

Excellence Center for Sleep Disorders, King Chulalongkorn Memorial Hospital, Thai Red Cross Society, Bangkok, Thailand
e-mail: naricha.c@chula.ac.th

W.-J. Cheng
National Center for Geriatrics and Welfare Research, National Health Research Institutes, Zhunan, Miaoli County, Taiwan

Department of Public Health, China Medical University, Taichung, Taiwan
e-mail: s871056@nhri.edu.tw

H.-Y. Li (ed.), *Snoring and Obstructive Sleep Apnea*,
https://doi.org/10.1007/978-981-95-2648-2_22

Abstract

Obstructive sleep apnea (OSA) presents distinct characteristics in women throughout their lifespan, notably influenced by physiological changes associated with pre-childbearing, childbearing, and peri-menopausal to post-menopausal stages. Recent advancements in the phenotyping and endotyping of OSA have enhanced our understanding of sex differences within the disorder, facilitating the development of more effective detection and treatment strategies specifically tailored for women. The growing recognition of OSA's implications during pregnancy has prompted the publication of consensus statements from relevant professional societies, emphasizing the importance of addressing OSA in maternal healthcare. This chapter comprehensively reviews OSA in women, focusing on critical aspects, such as physiological changes and upper airway endotypes, clinical manifestations, differential phenotyping, and detection methodologies. Moreover, it explores treatment options, including both positive airway pressure (PAP) and non-PAP therapies, to provide a holistic understanding of OSA management in female patients.

Keywords

Apnea-hypopnea index (AHI) · Arousal threshold · Collapsibility · Continuous positive airway pressure (CPAP) · Epworth Sleepiness Scale (ESS) · Endotype · Home sleep apnea testing (HSAT) · Hormonal replacement therapy · Loop gain · Machine learning (ML) · Mandibular advancement device (MAD) · Obstructive sleep apnea (OSA) · Phenotype · Polysomnography (PSG) · Pregnancy

Abbreviations

AHI	apnea–hypopnea index
BMI	body mass index
CO_2	carbon dioxide
CPAP	continuous positive airway pressure
MAD	mandibular advancement device
OSA	obstructive sleep apnea
PSG	polysomnography
PUP	Phenotyping Using Polysomnography
REM	rapid eye movement
SpO_2	oxygen saturation

22.1 Introduction

Obstructive sleep apnea (OSA) affects 9–38% of individuals with an apnea–hypopnea index (AHI) ≥5/h, with prevalence rates of 13–33% in men and 6–19% in women. The prevalence was 6–17% at AHI ≥15/h [1]. Despite its prevalence, OSA often goes undiagnosed and untreated.

Untreated OSA is linked to significant health risks, including motor vehicle accidents, reduced quality of life, depression, diabetes, metabolic syndrome, and cardiovascular diseases, such as hypertension, atrial fibrillation, and stroke [2, 3]. Severe OSA may double or triple the risk of all-cause mortality and major adverse cardiac and cerebrovascular events over 5–10 years compared to snorer, mild, or moderate OSA [4]. Early diagnosis and intervention are vital for reducing these risks.

Recent advancements in phenotyping and endotyping of OSA have enhanced our understanding of sex differences in the disorder [5]. OSA manifests uniquely in women across the lifespan, particularly influenced by pregnancy and hormonal changes [6]. A meta-analysis indicates that approximately 15% of pregnant women are affected by OSA, which is associated with adverse pregnancy outcomes, underscoring the need for focused research on female OSA [7]. In line with this, the Society of Anesthesia and Sleep Medicine and the Society for Obstetric Anesthesia and Perinatology recently published recommendations based on systematic reviews and expert opinions regarding the screening, diagnosis, and management of OSA in pregnancy [8].

This chapter examines OSA in women across the lifespan, addressing prechildbearing, childbearing, and peri-menopausal to post-menopausal stages. We focus on: 1) physiological changes and upper airway endotypes; 2) clinical manifestations and differential phenotyping; 3) detection methodologies, including screening and diagnosis; and 4) treatment options, encompassing positive airway pressure (PAP) and non-PAP therapies.

22.2 Pathophysiology

22.2.1 Physiologic Changes Across the Lifespan

22.2.1.1 Pre-child Bearing

There is an increase in central and peripheral chemosensitivity results in a lower CO_2 reserve in men, predisposing them to apneas [9]. Women typically exhibit peripherally distributed adipose tissue, whereas men have a greater central and visceral fat accumulation, reflecting the influence of sex hormones during reproductive years [10].

22.2.1.2 Child Bearing

The elevation of the diaphragm from the enlarged uterus can lead to tracheal shortening and a progressive reduction in functional residual capacity, expiratory reserve volume, and residual volume [11]. Elevated progesterone enhances ventilatory drive, leading to more negative inspiratory pressure [12]. Conversely, increased circulating progesterone may boost genioglossal activity, offering protection against airway collapse [13].

22.2.1.3 Peri-menopausal and Post-menopausal

As women age and ovarian reserves decline, estrogen and progesterone levels first fluctuate and then decrease significantly. This decline is generally accompanied by a relative increase in androgens due to decreased ovarian hormone levels and increased sex hormone-binding globulin [14], which can elevate the risk of OSA.

22.2.2 Upper Airway and Endotype

22.2.2.1 Collapsibility

Women with OSA exhibit a less collapsible upper airway compared to men, as evidenced by measurements such as passive critical pressure [15] and minimal ventilation (V_{min}) or ventilation at eupneic drive ($V_{passive}$) derived from polysomnographic signals [16, 17]. This difference is likely attributable to anatomical and physiological variations, with male airways generally being longer and thicker, resulting in increased susceptibility to collapse [18].

22.2.2.2 Arousal Threshold

Research indicates that women have a lower arousal threshold and experience shorter respiratory events than men, which may explain the higher prevalence of insomnia symptoms among women with OSA. However, these distinctions may lose significance after adjustment for AHI, suggesting that arousal threshold could be more closely related to OSA severity rather than sex alone [17].

22.2.2.3 Upper Airway Compensation

The evidence concerning gender differences in upper airway muscle compensation remains inconsistent. Some studies report similar pharyngeal resistance and retropalatal compliance in both sexes without OSA [19], while other polysomnography (PSG)-based studies indicate better pharyngeal compensation in women with OSA compared to men [5]. The role of sex hormones is significant; testosterone has been linked to impaired muscle compensation in men [16], while estrogen appears to enhance upper airway dilator muscle function, providing benefits for postmenopausal women with OSA [20]. Progesterone has also been shown to improve ventilatory responsiveness to hypoxia and hypercapnia [21]. These factors may contribute to the lower prevalence of OSA in premenopausal women. However, evidence linking changes in female sex hormones during menopause to increased prevalence of sleep-disordered breathing in midlife women remains limited.

22.2.2.4 Loop Gain

Loop gain, which measures the ventilatory response to disturbances like apnea and hypopnea, does not significantly differ between sexes [16]. In both genders, aging is associated with an increased incidence of central apneas [22], oscillatory breathing [23], and Cheyne–Stokes respiration in heart failure patients [24], reflecting a more unstable breathing pattern. However, the association between age and loop gain was inconsistent by measuring methods. The Phenotyping Using Polysomnography (PUP) method shows that loop gain increases with age [5, 16], while the continuous positive airway pressure (CPAP) drop method reveals lower loop gain in older patients compared to younger ones with OSA [25]. Although female sex hormones may enhance responses to hypoxia and hypercapnia, and exogenous progesterone combined with estrogen may amplify these effects in postmenopausal women, the evidence is still limited. Consequently, definitive conclusions regarding hormone replacement therapy in treating OSA in menopausal women remain inconclusive.

22.3 Clinical Manifestation

22.3.1 Differential Phenotyping

Women tend to present with nonspecific symptoms, including headache, fatigue, depression, anxiety, sleep-onset insomnia, and sleep disruption than the classic OSA presentations [26]. Recent cluster analysis identified four subtypes: (1) subtype A—young, obese males with severe OSA and excessive daytime sleepiness; [2] subtype B—older, obese males with severe OSA and minimal symptoms; (3) subtype C—middle-aged, mildly obese females with insomnia and moderate-to-severe OSA; and (4) subtype D—young, non-obese males with severe OSA and primary upper airway symptoms, displaying the least hypoxemia and comorbidity rates [27]. This analysis highlights gender differences in clinical and polysomnographic characteristics and associated comorbidities.

22.3.2 Consequences

22.3.2.1 Women's Vulnerability to Cardiovascular Consequences

Women with OSA more frequently report insomnia symptoms than excessive daytime sleepiness, correlating with increased cardiovascular mortality risk [28, 29]. The duration of SpO_2 <90% is more strongly associated with cardiac events in women than in men [30]. Premenopausal women appear hormonally protected, as estradiol can mitigate cardiopulmonary disorders and oxidative stress from chronic intermittent hypoxia in animal models [31]. In humans, OSA contributes less to hypertension in women than in men, although arrhythmia rates are similar across genders [32]. Future studies should evaluate menstrual status to clarify its role in these associations.

22.3.2.2 Non-pregnant Women Versus Pregnant Women

22.3.2.2.1 Non-pregnant Women

In a cohort with suspected OSA, the impact of nocturnal oxygen desaturation on composite outcomes—including all-cause mortality and hospitalization for myocardial infarction, stroke, heart failure, or atrial fibrillation—was greater in women than in men [30]. Additionally, the hypoxic burden during REM sleep, which is more pronounced in women, may contribute to greater hemodynamic consequences of OSA [33].

22.3.2.2.2 Pregnant Women

OSA during pregnancy is associated with higher risks for health complications and adverse outcomes, including gestational diabetes, gestational hypertension, preeclampsia, cesarean delivery, small-for-gestational-age infants, and longer hospital stays. Serious complications, such as pulmonary edema, congestive heart failure, admissions to intensive care units, and hysterectomy, have also been reported [7, 27].

22.4 OSA Detection

22.4.1 OSA Assessment and the Use of Screening Tools

The assessment of OSA starts with a detailed review of the patient's sleep history and a thorough examination of the upper airway. Key inquiries include the frequency of observed apneas, instances of choking or gasping during sleep, sleep disturbances, and levels of daytime sleepiness and fatigue [2, 3]. Screening methods may involve instruments such as questionnaires, predictive algorithms, and consumer or medical-grade devices, including smartphone applications, to facilitate the early identification of OSA [34].

22.4.1.1 Questionnaire Instruments

Historically, OSA screening has relied on questionnaires [35–37]. The questionnaires frequently utilized for evaluating OSA risk are the STOP-Bang [38], NoSAS [36], Berlin Questionnaire [39], and the Epworth Sleepiness Scale (ESS) [37]. A meta-analysis comparing the diagnostic performance of the STOP-Bang, Berlin Questionnaire, and ESS at AHI thresholds of ≥5/h, ≥15/h, and ≥30/h found that a STOP-Bang score of >3 had the highest sensitivity, whereas the ESS displayed the lowest accuracy across all thresholds [35]. Given that women often present with a lower body mass index (BMI), greater fat distribution in the lower body, diminished snoring, prolonged sleep-onset latency, and shorter sleep duration than men [40], utilizing a lower AHI cutoff may be more suitable.

22.4.1.2 Prediction Model

Numerous OSA prediction models have been developed over the years, utilizing various clinical features—such as demographics, patient-reported symptoms, sleep patterns, comorbidities, and anthropometrics. There is a notable shift toward advanced screening techniques that leverage artificial intelligence, machine learning, and deep learning algorithms to evaluate the likelihood of OSA. These models generally prioritize sensitivity over specificity to enhance early detection of moderate-to-severe OSA. However, this focus can lead to increased false-positive rates, resulting in unnecessary sleep testing prescriptions [34, 40–46] (Table 22.1).

22.4.2 OSA Diagnostic Tools

Diagnostic testing for OSA involves in-laboratory PSG, unattended PSG, or home sleep apnea testing, integrated into a comprehensive evaluation with follow-up [34, 47]. Testing is recommended for those with OSA symptoms and individuals in mandatory screening roles, such as pilots, commercial drivers, and patients with significant comorbidities like cardiovascular disease and diabetes [47]. Conversely, routine sleep testing is not recommended for asymptomatic individuals in the general population, as there is insufficient evidence of benefit [3, 48].

22.4.2.1 Polysomnographic Feature

Women with OSA typically exhibit distinct polysomnographic characteristics, including a lower overall AHI, shorter apneic episodes, more frequent subcriterion events, reduced supine OSA, increased REM-related OSA, and greater sleep fragmentation, particularly during pregnancy [49]. Recent interest in endotyping OSA using polysomnographic features has led to various methods. Invasive measures, such as epiglottic pressure catheters and diaphragm electromyography, have been utilized to assess ventilation [50]. Wellman et al. employed CPAP manipulation in controlled settings [51], but these methods are often limited to physiology

Table 22.1 Data profile and performance of OSA prediction models with clinical features

Author year	Subject No.	AHI cutoff (/h)	Preva (%)	Algorithm	AUROC	Accu (%)	Sen (%)	Spec (%)
Kirby [73]	150	≥10	65.00	GRNN	0.94	91.3	98.90	80.00
Caffo [45]	1383	≥7*	37.80	LR	0.75	N/A	66.00	70.00
Bouloukaki [74]	2690	≥15	79.00	LR	0.78	N/A	70.00	73.10
Zou [75]	784	≥5	83.80	LR	[a]1. 0.84	N/A	1. 86.91	1. 74.80
					[b]2. 0.98	N/A	2. 94.22	2. 85.83
Marti-Soler [36]	1042	>20	11.52	LR	0.74	N/A	85.00	77.00
Shah [76]	12,158	≥15	9.02	LR	0.83	N/A	77.00	75.00
Ustun [43]	1922	>5 or > 10*	76.90	SLIM	[c]1. 0.77	N/A	1. 82.80	1. 56.20
					[d]2. 0.79	N/A	2. 83.20	2. 58.90
Wang [77]	3343	<5, 5–15, 15–30, >30	85.68	SMOTE+FDT	N/A	81.82	N/A	N/A
Liu [42]	1154	1. ≥15	1. 61.87	SVM	N/A&	N/A&	1. 68.35	1. 82.55
		2. ≥30	2. 44.04		N/A		2. 68.32	2. 79.85
Shin [78]	108,781	>5	2.08	LR	0.82	N/A	72.30	76.20
Traxdorf [79]	100	1. ≥5	1. 70.00	LR	N/A	N/A	1. 94.30	1. 50.00
		2. ≥15	2. 55.50		N/A	N/A	2. 92.70	2. 33.30
		3. ≥30	3. 26.00		N/A	N/A	3. 92.30	3. 22.90
Duarte [80]	2035	1. ≥5	1. 76.40	LR	1. 0.78	1.77.20	1.83.10	2.58.20
		2. ≥15	2. 54.70		2. 0.76	2.69.00	2. 88.70	2. 45.30
		3. ≥30	3. 35.80		3. 0.75	3.56.40	3. 91.50	3. 36.80
Huang [40]	6875	1. ≥5	1. 82.45	SVM	1. 0.82	1.74.24	1. 74.14	1. 74.71
		2. ≥15	2. 61.25		2. 0.80	2.72.68	2. 75.18	2. 68.73
		3. ≥30	3. 40.60		3. 0.78	3.70.28	3. 70.26	3. 70.30
Kim [81]	279	≥5	76.3	SVM	0.87	83.33	80.33	86.96
Keshavarz [82]	250	≥15	65.8	NB	0.768	72.3	78.95	59.49

Zhang [83]	481	≥15	35.55	SVM	0.83	N/A	92	75
Holfinger [44]	17,448	≥15	53.22	ANN	0.68	N/A	74	51
Casal-Guisande [84]	5000	1. ≥10	1. 72.59	ANFIS	1. 0.88	N/A	1. 89	1. 67
		2. ≥15	2. 61.46		2. 0.88	N/A	2. 87	2. 74
		3. ≥20	3. 51.30		3. 0.83	N/A	3. 87	3. 65
		4. ≥25	4. 43.41		4. 0.83	N/A	4. 70	4. 81
		5. ≥30	5. 36.80		5. 0.86	N/A	5. 76	5. 80
Han [85]	803	1. ≥5	1. 22.42	LightGBM	1. 0.82	1. 87.82	1. 97.85	N/A
		2. ≥15	2. 24.16		2. 0.95	2. 87.84	2. 92.65	N/A
		3. ≥30	3. 35.87		3. 0.95	3. 91.06	3. 70.19	N/A
Huo [86]	1237	≥15	54.81	LR	0.78	N/A	77	68

Adapted from Lee et al. (34)

*Respiratory disturbance index (RDI); #OSA; &No whole-group data available. [a]Predction1 = 1/[1+ exp (0.0486 age +0.1356 BMI +0.0616 WC +0.1626 ESS +0.4546 sex –10.535)]; [b] Prediction2=1/[1+ exp (0.0296age +0.0596WC +0.1116ESS –0.3596minimum SaO2 +26.202)]; [c]Size 5, false positive rate goal of 20%; dSize 10, false positive rate goal of 20%

Abbreviations: *Acc* accuracy, *AHI* apnea–hypopnea index, *ANN* artificial neural network, *ANFIS* Adaptive Neuro-fuzzy Inference System, *AUROC* area under receiver operating characteristic, *Accu* accuracy, *FDT* Fuzzy decision tree, *GRNN* generalized regression neural network, *LR* logistic regression, N/A not available, *NB* Naïve Bayes, *Preva* prevalence, *Sen* sensitivity, *SLIM* Supersparse Linear Integer Model, *Spec* specificity, *SOMTE* synthetic minority oversampling technique, *SVM* Support Vector Machine, *SW* surgical ward laboratories. The PUP method provides a noninvasive means to estimate OSA endotypes using clinically scored data, quantifying traits such as arousal threshold, upper airway collapsibility, compensation, and loop gain. This approach can be incorporated to routine in-laboratory polysomnographic studies [52].

22.5 Intervention

22.5.1 PAP Therapy

22.5.1.1 Non-pregnant Women

Gender differences impact treatment outcomes. CPAP remains the gold standard for OSA; however, it may be less effective in women compared to men [53]. While men and women do not differ in CPAP treatment pressure for the same level of OSA severity during titration studies [54], women typically exhibit lower residual AHI. Cluster analysis reveals that subtype C (middle-aged, mildly obese females with insomnia, and moderate-to-severe OSA) may respond moderately to CPAP therapy [27]. Women in the "disturbed sleep" cluster of the Sleep Heart Health Study showed a reduced risk of stroke with CPAP use compared to other OSA subtypes [55].

Women, especially younger women, tend to have lower adherence to CPAP treatment [56]. This may also relate to a lower arousal threshold in women with OSA [57]. Clinical trials indicate that elevating the arousal threshold with benzodiazepines and other sedatives may serve as adjuncts to CPAP therapy in women with OSA [58].

22.5.1.2 Pregnant Women

Data on OSA treatment during pregnancy are limited. Auto-titrating positive airway pressure remains the primary treatment option. Studies have shown favorable effects, including significant blood pressure reduction [59], mitigated cardiac output decreases with reduced total peripheral resistance [60], increased fetal movement [60], improved insulin secretion [61], and better pregnancy outcomes, such as lower rates of preterm delivery, unplanned cesarean sections, and neonatal intensive care unit admissions [61].

22.5.2 Non-PAP Therapy

22.5.2.1 Non-pregnant Women

Weight loss is a recommended treatment for obese patients with OSA, as obesity is characterized by a collapsible upper airway and respiratory instability (loop gain) [62]. Studies indicate that women with OSA achieve greater weight loss through lifestyle modifications [63].

Adjustable mandibular advancement device (MAD) are recommended for patients intolerant to CPAP or bilevel positive airway pressure, and they may be used in conjunction with CPAP when pressure intolerance occurs [64]. Although MAD is less effective in reducing the overall respiratory disturbance index compared to CPAP, compliance tends to be better with MAD. Studies show that women respond better to MAD than men [65].

For other treatments, such as negative-pressure intraoral devices and multilevel surgical interventions, no significant differences in efficacy have been observed between genders [33, 66].

The effects of hormone replacement therapy in postmenopausal women with OSA are controversial. One study found no effect of medroxyprogesterone [67], while another reported slightly lower AHI in postmenopausal women receiving hormone replacement therapy compared to those not on therapy [68].

22.5.2.2 Pregnant Women

Adjustments to MAD require weeks to months, making this treatment impractical for pregnant women. Surgical modalities have not been studied in this population due to potential risks associated with anesthesia and surgical procedures [6].

22.5.2.3 Non-CPAP Treatment and Endotypic Traits

Limited research has examined sex differences in non-CPAP treatment response related to sex-specific pathology. In upper airway surgery, high collapsibility is corrected, and high loop gain predicts surgical success [69]. While more women are identified as surgical responders, differential changes in loop gain by sex remain unclear. Considering the effects of estrogen and progesterone on upper airway compensation and ventilatory response, hormone therapy shows potential as a treatment for postmenopausal women with OSA [70]. Women also demonstrate higher treatment success rates with MAD compared to men [71], likely due to less collapsible upper airways, as indicated by smaller neck circumference. A high arousal threshold predicts responses to hypoglossal nerve stimulation [72], with a higher response rate observed in women; however, the postmenopausal status of participants limits the generalizability of findings.

References

1. Senaratna CV, Perret JL, Lodge CJ, Lowe AJ, Campbell BE, Matheson MC, et al. Prevalence of obstructive sleep apnea in the general population: a systematic review. Sleep Med Rev. 2017;34:70–81.
2. Lee PL, Wu YW, Cheng HM, Wang CY, Chuang LP, Lin CH, et al. Recommended assessment and management of sleep disordered breathing in patients with atrial fibrillation, hypertension and heart failure: Taiwan Society of Cardiology/Taiwan Society of sleep Medicine/Taiwan Society of pulmonary and Critical Care Medicine joint consensus statement. J Formos Med Assoc. 2024;123(2):159–78.
3. Gottlieb DJ, Punjabi NM. Diagnosis and management of obstructive sleep Apnea: a review. JAMA. 2020;323(14):1389–400.

4. Marin JM, Carrizo SJ, Vicente E, Agusti AG. Long-term cardiovascular outcomes in men with obstructive sleep apnoea-hypopnoea with or without treatment with continuous positive airway pressure: an observational study. Lancet. 2005;365(9464):1046–53.
5. Hang LW, Tsai YC, Finnsson E, Agustsson JS, Sands SA, Cheng WJ. Sex-specific age-related worsening of pathological endotypic traits in patients with obstructive sleep apnea. Sleep. 2025;48(1):zsae185.
6. Chirakalwasan NR, Sirimon. Sleep disorders and gestational diabetes mellitus. In: Watson R, editor. Neurological modulation of sleep: mechanisms and function of sleep health. Elsevier; 2020. p. 285–302.
7. Liu L, Su G, Wang S, Zhu B. The prevalence of obstructive sleep apnea and its association with pregnancy-related health outcomes: a systematic review and meta-analysis. Sleep Breath. 2019;23(2):399–412.
8. Dominguez JE, Cantrell S, Habib AS, Izci-Balserak B, Lockhart E, Louis JM, et al. Society of anesthesia and sleep medicine and the society for obstetric anesthesia and perinatology consensus guideline on the screening, diagnosis, and treatment of obstructive sleep apnea in pregnancy. Obstet Gynecol. 2023;142(2):403–23.
9. Zhou XS, Shahabuddin S, Zahn BR, Babcock MA, Badr MS. Effect of gender on the development of hypocapnic apnea/hypopnea during NREM sleep. J Appl Physiol (1985). 2000;89(1):192–9.
10. Geer EB, Shen W. Gender differences in insulin resistance, body composition, and energy balance. Giend Med. 2009;6(Suppl 1):60–75.
11. Hirnle L, Lysenko L, Gerber H, Lesnik P, Baranowska A, Rachwalik M, et al. Respiratory function in pregnant women. Adv Exp Med Biol. 2013;788:153–60.
12. Remmers JE, Anch AM, deGroot WJ. Respiratory disturbances during sleep. Clin Chest Med. 1980;1(1):57–71.
13. Bourjeily G, Ankner G, Mohsenin V. Sleep-disordered breathing in pregnancy. Clin Chest Med. 2011;32(1):175–89.
14. Brzozowska M, Lewinski A. Changes of androgens levels in menopausal women. Prz Menopauzalny. 2020;19(4):151–4.
15. Kirkness JP, Schwartz AR, Schneider H, Punjabi NM, Maly JJ, Laffan AM, et al. Contribution of male sex, age, and obesity to mechanical instability of the upper airway during sleep. J Appl Physiol (1985). 2008;104(6):1618–24.
16. Sands SA, Alex RM, Mann D, Vena D, Terrill PI, Gell LK, et al. Pathophysiology underlying demographic and obesity determinants of sleep apnea severity. Ann Am Thorac Soc. 2023;20(3):440–9.
17. Hang LW, Huang CS, Cheng WJ. Clinical characteristics of Asian patients with sleep apnea with low arousal threshold and sleep structure change with continuous positive airway pressure. Sleep Breath. 2021;25(3):1309–17.
18. Malhotra A, Huang Y, Fogel RB, Pillar G, Edwards JK, Kikinis R, et al. The male predisposition to pharyngeal collapse: importance of airway length. Am J Respir Crit Care Med. 2002;166(10):1388–95.
19. Pillar G, Malhotra A, Fogel R, Beauregard J, Schnall R, White DP. Airway mechanics and ventilation in response to resistive loading during sleep: influence of gender. Am J Respir Crit Care Med. 2000;162(5):1627–32.
20. Zhang L, Ou X, Zhu T, Lv X. Beneficial effects of estrogens in obstructive sleep apnea hypopnea syndrome. Sleep Breath. 2020;24(1):7–13.
21. Regensteiner JG, Woodard WD, Hagerman DD, Weil JV, Pickett CK, Bender PR, et al. Combined effects of female hormones and metabolic rate on ventilatory drives in women. J Appl Physiol (1985). 1989;66(2):808–13.
22. Bixler EO, Vgontzas AN, Ten Have T, Tyson K, Kales A. Effects of age on sleep apnea in men: I. Prevalence and severity. Am J Respir Crit Care Med. 1998;157(1):144–8.
23. Hudgel DW, Devadatta P, Hamilton H. Pattern of breathing and upper airway mechanics during wakefulness and sleep in healthy elderly humans. J Appl Physiol (1985). 1993;74(5):2198–204.

24. Sin DD, Fitzgerald F, Parker JD, Newton G, Floras JS, Bradley TD. Risk factors for central and obstructive sleep apnea in 450 men and women with congestive heart failure. Am J Respir Crit Care Med. 1999;160(4):1101–6.
25. Edwards BA, Wellman A, Sands SA, Owens RL, Eckert DJ, White DP, et al. Obstructive sleep apnea in older adults is a distinctly different physiological phenotype. Sleep. 2014;37(7):1227–36.
26. Lozo T, Komnenov D, Badr MS, Mateika JH. Sex differences in sleep disordered breathing in adults. Respir Physiol Neurobiol. 2017;245:65–75.
27. Zinchuk A, Yaggi HK. Phenotypic subtypes of OSA: a challenge and opportunity for precision medicine. Chest. 2020;157(2):403–20.
28. Labarca G, Dreyse J, Salas C, Letelier F, Jorquera J. A validation study of four different cluster analyses of OSA and the incidence of cardiovascular mortality in a Hispanic population. Chest. 2021;160(6):2266–74.
29. Covassin N, Lu D, St Louis EK, Chahal AA, Schulte PJ, Mansukhani MP, et al. Sex-specific associations between daytime sleepiness, chronic diseases and mortality in obstructive sleep apnea. Front Neurosci. 2023;17:1210206.
30. Kendzerska T, Leung RS, Atzema CL, Chandy G, Meteb M, Malhotra A, et al. Cardiovascular consequences of obstructive sleep apnea in women: a historical cohort study. Sleep Med. 2020;68:71–9.
31. Laouafa S, Ribon-Demars A, Marcouiller F, Roussel D, Bairam A, Pialoux V, et al. Estradiol protects against cardiorespiratory dysfunctions and oxidative stress in intermittent hypoxia. Sleep. 2017;40(8) https://doi.org/10.1093/sleep/zsx104.
32. Parise BK, Ferreira NL, Drager LF. The cardiovascular impact of obstructive sleep apnea in women: current knowledge and future perspectives. Sleep Med Clin. 2023;18(4):473–80.
33. Nilius G, Farid-Moayer M, Lin CM, Knaack L, Wang YP, Dellweg D, et al. Multi-center safety and efficacy study of a negative-pressure intraoral device in obstructive sleep apnea. Sleep Med. 2024;119:139–46.
34. Lee P-L, Gu W, Huang W-C, Chiang AA. From screening at clinic to diagnosis at home: how AI/ML/DL algorithms are transforming sleep apnea detection. In: Berry RB, Pardalos PM, Xian X, editors. Handbook of AI and data sciences for sleep disorders, Springer Optimization and Its Applications. 216. Cham: Springer; 2024. p. 109–60.
35. Chiu HY, Chen PY, Chuang LP, Chen NH, Tu YK, Hsieh YJ, et al. Diagnostic accuracy of the Berlin questionnaire, STOP-BANG, STOP, and Epworth sleepiness scale in detecting obstructive sleep apnea: a bivariate meta-analysis. Sleep Med Rev. 2017;36:57–70.
36. Marti-Soler H, Hirotsu C, Marques-Vidal P, Vollenweider P, Waeber G, Preisig M, et al. The NoSAS score for screening of sleep-disordered breathing: a derivation and validation study. Lancet Respir Med. 2016;4(9):742–8.
37. Chen NH, Johns MW, Li HY, Chu CC, Liang SC, Shu YH, et al. Validation of a Chinese version of the Epworth sleepiness scale. Qual Life Res. 2002;11(8):817–21.
38. Chung F, Abdullah HR, Liao P. STOP-Bang questionnaire: a practical approach to screen for obstructive sleep apnea. Chest. 2016;149(3):631–8.
39. Luo J, Huang R, Zhong X, Xiao Y, Zhou J. STOP-Bang questionnaire is superior to Epworth sleepiness scales, Berlin questionnaire, and STOP questionnaire in screening obstructive sleep apnea hypopnea syndrome patients. Chin Med J. 2014;127(17):3065–70.
40. Huang WC, Lee PL, Liu YT, Chiang AA, Lai F. Support vector machine prediction of obstructive sleep apnea in a large-scale Chinese clinical sample. Sleep. 2020;43(7):zsz295.
41. Duarte RLM, Rabahi MF, Magalhães-da-Silveira FJ, de Oliveira-E-Sá TS, Mello FCQ, Gozal D. Simplifying the screening of obstructive sleep apnea with a 2-item model, no-apnea: a cross-sectional study. J Clin Sleep Med. 2018;14(7):1097–107.
42. Liu WT, Wu HT, Juang JN, Wisniewski A, Lee HC, Wu D, et al. Prediction of the severity of obstructive sleep apnea by anthropometric features via support vector machine. PLoS One. 2017;12(5):e0176991.
43. Ustun B, Westover MB, Rudin C, Bianchi MT. Clinical prediction models for sleep apnea: the importance of medical history over symptoms. J Clin Sleep Med. 2016;12(2):161–8.

44. Holfinger SJ, Lyons MM, Keenan BT, Mazzotti DR, Mindel J, Maislin G, et al. Diagnostic performance of machine learning-derived OSA prediction tools in large clinical and community-based samples. Chest. 2022;161(3):807–17.
45. Caffo B, Diener-West M, Punjabi NM, Samet J. A novel approach to prediction of mild obstructive sleep disordered breathing in a population-based sample: the Sleep Heart Health Study. Sleep. 2010;33(12):1641–8.
46. Bazoukis G, Bollepalli SC, Chung CT, Li X, Tse G, Bartley BL, et al. Application of artificial intelligence in the diagnosis of sleep apnea. J Clin Sleep Med. 2023;19(7):1337–63.
47. Kapur VK, Auckley DH, Chowdhuri S, Kuhlmann DC, Mehra R, Ramar K, et al. Clinical practice guideline for diagnostic testing for adult obstructive sleep apnea: an American Academy of sleep medicine clinical practice guideline. J Clin Sleep Med. 2017;13(3):479–504.
48. Jonas DE, Amick HR, Feltner C, Weber RP, Arvanitis M, Stine A, et al. Screening for obstructive sleep apnea in adults: evidence report and systematic review for the US preventive services task force. JAMA. 2017;317(4):415–33.
49. Bonsignore MR, Saaresranta T, Riha RL. Sex differences in obstructive sleep apnoea. Eur Respir Rev. 2019;28(154):190030.
50. Eckert DJ, White DP, Jordan AS, Malhotra A, Wellman A. Defining phenotypic causes of obstructive sleep apnea. Identification of novel therapeutic targets. Am J Respir Crit Care Med. 2013;188(8):996–1004.
51. Wellman A, Edwards BA, Sands SA, Owens RL, Nemati S, Butler J, et al. A simplified method for determining phenotypic traits in patients with obstructive sleep apnea. J Appl Physiol (1985). 2013;114(7):911–22.
52. Sands SA, Edwards BA, Terrill PI, Taranto-Montemurro L, Azarbarzin A, Marques M, et al. Phenotyping pharyngeal pathophysiology using polysomnography in patients with obstructive sleep apnea. Am J Respir Crit Care Med. 2018;197(9):1187–97.
53. Chang ET, Wang HM, Lai HL. Gender differences in obstructive sleep apnea syndrome. Eur J Intern Med. 2016;33:e9–e10.
54. Duarte RLM, Magalhaes-da-Silveira FJ, Gozal D. Are there sex-related differences in therapeutic CPAP levels in adults undergoing in-laboratory titration? J Clin Sleep Med. 2021;17(9):1815–20.
55. Mazzotti DR, Keenan BT, Lim DC, Gottlieb DJ, Kim J, Pack AI. Symptom subtypes of obstructive sleep apnea predict incidence of cardiovascular outcomes. Am J Respir Crit Care Med. 2019;200(4):493–506.
56. Patel SR, Bakker JP, Stitt CJ, Aloia MS, Nouraie SM. Age and sex disparities in adherence to CPAP. Chest. 2021;159(1):382–9.
57. Cheng WJ, Finnsson E, Agustsson JS, Sands SA, Hang LW. Continuous positive airway pressure and adherence in patients with different endotypes of obstructive sleep apnea. J Sleep Res. 2024;33(1):e13999.
58. Taranto-Montemurro L, Messineo L, Wellman A. Targeting Endotypic traits with medications for the pharmacological treatment of obstructive sleep apnea. A review of the current literature. J Clin Med. 2019;8(11):1846.
59. Edwards N, Blyton DM, Kirjavainen T, Kesby GJ, Sullivan CE. Nasal continuous positive airway pressure reduces sleep-induced blood pressure increments in preeclampsia. Am J Respir Crit Care Med. 2000;162(1):252–7.
60. Blyton DM, Sullivan CE, Edwards N. Reduced nocturnal cardiac output associated with preeclampsia is minimized with the use of nocturnal nasal CPAP. Sleep. 2004;27(1):79–84.
61. Chirakalwasan N, Amnakkittikul S, Wanitcharoenkul E, Charoensri S, Saetung S, Chanprasertyothin S, et al. Continuous positive airway pressure therapy in gestational diabetes with obstructive sleep apnea: a randomized controlled trial. J Clin Sleep Med. 2018;14(3):327–36.
62. Bokov P, Essalhi M, Delclaux C. Loop gain in severely obese women with obstructive sleep apnoea. Respir Physiol Neurobiol. 2016;221:49–53.

63. Newman AB, Foster G, Givelber R, Nieto FJ, Redline S, Young T. Progression and regression of sleep-disordered breathing with changes in weight: the Sleep Heart Health Study. Arch Intern Med. 2005;165(20):2408–13.
64. Ferguson KA, Cartwright R, Rogers R, Schmidt-Nowara W. Oral appliances for snoring and obstructive sleep apnea: a review. Sleep. 2006;29(2):244–62.
65. Cistulli PA, Sutherland K. Phenotyping obstructive sleep apnoea-bringing precision to oral appliance therapy. J Oral Rehabil. 2019;46(12):1185–91.
66. Su YY, Lin PW, Lin HC, Chang CT, Lin CY, Friedman M, et al. Systematic review and updated meta-analysis of multi-level surgery for patients with OSA. Auris Nasus Larynx. 2022;49(3):421–30.
67. Block AJ, Wynne JW, Boysen PG, Lindsey S, Martin C, Cantor B. Menopause, medroxyprogesterone and breathing during sleep. Am J Med. 1981;70(3):506–10.
68. Pickett CK, Regensteiner JG, Woodard WD, Hagerman DD, Weil JV, Moore LG. Progestin and estrogen reduce sleep-disordered breathing in postmenopausal women. J Appl Physiol (1985). 1989;66(4):1656–61.
69. Joosten SA, Leong P, Landry SA, Sands SA, Terrill PI, Mann D, et al. Loop gain predicts the response to upper airway surgery in patients with obstructive sleep apnea. Sleep. 2017;40(7) https://doi.org/10.1093/sleep/zsx094.
70. O'Halloran KD, Lewis P, McDonald F. Sex, stress and sleep apnoea: decreased susceptibility to upper airway muscle dysfunction following intermittent hypoxia in females. Respir Physiol Neurobiol. 2017;245:76–82.
71. Vecchierini MF, Attali V, Collet JM, d'Ortho MP, Goutorbe F, Kerbrat JB, et al. Sex differences in mandibular repositioning device therapy effectiveness in patients with obstructive sleep apnea syndrome. Sleep Breath. 2019;23(3):837–48.
72. Op de Beeck S, Wellman A, Dieltjens M, Strohl KP, Willemen M, Van de Heyning PH, et al. Endotypic mechanisms of successful hypoglossal nerve stimulation for obstructive sleep apnea. Am J Respir Crit Care Med. 2021;203(6):746–55.
73. Kirby SD, Eng P, Danter W, George CF, Francovic T, Ruby RR, et al. Neural network prediction of obstructive sleep apnea from clinical criteria. Chest. 1999;116(2):409–15.
74. Bouloukaki I, Kapsimalis F, Mermigkis C, Kryger M, Tzanakis N, Panagou P, et al. Prediction of obstructive sleep apnea syndrome in a large Greek population. Sleep Breath. 2011;15(4):657–64.
75. Zou J, Guan J, Yi H, Meng L, Xiong Y, Tang X, et al. An effective model for screening obstructive sleep apnea: a large-scale diagnostic study. PLoS One. 2013;8(12):e80704.
76. Shah N, Hanna DB, Teng Y, Sotres-Alvarez D, Hall M, Loredo JS, et al. Sex-specific prediction models for sleep apnea from the Hispanic community health study/study of Latinos. Chest. 2016;149(6):1409–18.
77. Wang KJ, Chen KH, Huang SH, Teng NC. A prognosis tool based on fuzzy anthropometric and questionnaire data for obstructive sleep apnea severity. J Med Syst. 2016;40(4):110.
78. Shin CH, Grabitz SD, Timm FP, Mueller N, Chhangani K, Ladha K, et al. Development and validation of a Score for Preoperative Prediction of Obstructive Sleep Apnea (SPOSA) and its perioperative outcomes. BMC Anesthesiol. 2017;17(1):71.
79. Traxdorf M, Tziridis K, Scherl C, Iro H, Haferkamp J. The Erlangen questionnaire: a new 5-item screening tool for obstructive sleep apnea in a sleep clinic population – a prospective, double blinded study. Eur Rev Med Pharmacol Sci. 2017;21(16):3690–8.
80. Duarte RLM, Rabahi MF, Magalhaes-da-Silveira FJ, de Oliveira ESTS, Mello FCQ, Gozal D. Simplifying the screening of obstructive sleep apnea with a 2-item model, no-apnea: a cross-sectional study. J Clin Sleep Med. 2018;14(7):1097–107.
81. Kim YJ, Jeon JS, Cho SE, Kim KG, Kang SG. Prediction models for obstructive sleep apnea in Korean adults using machine learning techniques. Diagnostics (Basel). 2021;11(4):612.
82. Keshavarz Z, Rezaee R, Nasiri M, Pournik O. Obstructive sleep apnea: a prediction model using supervised machine learning method. Stud Health Technol Inform. 2020;272:387–90.

83. Zhang L, Yan YR, Li SQ, Li HP, Lin YN, Li N, et al. Moderate to severe OSA screening based on support vector machine of the Chinese population faciocervical measurements dataset: a cross-sectional study. BMJ Open. 2021;11(9):e048482.
84. Casal-Guisande M, Torres-Durán M, Mosteiro-Añón M, Cerqueiro-Pequeño J, Bouza-Rodríguez JB, Fernández-Villar A, et al. Design and conceptual proposal of an intelligent clinical decision support system for the diagnosis of suspicious obstructive sleep apnea patients from health profile. Int J Environ Res Public Health. 2023;20(4):3627.
85. Han H, Oh J. Application of various machine learning techniques to predict obstructive sleep apnea syndrome severity. Sci Rep. 2023;13(1):6379.
86. Huo J, Quan SF, Roveda J, Li A. Coupling analysis of heart rate variability and cortical arousal using a deep learning algorithm. PLoS One. 2023;18(4):e0284167.

The Management of COMISA (Comorbid Insomnia and Sleep Apnea)

23

Ting-Wei Liao and Li-Pang Chuang

Contents

Abstract

The traditional approach to managing COMISA has focused on treating obstructive sleep apnea (OSA) and insomnia as separate entities, with OSA typically prioritized as the primary condition. However, emerging evidence underscores the bidirectional relationship between insomnia and OSA, where each condition can exacerbate or contribute to the development of the other.

Management of COMISA is particularly challenging due to the variability in treatment responses, as insomnia symptoms are often associated with poor adherence to positive airway pressure (PAP) therapy. Furthermore, certain treatments for insomnia can aggravate OSA or related symptoms, such as excessive

T.-W. Liao · L.-P. Chuang (✉)
Department of Pulmonary and Critical Care Medicine, Sleep Center, Chang Gung Memorial Hospital, Taoyuan, Taiwan

Department of Neurology, Sleep Center, Chang Gung Memorial Hospital, Taoyuan, Taiwan

H.-Y. Li (ed.), *Snoring and Obstructive Sleep Apnea*,
https://doi.org/10.1007/978-981-95-2648-2_23

daytime sleepiness. Given this interplay, managing COMISA requires an integrated and multifaceted approach that addresses both conditions simultaneously.

Keywords

Insomnia · Sleep Apnea · Comorbidity · Positive airway pressure · Cognitive behavioral therapy

23.1 Introduction of COMISA

Obstructive sleep apnea (OSA) and insomnia are the two most prevalent sleep disorders that were reported as high as 30% prevalence individually [1, 2]. Those two sleep disorders also have been described frequently to coexist [3]. OSA is characterized by repetitive apneas and hypopneas consequent by upper airway complete or partial collapse during sleep [4]. The participant pathophysiological processes are intermittent hypoxemia, hypercapnia, sympathetic activation, and arousal from occlusion of the upper airway, which lead to long-term comorbidities with cardiovascular disease, impair blood sugar control and neurocognitive impairment [5–8]. Chronic insomnia, which is defined as persistent for a minimum of 3 months and leads to daytime dysfunction, is characterized by frequent reports of difficulty initiating and maintaining sleep or undesired early-morning awakenings despite adequate sleep [9, 10]. Both OSA and insomnia are associated with increased risk of mental and physical problems, reduced quality of life, and increased economic costs of healthcare [11–13].

23.1.1 Prevalence

The term of comorbidity of insomnia and OSA was first described as a new syndrome by Professor Guilleminault in 1973 [14]. However, until 2017, the so-called "COMISA," comorbid insomnia and sleep apnea, was first proposed by Professor Alexander [3]. After that, clinical interests emerged along with published literature over the following years. OSA and insomnia co-occur commonly. About 30–40% of insomnia patients have confirmed OSA diagnosis and around 20–40% OSA patients report clinically insomnia symptoms [15–17]. Some evidence revealed that those two diseases, OSA and insomnia, are bidirectionally related [16]. The mechanisms of each condition will contribute to the development and progression of the other, leading to higher morbidity and mortality, and poor response to each treatment [18].

23.1.2 Comorbidity: Cardiovascular Disease

The cumulative impact of COMISA on pathophysiological mechanisms may contribute to the development of cardiovascular disease (CVD), but the evidence linking COMISA and CVD remains complex [19]. Untreated insomnia or OSA is separately and independently associated with adverse consequences of CVD [6, 20]. COMISA has found to be linked to increased cardiovascular morbidity in cross-sectional studies [18, 21, 22]. The possible mechanisms in between COMISA and high CVD risk include sleep disruption and oxygen desaturation [23]. Both insomnia and OSA disrupt sleep architecture, leading to increased levels of sympathetic nervous system activation and contribute to systemic inflammation, a key factor in the development of CVD. Sleep disturbances can also affect the regulation of hormones, such as renin–angiotensin–aldosterone, which is involved in blood pressure and metabolism. Besides sleep disruption, oxygen desaturation also plays an important role in the development of CVD in COMISA. In OSA, repeated episodes of interrupted breathing can lead to intermittent hypoxia, which activates the circulating blood cells with the enhancement of adhesion and chemotaxis to endothelial cells, leading to further atherosclerosis formation [24].

23.1.3 Comorbidity: Diabetes

Other than increased CVD risk, disturbance of blood sugar control was also affected by COMISA [25]. There is some evidence regarding the relationship between COMISA and diabetes [19, 26]. Compared with insomnia alone, COMISA was associated with a higher odds ratio of hypertension, but not with incidence of diabetes among a US Hispanic/Latino population [26]. An association between COMISA and diabetes was revealed in another cross-sectional study, with a relatively small sample size [19]. A recent published literature demonstrates a novel approach for defining a COMISA phenotype that could predict incident diabetes in older man [25]. Although CVD was a confounder in the prediction of incident diabetes, after adjustment for the main confounding factors associated with cardiovascular risk, unlike OSA syndrome or insomnia alone, only COMISA was associated with higher risk of CVD in type 2 diabetics [23].

23.2 Management for Patients with COMISA

Management should begin with lifestyle changes and interdisciplinary weight control in overweight or obese patients [27]. Cognitive behavioral therapy for insomnia (CBTi) remains the first-line treatment, with or without PAP [17]. Tirzepatide could be used in adult patients with obesity [28]. Non-Benzodiazepine hypnotics like eszopiclone can improve sleep quality [29], and hypoglossal nerve stimulation presents a promising option [30]. Further research is needed for other non-PAP therapies.

23.2.1 Cognitive Behavioral Therapy for Insomnia (CBTi)

Cognitive behavioral therapy for insomnia (CBTi) focuses on addressing the maladaptive thoughts and behaviors that perpetuate insomnia and is considered the first-line treatment for chronic insomnia [31]. CBTi has also been shown to effectively treat insomnia in patients with COMISA.

CBTi has demonstrated efficacy in reducing insomnia symptoms in patients with COMISA, irrespective of OSA treatment status, as highlighted by meta-analyses [32]. Higher apnea–hypopnea index (AHI) values have been associated with greater improvement in insomnia symptoms following CBTi in this population [33]. Additionally, randomized controlled trials have shown that CBTi not only reduces AHI and nocturnal awakenings but also enhances adherence to PAP therapy when initiated before or concurrently with PAP [34–37].

A potential concern with CBTi is that the bedtime restriction component may exacerbate daytime sleepiness or impair neurocognitive function in patients with COMISA [38, 39]. One study reported a 15% increase in Epworth sleepiness scale (ESS) scores during the first week of bedtime restriction, though these scores returned to baseline in subsequent weeks [38]. While CBTi combined with PAP shows promising results, neurocognitive performance may be suboptimal when CBTi is added to PAP therapy compared to PAP alone, likely because of sleep restriction [39]. These findings highlight the need for tailored CBTi protocols in COMISA management.

23.2.2 Continuous Positive Airway Pressure (CPAP)

Continuous positive airway pressure (CPAP) is the gold standard treatment for OSA, effectively maintaining airway patency during sleep. However, in patients with COMISA, CPAP therapy can sometimes exacerbate pre-existing insomnia symptoms, leading to decreased adherence [17].

Despite this, PAP therapy can still alleviate insomnia symptoms in a subset of patients with COMISA. A recent review reported a 20–50% improvement in insomnia severity among patients with COMISA undergoing PAP therapy [16]. Greater OSA severity and more frequent brief awakenings (middle insomnia) appear to predict better insomnia improvement with PAP treatment, suggesting that untreated OSA may contribute to insomnia symptoms in these cases [40, 41].

Unfortunately, it remains challenging to predict which patients with COMISA will respond to PAP therapy. Thus, combining CBTi with PAP is often recommended to alleviate insomnia symptoms, enhance PAP adherence, and foster a more positive treatment experience [17].

23.2.3 Pharmacological Interventions

Pharmacotherapy targeting OSA pathophysiology and specific phenotypes is a promising area of research. These medications aim to reduce apnea events (e.g., weight loss agents, ventilation activators, and muscle/nervous system stimulants), alleviate symptoms (e.g., wake-promoting agents for excessive daytime sleepiness), and improve adherence to therapies like CPAP (e.g., hypnotics) [42]. However, most medications remain understudied in patients with COMISA.

Tirzepatide is the first FDA-approved drug for moderate-to-severe OSA in adults with obesity. In two placebo-controlled trials, tirzepatide reduced AHI, weight, hypoxic burden, and systolic blood pressure while improving sleep-related and daytime functional outcomes [28]. These results suggest its potential role in managing COMISA in obese patients.

Noradrenergic and antimuscarinic therapies, such as atomoxetine plus oxybutynin, improve pharyngeal muscle tone and have shown significant improvements in AHI and oxygen saturation [43]. However, caution is needed in COMISA patients due to atomoxetine's wake-promoting effects, which may reduce total sleep time, sleep efficiency, and REM sleep, while increasing wake after sleep and N1 sleep, particularly at higher doses [44].

Combining atomoxetine with hypnotics is a potential alternative. Atomoxetine with trazodone reduces AHI and hypoxic burden but also decreases REM sleep, worsens sleep quality, and increases subjective sleepiness [45, 46]. Atomoxetine with lemborexant has minimal impact on AHI, but worsens self-reported sleep quality, while adding zolpidem to atomoxetine–oxybutynin improves sleep efficiency without affecting OSA severity [45, 47]. However, small sample sizes limit firm conclusions, warranting further studies.

Non-Benzodiazepine hypnotics (Z-Drugs) like eszopiclone improve sleep quality (e.g., latency, efficiency, and wake time after sleep) without worsening OSA severity [29]. Eszopiclone may reduce AHI in some studies and improve CPAP adherence, but results are inconsistent and warrant further investigation [48, 49].

Trazodone, commonly used as a hypnotic, has shown a significant reduction in AHI in limited studies, though further research is needed to confirm its effects and identify responders based on OSA endotypes, such as REM-predominant OSA [50, 51].

Melatonin has limited efficacy for primary insomnia but may benefit COMISA patients [52, 53]. A small, randomized trial demonstrated improved sleep quality, insomnia symptoms, and CPAP adherence with melatonin [53]. However, high dropout rates highlight the need for larger, robust studies.

Other agents, such as sodium oxybate and donepezil, have shown modest AHI reduction, but present risks like central apneas or decreased sleep efficiency [51, 54, 55]. Lemborexant, a dual orexin receptor antagonist, has a favorable safety profile for insomnia, but lacks evidence of significant benefits for COMISA management [56].

23.2.4 Other Interventions

Hypoglossal nerve stimulation (HGNS) is an effective alternative treatment for moderate-to-severe OSA, particularly in patients who cannot tolerate PAP therapy [57]. While HGNS primarily targets respiratory improvements, its effects on insomnia in patients with COMISA are noteworthy.

Data from the ADHERE registry revealed that HGNS therapy improves insomnia and daytime sleepiness scores in all patients, with greater benefits in those with moderate-to-severe baseline insomnia (ISI ≥15) [30]. Patients with major sleep maintenance issues showed the greatest improvement in insomnia and daytime sleepiness, while those with early-morning awakenings reported lower satisfaction rates. Sleep-onset insomnia was linked to longer device usage [58]. Smaller studies also show that HGNS reduces insomnia symptoms alongside respiratory improvements [59]. However, severe baseline insomnia may limit respiratory outcome gains, emphasizing the need for tailored management strategies. HGNS holds promises for patients with COMISA, but further research is needed to optimize outcomes and better understand its role in COMISA management.

Other non-PAP therapies, including upper airway surgery [60], nasal dilator strip therapy [61], and maxillomandibular advancement [62], have been shown to improve insomnia symptoms in OSA patients. However, patients with COMISA may demonstrate lower acceptance and therapeutic benefit from these interventions compared to those with OSA alone [17].

23.3 Conclusion

Managing COMISA is challenging due to the complex interplay between conditions and variability in treatment responses. While combining CBTi and CPAP is the most effective first-line treatment, limitations remain in predicting individual responses and addressing treatment-resistant cases.

Pharmacological therapies and non-PAP interventions, such as HGNS, show promise but require further research, particularly in combination with existing therapies. Future studies should focus on identifying clinical phenotypes and OSA endotypes to guide personalized treatments and evaluate long-term outcomes on quality of life, cardiovascular health, and neurocognitive function.

References

1. Lechat B, Naik G, Reynolds A, Aishah A, Scott H, Loffler KA, et al. Multinight prevalence, variability, and diagnostic misclassification of obstructive sleep apnea. Am J Respir Crit Care Med. 2022;205(5):563–9.
2. Zeng LN, Zong QQ, Yang Y, Zhang L, Xiang YF, Ng CH, et al. Gender difference in the prevalence of insomnia: a meta-analysis of observational studies. Front Psych. 2020;11:577429.
3. Sweetman AM, Lack LC, Catcheside PG, Antic NA, Chai-Coetzer CL, Smith SS, et al. Developing a successful treatment for co-morbid insomnia and sleep apnoea. Sleep Med Rev. 2017;33:28–38.

4. Malhotra RK. AASM scoring manual 3: a step forward for advancing sleep care for patients with obstructive sleep apnea. J Clin Sleep Med. 2024;20(5):835–6.
5. Gottlieb DJ, Punjabi NM. Diagnosis and management of obstructive sleep apnea: a review. JAMA. 2020;323(14):1389–400.
6. Yeghiazarians Y, Jneid H, Tietjens JR, Redline S, Brown DL, El-Sherif N, et al. Obstructive sleep apnea and cardiovascular disease: a scientific statement from the American Heart Association. Circulation. 2021;144(3):e56–67.
7. Reutrakul S, Mokhlesi B. Obstructive sleep apnea and diabetes: a state of the art review. Chest. 2017;152(5):1070–86.
8. Liguori C, Maestri M, Spanetta M, Placidi F, Bonanni E, Mercuri NB, et al. Sleep-disordered breathing and the risk of Alzheimer's disease. Sleep Med Rev. 2021;55:101375.
9. Furukawa Y, Sakata M, Furukawa TA, Efthimiou O, Perlis M. Initial treatment choices for long-term remission of chronic insomnia disorder in adults: a systematic review and network meta-analysis. Psychiatry Clin Neurosci. 2024;78(11):646–53.
10. Sateia MJ. International classification of sleep disorders-third edition: highlights and modifications. Chest. 2014;146(5):1387–94.
11. Sweetman A, Lack L, Van Ryswyk E, Vakulin A, Reed RL, Battersby MW, et al. Co-occurring depression and insomnia in Australian primary care: recent scientific evidence. Med J Aust. 2021;215(5):230–6.
12. Natsky AN, Vakulin A, Chai-Coetzer CL, Lack L, McEvoy RD, Lovato N, et al. Economic evaluation of cognitive behavioural therapy for insomnia (CBT-I) for improving health outcomes in adult populations: a systematic review. Sleep Med Rev. 2020;54:101351.
13. Li L, Gan Y, Zhou X, Jiang H, Zhao Y, Tian Q, et al. Insomnia and the risk of hypertension: a meta-analysis of prospective cohort studies. Sleep Med Rev. 2021;56:101403.
14. Guilleminault C, Eldridge FL, Dement WC. Insomnia with sleep apnea: a new syndrome. Science. 1973;181(4102):856–8.
15. Ong JC, Crawford MR, Wallace DM. Sleep apnea and insomnia: emerging evidence for effective clinical management. Chest. 2021;159(5):2020–8.
16. Sweetman A, Lack L, McEvoy RD, Smith S, Eckert DJ, Osman A, et al. Bi-directional relationships between co-morbid insomnia and sleep apnea (COMISA). Sleep Med Rev. 2021;60:101519.
17. Sweetman A, Lack L, Crawford M, Wallace DM. Comorbid insomnia and sleep apnea: assessment and management approaches. Sleep Med Clin. 2022;17(4):597–617.
18. Lechat B, Appleton S, Melaku YA, Hansen K, McEvoy RD, Adams R, et al. Comorbid insomnia and sleep apnoea is associated with all-cause mortality. Eur Respir J. 2022;60(1):2101958.
19. Meira ECM, Salles C, Gozal D. A reappraisal on the associations between sleep-disordered breathing, insomnia, and cardiometabolic risk. Am J Respir Crit Care Med. 2021;203(12):1583–4.
20. Javaheri S, Redline S. Insomnia and risk of cardiovascular disease. Chest. 2017;152(2):435–44.
21. Sweetman A, Melaku YA, Lack L, Reynolds A, Gill TK, Adams R, et al. Prevalence and associations of co-morbid insomnia and sleep apnoea in an Australian population-based sample. Sleep Med. 2021;82:9–17.
22. Gupta MA, Knapp K. Cardiovascular and psychiatric morbidity in obstructive sleep apnea (OSA) with insomnia (sleep apnea plus) versus obstructive sleep apnea without insomnia: a case-control study from a Nationally Representative US sample. PLoS One. 2014;9(3):e90021.
23. Hein M, Lanquart JP, Mungo A, Loas G. Cardiovascular risk associated with co-morbid insomnia and sleep apnoea (COMISA) in type 2 diabetics. Sleep Sci. 2022;15(Spec 1):184–94.
24. Chuang LP, Wu HP, Lee LA, Chiu LC, Lin SW, Hu HC, et al. Elevated Monocytic Interleukin-8 expression under intermittent hypoxia condition and in obstructive sleep apnea patients. Int J Mol Sci. 2021;22(21):11396.
25. Guo J, Redline S, Stone KL, Xiao Y. Redefining comorbid insomnia and sleep apnea: the Association of Sleep Breathing Impairment and Insomnia with incident diabetes. Ann Am Thorac Soc. 2023;20(12):1791–800.

26. Li X, Sotres-Alvarez D, Gallo LC, Ramos AR, Aviles-Santa L, Perreira KM, et al. Associations of sleep-disordered breathing and insomnia with incident hypertension and diabetes. The Hispanic community health study/study of Latinos. Am J Respir Crit Care Med. 2021;203(3):356–65.
27. Carneiro-Barrera A, Amaro-Gahete FJ, Guillen-Riquelme A, Jurado-Fasoli L, Saez-Roca G, Martin-Carrasco C, et al. Effect of an interdisciplinary weight loss and lifestyle intervention on obstructive sleep apnea severity: the INTERAPNEA Randomized Clinical Trial. JAMA Netw Open. 2022;5(4):e228212.
28. Malhotra A, Grunstein RR, Fietze I, Weaver TE, Redline S, Azarbarzin A, et al. Tirzepatide for the treatment of obstructive sleep apnea and obesity. N Engl J Med. 2024;391(13):1193–205.
29. Zhang XJ, Li QY, Wang Y, Xu HJ, Lin YN. The effect of non-benzodiazepine hypnotics on sleep quality and severity in patients with OSA: a meta-analysis. Sleep Breath. 2014;18(4):781–9.
30. Dhanda Patil R, Ishman SL, Chang JL, Thaler E, Suurna MV. Impact of insomnia on hypoglossal nerve stimulation outcomes in the ADHERE registry. Laryngoscope. 2024;134(1):471–9.
31. Sateia MJ, Buysse DJ, Krystal AD, Neubauer DN, Heald JL. Clinical practice guideline for the pharmacologic treatment of chronic insomnia in adults: an American Academy of sleep medicine clinical practice guideline. J Clin Sleep Med. 2017;13(2):307–49.
32. Sweetman A, Farrell S, Wallace DM, Crawford M. The effect of cognitive behavioural therapy for insomnia in people with comorbid insomnia and sleep apnoea: a systematic review and meta-analysis. J Sleep Res. 2023;32(6):e13847.
33. Sweetman A, Lechat B, Catcheside PG, Smith S, Antic NA, O'Grady A, et al. Polysomnographic predictors of treatment response to cognitive behavioral therapy for insomnia in participants with co-morbid insomnia and sleep apnea: secondary analysis of a randomized controlled trial. Front Psychol. 2021;12:676763.
34. Alessi CA, Fung CH, Dzierzewski JM, Fiorentino L, Stepnowsky C, Rodriguez Tapia JC, et al. Randomized controlled trial of an integrated approach to treating insomnia and improving the use of positive airway pressure therapy in veterans with comorbid insomnia disorder and obstructive sleep apnea. Sleep. 2021;44(4):zsaa235.
35. Sweetman A, Lack L, McEvoy RD, Antic NA, Smith S, Chai-Coetzer CL, et al. Cognitive behavioural therapy for insomnia reduces sleep apnoea severity: a randomised controlled trial. ERJ Open Res. 2020;6(2):00161.
36. Sweetman A, Lack L, Catcheside PG, Antic NA, Smith S, Chai-Coetzer CL, et al. Cognitive and behavioral therapy for insomnia increases the use of continuous positive airway pressure therapy in obstructive sleep apnea participants with comorbid insomnia: a randomized clinical trial. Sleep. 2019;42(12):zsz178.
37. Richards D, Bartlett DJ, Wong K, Malouff J, Grunstein RR. Increased adherence to CPAP with a group cognitive behavioral treatment intervention: a randomized trial. Sleep. 2007;30(5):635–40.
38. Sweetman A, McEvoy RD, Smith S, Catcheside PG, Antic NA, Chai-Coetzer CL, et al. The effect of cognitive and behavioral therapy for insomnia on week-to-week changes in sleepiness and sleep parameters in patients with comorbid insomnia and sleep apnea: a randomized controlled trial. Sleep. 2020;43(7):zsaa002.
39. Turner AD, Ong JC, Jones AL, Tu A, Salanitro M, Crawford MR. Neurocognitive functioning in comorbid insomnia and sleep apnea patients is better after positive airway pressure therapy, but worse after cognitive behavioral therapy for insomnia: exploratory analysis of cognitive outcomes from the Multidisciplinary Approach to the Treatment of Insomnia and Comorbid Sleep Apnea study. Sleep. 2023;46(8):zsad128.
40. Glidewell RN, Renn BN, Roby E, Orr WC. Predictors and patterns of insomnia symptoms in OSA before and after PAP therapy. Sleep Med. 2014;15(8):899–905.
41. Bjornsdottir E, Janson C, Sigurdsson JF, Gehrman P, Perlis M, Juliusson S, et al. Symptoms of insomnia among patients with obstructive sleep apnea before and after two years of positive airway pressure treatment. Sleep. 2013;36(12):1901–9.

42. Harding CD, Fuentes AL, Malhotra A. Tackling obstructive sleep apnea with pharmacotherapeutics: expert guidance. Expert Opin Pharmacother. 2024;25(8):1019–26.
43. Lee YC, Lu CT, Chuang LP, Lee LA, Fang TJ, Cheng WN, et al. Pharmacotherapy for obstructive sleep apnea – a systematic review and meta-analysis of randomized controlled trials. Sleep Med Rev. 2023;70:101809.
44. Rosenberg R, Abaluck B, Thein S. Combination of atomoxetine with the novel antimuscarinic aroxybutynin improves mild to moderate OSA. J Clin Sleep Med. 2022;18(12):2837–44.
45. Corser B, Eves E, Warren-McCormick J, Rucosky G. Effects of atomoxetine plus a hypnotic on obstructive sleep apnea severity in patients with a moderately collapsible pharyngeal airway. J Clin Sleep Med. 2023;19(6):1035–42.
46. Shahbazi M, Heidari R, Tafakhori A, Samadi S, Nikeghbalian Z, Amirifard H, et al. The effects of atomoxetine and trazodone combination on obstructive sleep apnea and sleep microstructure: a double-blind randomized clinical trial study. Sleep Med. 2024;113:13–8.
47. Messineo L, Carter SG, Taranto-Montemurro L, Chiang A, Vakulin A, Adams RJ, et al. Addition of zolpidem to combination therapy with atomoxetine-oxybutynin increases sleep efficiency and the respiratory arousal threshold in obstructive sleep apnoea: a randomized trial. Respirology. 2021;26(9):878–86.
48. Nigam G, Camacho M, Riaz M. The effect of nonbenzodiazepines sedative hypnotics on apnea-hypopnea index: a meta-analysis. Ann Thorac Med. 2019;14(1):49–55.
49. Wang D, Tang Y, Chen Y, Zhang S, Ma D, Luo Y, et al. The effect of non-benzodiazepine sedative hypnotics on CPAP adherence in patients with OSA: a systematic review and meta-analysis. Sleep. 2021;44(8):zsab077.
50. Chen CY, Chen CL, Yu CC. Trazodone improves obstructive sleep apnea after ischemic stroke: a randomized, double-blind, placebo-controlled, crossover pilot study. J Neurol. 2021;268(8):2951–60.
51. Nobre ML, Sarmento ACA, de Oliveira PF, Wanderley FF, Diniz Junior J, Goncalves AK. Pharmacological treatment for obstructive sleep apnea: a systematic review and meta-analysis. Clinics (Sao Paulo). 2024;79:100330.
52. De Crescenzo F, D'Alo GL, Ostinelli EG, Ciabattini M, Di Franco V, Watanabe N, et al. Comparative effects of pharmacological interventions for the acute and long-term management of insomnia disorder in adults: a systematic review and network meta-analysis. Lancet. 2022;400(10347):170–84.
53. Motlaq TM, Rahimi B, Amini S. Effect of melatonin on insomnia and daytime sleepiness, in patients with obstructive sleep apnea and insomnia (COMISA): a randomized double-blinded placebo-controlled trial. J Pharm Health Care Sci. 2024;10(1):25.
54. George CF, Feldman N, Inhaber N, Steininger TL, Grzeschik SM, Lai C, et al. A safety trial of sodium oxybate in patients with obstructive sleep apnea: acute effects on sleep-disordered breathing. Sleep Med. 2010;11(1):38–42.
55. Li Y, Owens RL, Sands S, Orr J, Moraes W, DeYoung P, et al. The effect of donepezil on arousal threshold and apnea-hypopnea index. A randomized, double-blind, cross-over study. Ann Am Thorac Soc. 2016;13(11):2012–8.
56. Cheng JY, Lorch D, Lowe AD, Uchimura N, Hall N, Shah D, et al. A randomized, double-blind, placebo-controlled, crossover study of respiratory safety of lemborexant in moderate to severe obstructive sleep apnea. J Clin Sleep Med. 2024;20(1):57–65.
57. Certal VF, Zaghi S, Riaz M, Vieira AS, Pinheiro CT, Kushida C, et al. Hypoglossal nerve stimulation in the treatment of obstructive sleep apnea: a systematic review and meta-analysis. Laryngoscope. 2015;125(5):1254–64.
58. Dhanda Patil R, Suurna MV, Steffen A, Soose R, Coxe J, Chan T, et al. Relationship of nocturnal insomnia symptoms and outcomes after hypoglossal nerve stimulation. OTO Open. 2024;8(2):e134.
59. Pordzik J, Ludwig K, Seifen C, Huppertz T, Bahr-Hamm K, Matthias C, et al. Insomnia in patients undergoing hypoglossal nerve stimulation therapy for obstructive sleep apnea. Biology (Basel). 2023;12(1):98.

60. Guilleminault C, Davis K, Huynh NT. Prospective randomized study of patients with insomnia and mild sleep disordered breathing. Sleep. 2008;31(11):1527–33.
61. Krakow B, Melendrez D, Sisley B, Warner TD, Krakow J, Leahigh L, et al. Nasal dilator strip therapy for chronic sleep-maintenance insomnia and symptoms of sleep-disordered breathing: a randomized controlled trial. Sleep Breath. 2006;10(1):16–28.
62. Proothi M, Grazina VJR, Gold AR. Chronic insomnia remitting after maxillomandibular advancement for mild obstructive sleep apnea: a case series. J Med Case Rep. 2019;13(1):252.

MIX
Papier aus verantwortungsvollen Quellen
Paper from responsible sources
FSC® C105338

If you have any concerns about our products,
you can contact us on
ProductSafety@springernature.com

In case Publisher is established outside the EU,
the EU authorized representative is:
Springer Nature Customer Service Center GmbH
Europaplatz 3, 69115 Heidelberg, Germany

Printed by Libri Plureos GmbH
in Hamburg, Germany